T0201174

Successful Construction Supply Chain Management

Successful Construction Supply Chain Management

Concepts and Case Studies

Second Edition

Edited by
Stephen Pryke
Bartlett School of Construction and Project Management
University College London

This edition first published 2020
© 2020 John Wiley & Sons Ltd

Edition History
First published in 2009 by Blackwell Publishing Ltd as *Construction Supply Chain Management - Concepts and Case Studies*
Second retitled edition published in 2020 by John Wiley & Sons Ltd

Registered Offices
John Wiley & Sons, Inc., 111 River Street, Hoboken, NJ 07030, USA
John Wiley & Sons Ltd, The Atrium, Southern Gate, Chichester, West Sussex, PO19 8SQ, UK

Editorial Office
9600 Garsington Road, Oxford, OX4 2DQ, UK

For details of our global editorial offices, customer services, and more information about Wiley products visit us at www.wiley.com.

Wiley also publishes its books in a variety of electronic formats and by print-on-demand. Some content that appears in standard print versions of this book may not be available in other formats.

Library of Congress Cataloging-in-Publication Data applied for
Hardback ISBN - 9781119450689

Cover Design: Wiley
Cover Image: Karen Rubin

Set in 10/12pt Warnock by SPi Global, Chennai, India

Printed and bound by CPI Group (UK) Ltd, Croydon, CR0 4YY

10 9 8 7 6 5 4 3 2 1

Contents

List of Contributors *xv*
Preface *xxi*
Acknowledgements *xxiii*

1   **Introduction** *1*
        Stephen Pryke
1.1   Overview: Part A *2*
1.1.1   IT, Digital, and BIM *2*
1.1.2   Self-Organising Networks in Supply Chains *2*
1.1.3   Green Issues *3*
1.1.4   Demand Chains and Supply Chains *4*
1.1.5   Lean *5*
1.1.6   Power Structures and Systemic Risk *5*
1.1.7   Decision-Making Maturity *6*
1.2   Overview: Part B *7*
1.2.1   Lessons from Megaprojects *7*
1.2.2   Collaboration and Integration *8*
1.2.3   Lesson Learned and Findings from Tier 1 Contractors *8*
1.2.4   Lean Practices in The Netherlands *9*
1.2.5   Knowledge Transfer in Supply Chains *10*
1.2.6   The Role of Trust in Managing Supply Chains *10*
1.3   Summary *11*
        References *11*

Part I   Chapters that Principally, but not Exclusively, Deal with Concepts and the Development of Theory *13*

2   **The Digital Supply Chain: Mobilising Supply Chain Management Philosophy to Reconceptualise Digital Technologies and Building Information Modelling (BIM)** *15*
        Eleni Papadonikolaki
2.1   Introduction *15*
2.2   The Nature of Construction *17*

2.2.1   Addressing Existing Complexity and Fragmentation in Construction   *17*
2.2.2   Advancements from Other Industries Applicable to Construction   *17*
2.2.3   Potential Synergies Between Supply Chain Management and
            Digitisation   *19*
2.3   Origins and Development of Supply Chain Thinking in AEC   *20*
2.3.1   The Emergence of Supply Chain Thinking from Operations and Logistics
            Research   *20*
2.3.2   The Adaptation of Supply Chain Management Concepts in AEC   *21*
2.4   Pragmatic Impact of Supply Chain Thinking in Construction   *22*
2.4.1   Supply Chain Thinking Schools   *22*
2.4.2   Supply Chain Concepts and Varying Interpretations   *23*
2.5   Origins and Development of Digitisation in the Built Environment   *23*
2.5.1   Development of Digital Capabilities in the Built Environment   *23*
2.5.2   From Building Product Models to Building Information Modelling (BIM)   *25*
2.5.3   Importance of Standards in a Digital Built Environment   *25*
2.5.4   Pluralism of Digital Artefacts and BIM Maturity Assessment Methods   *26*
2.6   Pragmatic Impact of Digitisation and BIM   *28*
2.6.1   BIM and the Enterprise: Bottom-Up Adoption   *28*
2.6.2   BIM and the Institutional Setting: Top-Down Diffusion   *28*
2.6.3   Mismatch Between Top-Down and Bottom-Up Strategies   *29*
2.7   Synthesis of Digital Technologies Construction Supply Chain   *30*
2.7.1   Potential and Outlook of Digital Technologies to Support Supply Chains   *30*
2.7.2   Co-Evolution of Supply Chain Management and Digital in AEC   *31*
2.8   Conclusion   *32*
            References   *34*

3   **At the Interface: When Social Network Analysis and Supply Chain
            Management Meet**   *43*
            Huda Almadhoob
3.1   Introduction   *43*
3.2   Reconceptualising Supply Chains   *44*
3.3   Supply Networks as Complex Adaptive Systems   *45*
3.4   What Is Social Network Analysis?   *50*
3.5   Rationale for a Network Approach   *52*
3.6   Key Challenges in Conducting Social Network Analysis   *54*
3.7   Conclusions and Directions for Future Research   *55*
3.8   Managerial Implications   *56*
            References   *57*

4   **Green Supply Chain Management in Construction: A Systematic
            Review**   *63*
            Niamh Murtagh and Sulafa Badi
4.1   Introduction   *63*
4.1.1   Environmental Impact of Construction   *64*
4.1.2   Definition   *65*
4.2   Research Methodology   *66*
4.2.1   Stage 1: Define Eligibility Criteria   *66*

4.2.2   Stage 2: Define Search Terms   *67*
4.2.3   Stage 3: Search, Screen, and Compile List of Included Papers   *67*
4.2.4   Stage 4: Code and Critically Evaluate Included Studies   *67*
4.2.5   Stage 5: Formulate Synthesis   *68*
4.3   Analysis   *68*
4.3.1   Research Interest over Time   *68*
4.3.2   Source Journals   *68*
4.3.3   Geographic Spread   *69*
4.3.4   Methods   *69*
4.3.5   Tools and Techniques   *72*
4.3.6   Stakeholders   *73*
4.3.7   Definitions of Green Supply Chain Management   *74*
4.4   Discussion   *75*
4.4.1   Overview   *75*
4.4.2   Definition   *75*
4.4.3   Nature of Construction   *76*
4.4.4   Stakeholder Roles   *77*
4.4.5   Practical Recommendations   *77*
4.5   Looking to the Future   *78*
4.6   Conclusion   *80*
           References   *81*

5   **Connecting the 'Demand Chain' with the 'Supply Chain': (Re)creating
           Organisational Routines in Life Cycle Transitions**   *87*
           Simon Addyman
5.1   Introduction   *87*
5.1.1   The Temporal Paradox in Temporary Organising   *89*
5.2   The Construction Industry – Procurement and Relational Difficulties   *90*
5.3   Temporary Organisations and the Project Life Cycle   *92*
5.4   Routines and the Capability of Projects   *95*
5.5   A Recursive Process Model of Transitioning   *98*
5.6   Discussion   *101*
5.7   Summary   *103*
           References   *104*

6   **Construction Supply Chain Management through a Lean Lens**   *109*
           Lauri Koskela, Ruben Vrijhoef and Rafaella Dana Broft
6.1   Introduction   *109*
6.2   Theoretical and Philosophical Grounding of Lean   *110*
6.2.1   Theoretical and Philosophical Grounding of the Mainstream Approach to
           Production Management   *110*
6.2.2   Theoretical and Philosophical Grounding of Lean   *111*
6.2.2.1   Theory of Production   *111*
6.2.2.2   Epistemology of the Lean Concept   *112*
6.2.2.3   Ontology of the Lean Concept   *112*
6.2.3   Implications for Management and Organising   *113*

6.3   Theoretical Background and Characterisation of Supply Chain
          Management   *114*
6.3.1   Production Perspective   *114*
6.3.2   Economic Perspective   *115*
6.3.3   Organisational Perspective   *116*
6.3.4   Social Perspective   *116*
6.4   Analysis of Supply Chain Approaches and Conceptualisations through a Lean
          Versus Mainstream Lens   *117*
6.5   Contingency of Supply Chain Management in Construction through a Lean
          Lens   *118*
6.5.1   Construction from a Production Perspective   *119*
6.5.2   Construction from an Economic Perspective   *119*
6.5.3   Construction from an Organisational Perspective   *119*
6.5.4   Construction from a Social Perspective   *121*
6.5.5   A Crossover of Supply Chain Management and Lean in the Context of
          Construction   *121*
6.6   Discussion   *121*
6.7   Conclusion   *122*
          References   *122*

**7   Supply Chain Management and Risk Set in Changing Times: Old Wine
          in New Bottles?**   *127*
          Andrew Edkins
7.1   Introduction and Overview   *127*
7.2   The Collapse of Carillion: Consequences for Consideration – Implications
          for Construction Supply Chains   *129*
7.3   Risk, Power Structures, and Supply Chains   *132*
7.3.1   Commercial Power and the Role of Law and Regulation   *133*
7.3.2   Technology-Based Power Structures: Cases of Construction Waste and
          BIM   *135*
7.4   Conclusions   *139*
          References   *140*

**8   Linkages, Networks, and Interactions: Exploring the Context for Risk
          Decision Making in Construction Supply Chains**   *143*
          Alex Arthur
8.1   Introduction   *143*
8.2   The Evolution of the UK Construction Industry and Supply Chain
          Relationships   *144*
8.3   The Concept of Risk   *147*
8.3.1   Uncertainty   *149*
8.3.2   Probability   *150*
8.3.3   Risk as a Potential Future Event   *150*
8.3.4   The Impact of a Risk Event on an Objective or Interest   *150*
8.4   The Construction Risk Management System   *150*
8.4.1   Risk Identification Subsystem   *152*
8.4.2   Risk Analysis Subsystem   *153*

8.4.3   Risk Response Subsystem   *153*
8.5   Risk Generation in Construction Supply Chain Relationships   *154*
8.5.1   Project Risk Events Generated through the Project Delivery Processes   *154*
8.5.2   Project Risk Events Generated through the Network and Interactions within
         Construction Supply Chain Relationships   *155*
8.6   Risk Management Decision-Making Systems in Construction Supply Chain
         Relationships   *156*
8.7   Conclusion   *159*
         References   *161*

9   Culture in Supply Chains   *167*
           Richard Fellows and Anita Liu
9.1   Introduction – Context   *167*
9.2   Culture   *170*
9.3   Dimensions of Culture   *173*
9.3.1   National Culture   *174*
9.3.2   Organisational Culture   *176*
9.3.3   Fitting with Other Cultures   *180*
9.3.4   Organisational Climate   *182*
9.3.5   Project Atmosphere   *182*
9.3.6   Behaviour Modification   *183*
9.4   Values and Value   *183*
9.5   Ethics   *185*
9.6   Organisational Citizenship Behaviour (OCB) and Corporate Social
         Responsibility (CSR)   *187*
9.7   Teams and Teamwork   *188*
9.8   Sensemaking   *189*
9.9   Motivated Reasoning   *190*
9.10   (Strategic) Alliances   *192*
9.11   Supply Chain Participants and Behaviour   *194*
9.12   Conclusion   *199*
         References   *201*

**Part II   Chapters that Principally, but not Exclusively, Deal with Case
Study Material**   *211*

**10   Managing Megaproject Supply Chains: Life After Heathrow
            Terminal 5**   *213*
            Dr Juliano Denicol
10.1   Motivation for the Research   *213*
10.2   Construction Supply Chain Management   *214*
10.2.1   Temporary vs Permanent Supply Chains (ETO vs MTS)   *217*
10.3   Why Are Megaprojects So Important?   *221*
10.4   Megaproject Supply Chain Management   *223*
10.5   Conclusion   *228*
         References   *231*

11 Anglian Water @one Alliance: A New Approach to Supply Chain Management *237*
Grant Mills, Dale Evans, and Chris Candlish
11.1 Introduction *237*
11.2 Supply Chain Management *238*
11.3 Alliance Supply Chain Management *239*
11.4 Anglian Water Alliance Case Study *240*
11.4.1 Strategic Approach to Alliance Supply Chain Management *240*
11.4.2 Alliance Supply Chain Work Clusters *241*
11.4.3 Alliance Supply Chain Early Involvement and Collaboration *242*
11.5 Evaluation of the Value of Alliance Supply Chain Management *244*
11.5.1 Strategic Approach to Alliance Supply Chain Management *244*
11.5.2 Alliance Supply Chain Management Provides an Effective Environment for the Early Engagement of Specialist Suppliers *244*
11.5.3 Alliance Supply Chain Management Can Create a Win-Win-Win Reciprocal Relationship *245*
11.5.4 Alliance Supply Chain Management Can Drive Team Innovation and Create New Service Relationships *245*
11.5.5 Long-Term Approaches to Alliance Supply Chain Management Can Drive Strategic Business Benefits *246*
11.5.6 Alliance Supply Chain Management that Uses Advanced Production Systems Can Deliver Tactical Benefits *246*
11.6 Conclusions *246*
References *247*

12 Understanding Supply Chain Management from a Main Contractor's Perspective *251*
Emmanuel Manu and Andrew Knight
12.1 Introduction *251*
12.2 Multilayered Subcontracting in the Construction Industry *252*
12.3 Supply Chain Management: Principles and Practices *254*
12.4 Supply Chain Management Practices from a Contractor's Perspective *256*
12.5 Case Study of a Large UK Main Contractor *257*
12.5.1 Supply Chain Management Goals *258*
12.5.2 Supply Chain Management Team *259*
12.5.3 Supply Chain Management Classification *260*
12.5.4 Supply Chain Management Practices *261*
12.5.4.1 Audit Supply Chain Firms *261*
12.5.4.2 Use Collaborative ICT Systems *263*
12.5.4.3 Measure Performance of Supply Chain Firms *263*
12.5.4.4 Engage in Continuous Performance Improvement Activities *264*
12.5.4.5 Develop Long-Term Collaborative Relationships *264*
12.5.4.6 Motivate and Incentivise the Supply Chain *265*
12.6 Conclusion *265*
References *267*

13 **Lean Supply Chain Management in Construction: Implementation at the 'Lower Tiers' of the Construction Supply Chain** *271*
Rafaella Dana Broft
13.1 Supply Chain Management in a Project-Based Environment *271*
13.1.1 The Supply Chain Management Concept *271*
13.1.2 The Project Focus in Construction *272*
13.1.3 The Lower Tiers of the Construction Supply Chain *273*
13.1.4 A Main Contractor's Position and Role in the Construction Supply Chain *274*
13.2 The Characteristics of Construction *275*
13.2.1 Construction from a Production Perspective *275*
13.2.2 Construction: True Peculiarities?! *277*
13.3 Lean Supply Chain Management in Construction *279*
13.3.1 An Introduction to Lean *279*
13.3.2 The Role of Lean in Combination with Supply Chain Management *280*
13.3.3 Lean and Supply Chain Management in Construction *281*
13.4 Conclusion *283*
References *283*

14 **Knowledge Transfer in Supply Chains** *289*
Hedley Smyth and Meri Duryan
14.1 Introduction *289*
14.1.1 The Supply Chain Issue *290*
14.1.2 Learning and Knowledge Transfer *291*
14.2 What Is Known – A Summary Review of the Literature *292*
14.2.1 The Supply Chain Ecosystem *292*
14.2.2 Supply Chain Learning and Knowledge Management *293*
14.2.3 Prequalification and Bidding Processes *294*
14.3 Methodology and Methods *295*
14.4 Findings *296*
14.5 Conclusions *301*
References *302*

15 **Understanding Trust in Construction Supply Chain Relationships** *307*
Jing Xu
15.1 Introduction *307*
15.2 Towards an Understanding of Trust in Construction Supply Chains *308*
15.2.1 Towards a Service-Dominant Logic View *308*
15.2.2 Towards a Process-Based View *311*
15.3 Methodology and Methods *314*
15.4 Case Study *315*
15.4.1 Context *316*
15.4.1.1 Assessing the Shadow of the Past *316*
15.4.1.2 Organisational Structure and Policy: Forming a Sense of Unfairness *316*

15.4.2   Procurement and Preconstruction Stage   *318*
15.4.2.1 Early Involvement: Forming a Sense of Security and Familiarity   *318*
15.4.2.2 Two-stage Procurement: Creating a Sense of Equity   *318*
15.4.2.3 The Value of Trust   *319*
15.4.3   Execution Stage   *320*
15.4.3.1 Structuring the Project: Maintaining Security and Familiarity   *320*
15.4.3.2 Joint Activities: Forming the Interpretations of Trustworthiness   *320*
15.4.3.3 Using Trust Relations in Resource Coordination: Bounded Solidarity and
          Economic Reciprocity   *321*
15.4.3.4 The Value of Trust   *322*
15.4.4   Completion Stage   *323*
15.4.4.1 Stabilising the Relationship: Trust as a Rule of Legitimation   *323*
15.4.4.2 The Shadow of the Future: Social Reciprocity   *323*
15.4.4.3 The Value of Trust   *323*
15.5   Discussion   *324*
15.5.1   The Constitution of Trust   *324*
15.5.2   The Value of Trust   *326*
15.5.3   Conditions of Trust: Influences of Ecosystems and Time   *326*
15.6   Conclusions and Recommendations   *328*
          References   *329*

16   **Summary and Conclusions**   *335*
          Stephen Pryke
16.1   Context – What's the Problem?   *335*
16.2   A Summary of the Contributions   *336*
16.2.1   IT, Digital, and BIM   *336*
16.2.2   Self-Organising Networks in Supply Chains   *336*
16.2.3   Green Issues   *337*
16.2.4   Demand Chains and Supply Chains   *337*
16.2.5   Lean   *337*
16.2.6   Power Structures and Systemic Risk   *337*
16.2.7   Decision-Making Maturity   *338*
16.2.8   Culture   *338*
16.2.9   Lessons from Megaprojects   *338*
16.2.10   Collaboration and Integration   *339*
16.2.11   Lesson Learned and Findings from Tier 1 Contractors   *339*
16.2.12   Lean Practices in The Netherlands   *340*
16.2.13   Knowledge Transfer   *340*
16.2.14   The Role of Trust in Managing Supply Chains   *341*
16.3   Key Themes and Agendas for Research and Practice   *341*
16.3.1   Complexity and Interdependence   *341*
16.3.2   Work Packages   *341*
16.3.3   Resistance to Change   *342*
16.3.4   Risk   *342*
16.3.5   Communications and Integration of Systems and the Green Agenda   *343*

16.3.6   The Role of the Contractor   *343*
16.3.7   The Role of the Client   *343*
16.3.8   Lean Construction   *343*
16.3.9   Collaborative Behaviour and Quality of Relationships   *344*
16.4   Final Remarks   *344*
          References   *344*

Index   *347*

List of Contributors

Professor Stephen Pryke is Professor of Supply Chain and Project Networks in the Bartlett School of Construction and Project Management, University College London (UCL). He is founder and Managing Director of CONA – The Centre for Organisational Network Analysis at UCL (CONA@UCL). For many years he has acted as the Programme Leader for the MSc Project and Enterprise Management programme at UCL. Stephen sits on the Infrastructure Supply Chain Management Roundtable for the Royal Institution of Chartered Surveyors (RICS) and has carried out research commissioned by RICS on a number of occasions. Stephen has written five books for Wiley-Blackwell and his work in the fields of supply chain management, the application of social network analysis to projects, and project management, has been disseminated in a wide range of academic journals.

Dr Simon Addyman has over 30 years' experience in the management of construction projects, predominantly public infrastructure projects, both internationally and in the UK. He specialises in the design of procurement and delivery models, focusing on development of interorganisational project capabilities. Simon was awarded Project Professional of the Year at the UK's Association for Project Management Awards 2013 for his work on London Underground's Bank Station Capacity Upgrade Project. Simon's research explores the (re)creation of organisational routines in temporary organisations, exploring the concept of 'transitioning' through project life cycle stages. Simon is a visiting lecturer at University College London and Kings College London, teaching project management and organisational design to postgraduate students.

Huda Almadhoob is a PhD researcher in the Bartlett School of Construction and Project Management at University College London (UCL). Prior to this, she worked as a lecturer in the Architectural department at the University of Bahrain after finishing her MSc degree in Project and Enterprise Management (with distinction) from the Bartlett School of Construction and Project Management at UCL. Currently she is working on her PhD research to establish a new theoretical framework for managing large construction projects, grounded in complexity theory with reference to social network theory and its analytical techniques. Her main research interests involve organisational behaviours, supply chain management, and the application of social network analysis in the study of construction project networks.

Dr Alex Arthur is a senior consultant at a London-based construction practice, and an assessor for the Royal Institution of Chartered Surveyors. His PhD at University College London focused on the behavioural patterns of intuitive risk management systems. He has published a number of papers on construction risk management systems, intuitive decision making, and planning systems.

Dr Sulafa Badi is Associate Professor of Management and Organizational Behaviour at the British University in Dubai (BUiD) and Honorary Senior Fellow at the Bartlett School of Construction and Project Management, University College London (UCL). She started her career as an Architect before joining academia and holds an MSc in Construction Economics and Management and a PhD in Project Management, both from UCL. Sulafa's research interests include sustainability innovation in public sector procurement, supply chain management, and social networks in project and enterprise organisations. Her research has had both academic and practical impact and involved studies in the United Kingdom, China, India, and the Middle East. She examined a variety of contexts including large infrastructure projects, service ecosystems, small and medium-sized enterprises (SMEs), and community stakeholders.

Rafaella Dana Broft started her career as Project Engineer to the Pavilion of the Netherlands at the World EXPO 2010 in Shanghai. After the project had finished, she joined University College London for an MSc in Construction Economics and Management and developed an interest in supply chain management (SCM), partly due to her part-time job as a Procurement Manager at the London 2012 Olympic Site. Rafaella is currently working as an SCM expert for the Dutch construction industry, guiding representatives of both main contractor and subcontractor/supplier organisations on the implementation of Lean and SCM principles with the aim of creating successful construction supply chains. In 2016 she successfully applied for registration to study for a PhD at the Bartlett School of Construction and Project Management, increasing the connection between theory and practice.

Chris Candlish has worked in the water and wastewater treatment industry for nearly 40 years, starting as a process engineer and working as project, contracts, and commercial manager for an international water treatment company, and has been involved in many different types of project using a range of contract forms. He was involved in the bidding and negotiation of several major private finance initiative schemes for treatment works in Scotland. Chris has a degree in chemical engineering and has been responsible for the Supply Chain Business Unit in the @one Alliance which is delivering approximately £1.2 billion of Anglian Water's AMP6 capital delivery programme. He worked for Sweco, one of the @one partners, and was responsible for the '@one' supply chain strategy and managing the procurement of goods and services for the alliance. A key aspect of this role was developing and integrating the supply chain within the @one Alliance into a collaborative 'one team delivery network' to contribute to the @one's successful outperformance. Chris had been working with University College London on developing an understanding of supply chain relationships and behaviours using econometric analysis and on developing collaborative relationships and working practices. Chris has also supported certain Infrastructure Client Group initiatives including Project 13.

Dr Juliano Denicol is Assistant Professor in Project Management at the School of Construction and Project Management, the Bartlett Faculty of the Built Environment, UCL. Juliano holds a Ph.D. in Megaproject Management from University College London. He also holds a Bachelor of Architecture and Urbanism (B.Arch.) and a Master of Research (M.Res.) in Industrial Engineering from the Federal University of Rio Grande do Sul (UFRGS). Juliano's research explores the management of interorganisational structures in major and megaprojects. Previous research included several iconic UK megaprojects: High Speed 1, Heathrow Airport Terminal 5, London 2012 Olympics, Crossrail, Thames Tideway Tunnel, and High Speed 2. His work on megaprojects has been regarded of high global impact, receiving research awards from the Project Management Institute (PMI), the Major Projects Association (MPA), and the International Project Management Association (IPMA). Juliano has received the 2019 Global Young Researcher Award from the IPMA and the 2019 Best Reviewer Award from the Project Organising Special Interest Group of the European Academy of Management (EURAM). He is part of Project X, a major research network that aims to improve major project delivery in the UK, established by nine universities in collaboration with the Infrastructure and Projects Authority (IPA), and the Cabinet Office. Juliano has worked as a supply chain management consultant at High Speed 2, the largest infrastructure project in Europe, and advisor to the European Commission on public procurement policies.

Dr Meri Duryan is a Lecturer in Enterprise Management at the Bartlett School of Construction and Project Management, University College London. She also leads optional MSc modules on Knowledge Management in a Project Environment and Change Management in Organisations. Systems thinking, value co-creation, occupational well-being, and knowledge and change management are the areas of her research focus. In addition to her academic and publishing background, she has wide-ranging practitioner experience in delivering change and knowledge projects across different industries including IT, Construction, Healthcare, Transportation and Education. She also provided consultancy services and executive training working with the United Nations, World Bank, and the US International Development Agency.

Professor Andrew Edkins is Director of the Bartlett Real Estate Institute and Professor of the Management of Complex Projects in the Bartlett School of Construction and Project Management, University College London (UCL). Andrew's background is in the built environment and spans real estate, infrastructure, and the project delivery process. Andrew has worked in the industry with experience that ranges from complex commercial refurbishments through being part of the construction management team that built Chelsea and Westminster Hospital, to the delivery of complex private finance initiative (PFI) and public private partnerships projects. As a UCL academic, Andrew's first research projects led to two books on PFI. This success, coupled with experience gained working in the PFI sector, led to an extremely successful project with KPMG using the Knowledge Transfer Partnership scheme. Among many other research interests, a constant and principal research focus has been on projects and their management. This has included looking at typically strategic level issues that have been concentrated on the very earliest 'front end' of the project lifecycle as well as more recently in the transition to operations phase of the tail end of the project.

Dale Evans is Managing Director of the @one Alliance, the collaborative organisation formed between client Anglian Water and key supply chain partners. He is responsible for the delivery of a multibillion pound asset management plan (AMP) investment programme to upgrade and enhance water supply and recycling to more than six million domestic customers. Dale is chair of the infrastructure client group (ICG), a collaboration between the Institution of Civil Engineers and Infrastructure and Projects Authority (IPA), which represents over 20 key infrastructure clients. He has significant experience in operational management, programme management, and change management and has provided strategic board level advice on the set up and operation of utilities and construction sector alliances across the world. He was involved in the development of Project 13 which has been reshaping and improving productivity in UK infrastructure delivery. Within the @one Alliance, Dale set stretching targets that have driven industry-leading improvements in project delivery, performance, health, safety and wellbeing, and in carbon and is a strong advocate for standardisation, offsite manufacture, building information modelling, fair procurement practices, and the stimulation of early supply chain innovation.

Professor Richard Fellows is an Emeritus Professor at Loughborough University and an editor of *Construction Management and Economics*. Richard has taught at a number of universities in the UK and other countries and was coordinator for research in construction management for the Engineering and Physical Sciences Research Council in the UK. He has obtained many competitive research grants in the UK and Hong Kong and has supervised a large number of postgraduate research degrees. He has been external examiner for various programmes and research degrees in universities and has served on boards and committees of leading professional institutions. His research interests concern economics, contracts and law, and management of people in construction – especially cultural issues. He was a founder and for many years was joint coordinator of the CIB international research group, W112 – 'Culture in Construction'. Richard has published widely in books, journals, and international conferences and is qualified as a mediator.

Dr Andrew Knight is Dean of the School of Architecture, Design and the Built Environment, Nottingham Trent University. He has previously published a book and several academic papers in the field of construction procurement and supply chain management. He is a Chartered Quantity Surveyor and continues to engage with professional body activities especially in the field of professional ethics.

Professor Lauri Koskela is a Professor of Construction and Project Management at the University of Huddersfield. Previously he worked at the University of Salford as Professor of Lean, Theory Based Project and Production Management. Prior to that he was involved in applied research at VTT Technical Research Centre of Finland. Since 1991, Lauri has been involved in research on lean construction. His research has focused especially on theories of production management as well as project management, underlying lean construction. Lauri is a founding and continuously active member of the International Group for Lean Construction. He is Director and Trustee at Lean Construction Institute, UK.

Dr Anita Liu was Professor at The University of Hong Kong and, previously, at Loughborough University. She has taught at a number of leading universities in Hong Kong, the UK, and China. Anita has obtained many competitive research grants in Hong Kong and has supervised a large number of doctoral degrees. She has been external examiner for various programmes and research degrees in universities around the world and has served on the boards and committees of various professional institutions. Her research interests focus on research methods, and management of people in construction – especially cultural issues. She was a founder and was joint coordinator of the CIB international research group, W112 – 'Culture in Construction'. Anita has published extensively in international journals and conferences and has authored several books and chapters.

Dr Emmanuel Manu is a Senior Lecturer in Construction and Project Management at the School of Architecture, Design and the Built Environment, Nottingham Trent University. He has researched, taught, and been involved in the management of projects in various capacities. Emmanuel is a reviewer for several leading construction management journals and continues to undertake and publish research in the areas of procurement, supply chain management, and sustainability in the built environment.

Dr Grant Mills is a Senior Lecturer in Enterprise Management in the Bartlett School of Construction and Project Management, University College London. His research is in the integration of the planning, design, construction, and operation supply chains to deliver value. He has undertaken extensive action and translational research and consultancy with world-leading infrastructure alliances (including health and education social infrastructure). He has collaborated with organisations such as the European Investment Bank, Department of Health, Manchester City Council, Heathrow, Anglian Water, London Underground, Chartered Quality Institute (CQI) and the Environment Agency. He has been coinvestigator on various Engineering and Physical Sciences Research Council (EPSRC) research projects (~£4.6 M) and knowledge transfer initiatives (~196 K) and led extensive multisector infrastructure research into: efficient design and production management (e.g. as a design process management consultant to the construction sector); NHS infrastructure productivity (e.g. developing tools for the Department of Health); innovation in new supply and delivery model integration (e.g. as advisor to the Infrastructure Client Group); policy and operational productivity failure (e.g. through action research with leading infrastructure clients such as Heathrow and Department of Health).

Dr Niamh Murtagh is a Senior Research Fellow at the Bartlett School of Construction and Project Management, University College London. She is an environmental psychologist whose research examines sustainable behaviour, including energy, transport, and the built environment. Niamh's focus has been on applying insights from psychological theory to sustainable construction, investigating the psychological underpinnings of pro-environmental behaviour in construction professionals. She has a particular interest in methods in construction research, including systematic literature reviews. Niamh has published in, and is a regular reviewer for, journals including *Journal of Environmental Psychology, Construction Management and Economics, Frontiers,* and *Environment and Behavior.*

Dr Eleni Papadonikolaki is a Lecturer in Building Information Modelling (BIM) and Management in the Bartlett School of Construction and Project Management, University College London (UCL), and a consultant in the area of digital innovation and management. She holds a PhD on the 'Alignment of Partnering with Construction IT' from Delft University of Technology, Netherlands, a MSc degree in Digital Technologies, also from Delft University of Technology, and an Engineering Diploma in Architectural Engineering from the NTUA, Greece. Bringing practical experience of working as an architect engineer and design manager on a number of complex and international projects in Europe and the Middle East, Eleni is researching and helping teams manage the interfaces between digital technology and management. She is teaching at undergraduate, postgraduate, doctoral, and executive levels. Currently she is a steering committee member of the UCL Construction Blockchain Consortium (CBC).

Professor Hedley Smyth is Professor of Project Enterprises at the Bartlett School of Construction and Project Management, University College London, where he has been a full-time staff member since 2001. He has led teaching and research in the School regarding project-based enterprises – how you manage the firms that manage projects. His research interests are organisational behaviour, especially regarding relationship management, trust, and emotional intelligence. Marketing and knowledge management are two functional areas of research focus. He has published on these topics in journal articles, book chapters, and sole-authored books.

Dr Ruben Vrijhoef is a senior researcher at the Department of Management in the Built Environment at the Delft University of Technology. In addition he is a Professor at the Utrecht University of Applied Sciences in the field of urban construction. He received his PhD degree from the Delft University of Technology on supply chain integration in construction. His research expertise includes supply chain management and lean construction. In the field of construction supply chain management he has written multiple publications and edited the *Construction Supply Chain Management Handbook* with CRC Press, an imprint of Taylor & Francis Group.

Jing Xu is currently a PhD candidate at the Bartlett School of Construction and Project Management, University College London. Her PhD topic is about the value of trust in construction supply chains. Her research interests are mainly relationship-related issues, e.g., trust, relational norms, and social capital, within and between project-based organisations, particularly in the construction industry.

Preface

This book was prompted initially by a phone call from Madeleine Metcalfe – Senior Commissioning Editor, at that time, at Wiley-Blackwell. Madeleine raised the issue that the book which I had previously edited, published in 2009 – *Construction Supply Chain Management: Concepts and Case Studies* – was due for a second edition. I was leading a group looking at supply chains and project networks at University College London at the time and the group discussed next steps for a possible second edition. We quickly came to the conclusion that there were a lot of new ideas and that while some of the content of the original book might benefit from updating, the second book would be mostly new material. Hence the book presented here appears as a new book alongside the original, rather than replacing it simply as a second edition of the original book.

My fascination with the idea of conceptualising construction activity as a supply chain that needs to be actively managed grew out of working alongside Slough Estates plc and British Airports Authority (BAA) in the early 'noughties' when both organisations were beacons of effective supply chain management. At that time Slough Estates were building design-award-winning buildings in 30% less time than their competitors and at a construction cost of 15–20% less than the current 'Spons' price levels. Pre-Ferovial BAA, in their development of design standardisation and process, were also doing important work in this area. The Ministry of Defence was carrying out some ground-breaking work in the public sector through its partnership with The Tavistock Institute. That so much of this innovative work was not sustainable through some difficult economic conditions subsequently and a good measure of politics, should not detract from the really important progress made. I am encouraged to hear that some of our major infrastructure client organisations are revisiting Holti, Nicolini and Smalley's (2001) excellent little book – *The Handbook of Supply Chain Management*.

More recently, I have been involved in the work of St George plc, the property developer. Through this I gained an intimate knowledge of their inspiring work on prefabrication and standardisation developed through their development of the St George Wharf site on the River Thames. I have also recently been taking an interest in the work of IKEA, both as a retailer and more recently as a developer of hotels. IKEA are not reinventing supply chain management (SCM) in construction, they are simply using the knowledge and expertise in SCM that they have gained as the most successful furniture retailer in the world, to produce excellent, high quality, stylish hotels at low cost. Watch this space!!

At its best the UK construction industry is world class and it is a testament to this that that I have been approached by supply chain managers from other sectors (most notably

IT and manufacturing recently) who regard construction firms as experts in outsourcing and managing outsourced activities. IT professionals express the same concerns as construction professionals do in terms of dealing with the effective management of risk and lack of information and certainty.

Working on this most recent book on SCM has been a great opportunity to engage with some well-established academic friends and acquaintances writing in the field of SCM. It has also been a really good opportunity to engage with some of the rising stars of the SCM field.

Whether you are a student of SCM or a 'thinking practitioner', I hope that you find something to stimulate thought in the pages of this book. I feel really privileged to have worked with the group of talented academics which came forward to work on this book.

London, April 2019   *Stephen Pryke*

Acknowledgements

This book has benefitted from the authors' exposure to some really talented supply chain management professionals in industry. The subject has been kept alive and vibrant by the activities of some very talented PhD students at UCL and I am proud to have been working with these capable young individuals, many of whom have contributed chapters to this book.

I would like to thank each of the chapter authors for their hard work and determination in producing chapters of the very highest quality. As always, I want to acknowledge the efforts of Karen Rubin, who encouraged me to complete the book through ill-health and supported me in editing the book, as well as organising some of the graphics.

I want to acknowledge the support of the Bartlett School of Construction and Project Management, which effectively sponsored this book.

Lastly, thank you dear reader for taking an interest in our contribution to supply chain management knowledge. We hope that you enjoy it!

1

Introduction

Stephen Pryke

The aim of the book is to present evidence in support of some innovative supply chain management theory and to present case study findings that might improve the practice of supply chain management in construction. In Part A of the book, the intention is to draw on a diverse range of theoretical perspectives on supply chain management and to demonstrate application for the industry. The chapters in Part B set out to apply supply chain management principles to practice and to present some applied and conceptual findings. It is not the intention in this book to explain first principles of supply chain management because the previous book in this series – *Construction Supply Chain Management: Concepts and Case Studies* (Pryke 2009) – deals with these principles in some detail. However, a very brief explanation follows.

Our increasingly technically (not to say, organisationally) complex projects require ever higher levels of very specialised knowledge and expertise to deliver. The Tier 1 or main contractor may respond to this by investing in training and continually developing new expertise. This is a model preferred by some of the very large mainland European contractors – Bouygues and Vinci, for example. However, the UK approach tends to respond to this need for increasing levels of knowledge and expertise by outsourcing. Therefore, either because the client requires subcontracting to a particular specialist subcontractor, or the Tier 1 contractor deems it necessary to manage the risk of some specialist work by appointing a specialist Tier 2 subcontractor, a large proportion of the work contracted to Tier 1 contractors (perhaps 70% or more), is subcontracted. Although many bemoan this as *fragmentation*, it is argued here that this is inevitable in any area of work where very high levels of specialist expertise are required. In some cases, the fragmentation is actually desirable. The reason for managing the supply chain is to harness the benefit of long-term collaborative relationships and through these relationships create better value for clients and end-users. Firms entering the markets are acutely aware of the way in which their supply chain relationships can harness increased value and improve competitiveness for the supply chain. Unfortunately, all too often, firms *not* entering new markets become complacent about supply chain matters – the pursuit of continuously improving value and lowering costs to improve services and profitability are simply not a focus. Supply chain management is therefore the mark of a firm behaving strategically rather than reactively in the market place.

The book is divided into two distinct but overlapping parts. Part A comprises chapters dealing with the theoretical aspects of supply chain management and its application to construction. It is an opportunity to introduce some innovative ideas to help us understand supply chain management and how it might be increasingly exploited. All chapters have clearly identifiable applications to practice. The second half of the book (Part B), in six chapters, provides a series of discussions that begin with a case study scenario and then explore the implications for supply chain management. It is hoped that in this way there is material for all interests: setting agendas for industry; exploring and challenging published theory; providing some innovative frameworks for discussion; stimulating further research and case study work by students and practitioners alike.

1.1   Overview: Part A

1.1.1   IT, Digital, and BIM

The book opens with Eleni Papadonikolaki's chapter – *The Digital Supply Chain: Mobilising Supply Chain Management Philosophy to Reconceptualise Digital Technologies and Building Information Modelling (BIM)*. Papadonikolaki explores the way in which the transition from analogue to digital processes, or digitisation, affects construction. She reflects on the greatly improved computing infrastructure sitting alongside improving mobile devices and the fact that manufacturing has perhaps made an earlier start with leveraging what she describes as the 'digital thread'. Ironically, construction, with its context of complex, ill-defined and evolving information flows coupled with high levels of risk flowing from the business environment, has a greater need for rapid implementation of digitisation. Papadonikolaki looks at the advancements in other industries that might, in the future, have applications for construction and suggests that construction focuses too heavily on *projects* and places insufficient emphasis on *products*.

The synergies between supply chain management and digitisation are explored and she stresses the importance of using supply chain management strategically over the long term and within the context of real collaboration rather than simply adopting some of the terminology and classifying existing practices as supply chain management. BIM is discussed at some length, along with the importance of establishing appropriate standards in a digital built environment, through Industry Foundation Classes. Papadonikolaki reflects on the tension between 'top-down' and 'bottom-up' strategies in relation to the uptake of digital technologies and in particular BIM. She ponders how small and medium enterprises fare alongside the small number of very large organisations that the construction industry comprises. Finally, the chapter deals with the way in which digital technology might usefully address processual, technical and relationship complexities.

1.1.2   Self-Organising Networks in Supply Chains

Papadonikolaki's discussion on digital technologies in construction links neatly to Huda Almadhoob's development of one particular aspect of technology and theory associated with social network analysis in the context of construction supply management. Almadhoob's chapter: *At the Interface*: *When Social Network Analysis and Supply*

Chain Management Meet, builds some theory linking supply chains and networks based upon a large infrastructure case study stretching over a period of four years. Almadhoob explores the proposition that supply networks in construction are complex adaptative systems. She argues that a network perspective can be linked to complexity theory and refers to the nonlinear, self-organising, and emergent properties that we can see in operation. The chapter moves on to look at the key elements of supply networks in construction. Attention is turned to the idea that an organisation has a formal visible form based upon contractual relationships, formal rules and relationships, and dealing with contract and remedies. This lies alongside the possibly larger informal organisation that is essentially self-organising and, most fundamentally, is the part of the organisation that is *delivering the project.*

There is a little descriptive material dealing with social network analysis and a rationale for the use of network analysis in the context of construction supply chains. Part of this rationale rests upon the lack of rigour in some of the analysis that we have tried to perform in the construction sector. There has been a tendency towards descriptive rather than more structured analysis and communication of lessons learned from supply chains and projects. The importance of dealing with interdependence when studying construction supply chains is emphasised. The author finally reflects upon the need for the management of supply chain relationships to move away from a focus upon controlling and towards an emphasis on complexity and relationships. Almadhoob ends by noting that supply chain networks co-evolve in a nonlinear and dynamic manner where autonomy of action is promoted.

1.1.3 Green Issues

Chapter 4 is written by Niamh Murtagh and Sulafa Badi and the subject matter moves to sustainability. *Green Supply Chain Management in Construction: A Systematic Review* makes the point that environmental sustainability in construction is a pressing concern for the industry worldwide. The authors propose green supply chain management as a potential solution to the problems. The chapter provides a systematic literature review of peer - reviewed papers up to mid-2017. It deals with the status of the research domain and covers research methods and the tools and techniques employed. The chapter then offers some practical recommendations based upon the findings of the literature review. Murtagh and Badi provide a comprehensive definition of green supply chain management as applied to construction. Some future areas of research are suggested, including end-to-end and subdomain specific studies, and perhaps comparative research by project type. Pragmatic tools for decision making are discussed. This chapter includes discussion about:

- *Green design.* Limiting the impact on the natural world through consideration of material, production and resource consumption in use
- *Green manufacturing.* Dealing with the processes behind the *production of components* for construction, which can be overlooked when considering sustainability in construction
- *Green transportation.* At present there is very little consideration of the sustainably of transportation in relation to construction. This affects Tier 1, but the majority of the emissions occur in Tier 2 and lower tiers

- *Waste management.* Others will also look at this and at Lean Construction elsewhere in the book.
- *Green operation.* This is one of the areas that has recently received attention
- *End-of-life management.* Reverse logistics and recycling of building components.

Continuing the theme of looking to the future in the first two chapters, Murtagh and Badi finish with a section considering the future direction of research in the area of green supply chain management. Among the areas that they put forward are:

- Further work on complex decision making in practice
- A need to find more pragmatic methods to bridge the gap between academics and practitioners
- Looking at 'hot-spots' in supply chains – those links that have high impact in sustainability terms
- A better articulated model for innovation in sustainability through contract conditions
- A better understanding of the unique nature of the construction industry in green terms
- A look at the green agenda and the context of each of the myriad of interdependent roles in the supply chain
- The role of (and perhaps effectiveness) of industry bodies.

1.1.4   Demand Chains and Supply Chains

Simon Addyman's chapter *Connecting the 'Demand Chain' with the 'Supply Chain': (Re)creating Organisational Routines in Life Cycle Transitions* follows, arguing that although there is much reference to supply chains, there also exists a demand chain impacting upon the supply chain which is worthy of some consideration. Although this chapter is located in the 'theory building' part of the book, the chapter commences with the description of a major rail sector project within which Addyman played a senior role. A description of the physical and organisational attributes of the case study project provides a context for the theory which is derived later in the chapter. The chapter is based upon research carried out by a senior manager while employed on the project; uniquely the chapter benefits from a circular flow of: research findings; discussion with practitioners; theory development; publication of findings; and discussion with other practitioners and academics. Addyman examines the ongoing interdependent actions that comprise the activities of both the supply chain and their corresponding demand chains and the need for these interdependent relationships to be recreated at the commencement of each project undertaken (not necessary where supply chains or demand chains are 'standing' or identical over time). The challenges of 'temporary organising' are identified and how the industry searches for ways to manage the paradox through both long-term and short-term relationships are reviewed. This detailed study of a major, long-term rail supply chain concluded that there are five mains stages in dealing with the transition from procured project to active supply chain and these are:

Realising. Drawing attention and defining emergent problems and tasks in order to connect the routines of project design and delivery

Informing and assuming. Associated with the search for information, and the limited time available for information search

Turning and preparing. A process of formal validation, of defining governance

Validating. Gaining approval of the demand chain for the actions proposed by the supply chain

Enacting. Carrying out the adapted practices derived from the points above; and relating to time and space.

Addyman's chapter is located within Part A of the book, but delivers a descriptive case study, based upon a live construction supply chain with which he was associated.

1.1.5 Lean

Lauri Koskela's chapter with Ruben Vrijhoef and Rafaella Broft looks at *Construction Supply Chain Management Through a Lean Lens*. In the Netherlands, where two of the chapter authors are based, academics and practitioners tend to regard the subject of supply chain management as falling within the area of 'lean'. This chapter sets out to distinguish the subject of lean from supply chain management and to explain the relationship between the two topics. The roots of both supply chain management and lean are used to illustrate how these two topics relate to each other. The authors aim to:

- Look at theoretical, epistemological, and ontological aspects of both lean and supply chain management
- Establish the key characteristics of supply chain management
- Unpack the key theoretical and applied aspects of lean and supply chain management
- Identify the characteristics of construction and conceptualise construction as a supply chain within which lean can be applied.

The chapter starts with an exploration of the concept of *production management* – the idea that production (in our case typically of a building) comprises the transformation of inputs into outputs. *Decomposition* enables the task (building) to be packaged into sensible parcels of work and then delegated to an appropriate group or individual for completion. The authors make the link between production of the end product and the production of waste, discussing the 'just-in-time' and Toyota Production System along the way. There is a link to the previous chapter by Addyman through the discussion of production as a value generation function involving the engagement between the customers and the supply chain. Koskela, Vrijhoef, and Broft bring the chapter to a close with a discussion on four main perspectives in relation to supply chain management and lean: the production perspective, the economic and organisational perspectives, and the social perspectives. Finally, a case is made that lean has a broader base than supply chain management and that this broader base is beneficial in terms of its application to construction.

1.1.6 Power Structures and Systemic Risk

Towards the end of Part A of the book, Andrew Edkins presents *Supply Chain Management and Risk Set in Changing Times: Old Wine in New Bottles?* This first of two chapters looking at risk in the supply chain starts with the premise that achieving client satisfaction must essentially involve more than simply sourcing and procuring products and services. Edkins contends that any discussion about risk needs to broaden the definition away from negative uncertainties. There is also risk associated with a range of potential benefits. Edkins discusses the highly publicised collapse of Carillion plc in the UK in 2018. He argues that fragmentation, albeit on a very grand scale, had led to a cascade effect throughout multiple UK construction supply chains as a result of the nexus of contracts that major projects comprise and the associated interorganisational supply chain network relationships. Edkins makes the point that for every interorganisational relationship there is a power dynamic which depends on the relative prominence of the individual actors (firms) both within an individual supply chain and the market for a particular product or service within the market place.

Finally, Edkins concludes that supply chain actors need to recognise the systemic risk associated with power structures. These solution systems, comprising diverse organisational entities which are driven by different forces and operate in quite different ways, counter the difficulties and obstacles by combining careful selection of supply chain members with appropriate governance through contract and the management of relationships.

1.1.7 Decision-Making Maturity

The penultimate chapter in Part A is written by Alex Arthur: *Linkages, Networks and Interactions – Exploring the Context for Risk Decision Making in Construction Supply Chains*. This chapter begins with a look at the evolution of the construction industry to give some context for the discussion that follows. The organisational evolution that accompanied this is related to the relational aspects of supply chain management. Arthur considers the socially constructed nature of risk and its assessment and management: objectively and subjectively biased risk; socially mediated and transformed risk. The important relationship between uncertainty and risk is examined. This developing framework is supplemented by a review of the systems within which construction risk management operates. Risk identification analysis and response systems are outlined and appraised. The findings of this chapter draw upon a five-year research programme carried out with industry. The author finds that risk management systems in construction are ideologically biased even though the decision-making context is highly rational.

Arthur argues that, although there is space for both rationality and intuition in the risk management decision-making processes, there is a strong bias in practice for intuitive, nonevidence based, irrational assessment. The point of course, is that supply chain management operates within a context of risk assessment and response, and this affects the relationships between the supply chain actors. Where the client creates a context with high levels of risk being transferred, legitimately or otherwise, from the client to the supply chain actors, supply chain actors set about transferring some or all of that risk to other supply chain actors. This fundamentally affects the way in which the supply chain operates. Indeed, where a client sets out to internalise risk, rather than dispose of it, the project as a whole is subject to lower overall levels of risk and lower levels of cost in transferring, managing, and protecting against such risk. Elsewhere in the book we mention London Heathrow Terminal 5 and the way in which the client understood their role as supply chain manager. They understood that the client supply chain manager is uniquely placed to deal with risk across the whole supply chain, rather than have it as a matter for adversarial behaviour and cost premiums.

1.1.8 Culture

Part A of the book closes with a substantial chapter by Richard Fellows and Anita Liu entitled *Culture in Supply Chains*. Fellows and Liu start with a reference to the purpose of supply chains – to create value for the client and end-users. Value is created through the relationships that are established in the supply chain. Their point is essentially that success in supply chain management hinges around the ability to create and maintain a culture of value creation in the supply chain. The chapter defines culture and the main dimensions of culture. The authors discuss how culture in supply chains might modify the behaviour of the supply chain actors; the effects on value and values within the supply chain. There is some discussion on the subject of sense-making and this is linked to motivated reasoning – the way in which actors might take new data and information into account in developing their opinions and actions. The chapter concludes by recognising that individuals bring culture to the supply chain, through their background, education and experience, and their personal background. They enter an arena where the project perhaps has a culture of its own, based on history and context. Within these contexts, for the supply chain to be managed there needs to be a sense of identifying, harnessing, and perhaps modifying the culture within the supply chain. The supply chain needs to focus upon long-term value creation, even if the influential cultures from firms and individuals do not necessarily place this as a priority.

Part A comprises chapters that are primarily seeking to influence and attract the reader's attention through the development initially of theoretical principles, followed by discussion of application. Each chapter is evidentially based but does not necessarily present the evidence in case study form.

Part B consists of chapters that are essentially presenting case study matter to support their discussion and recommendations.

1.2   Overview: Part B

1.2.1   Lessons from Megaprojects

Juliano Denicol opens the second half of the book with a chapter based on case study research carried out over several years at University College London. In *Managing Megaproject Supply Chains: Life after Heathrow Terminal 5*, Denicol looks at the specific case of megaprojects (contract sum in excess of $1 billion) which followed from lessons learned from Heathrow Terminal 5. The chapter draws upon four megaprojects. These projects comprise in summary:

London Olympics 2012. A programme of projects delivering the facilities required to host the 2012 Olympics in London

Crossrail (the Elizabeth Line). A 118 km railway line in London and the counties of Berkshire, Buckinghamshire and Essex, England

Thames Tideway Tunnel. A 25 km tunnel running under the River Thames through central London, which captures, stores, and disposes of the sewage and rainwater discharges that currently overflow into the river at various points along its length

High Speed Two (HS2). At the time of writing this was the largest infrastructure project in Europe. A new high speed overground railway connecting London to Birmingham in excess of 200 km long (with subsequent phases to follow).

Denicol was able to conclude after his study of these four massive projects that:

- Frequency and consistency of demand from clients to construction supply chains are significant factors influencing the ability to exploit good supply chain management practice

- Although we see megaproject clients recognising the need to manage supply chains and employing organisations to act as supply chain management on their behalf, on much of the remainder of the workload of the construction industry, clients do not necessarily recognise the need to allocate resources to supply chain management
- The construction sector is resistant to change in a way that much newer industries are not
- The idea of having metrics and indicators to support effective supply chain management is relatively immature in UK construction. The industry needs to develop metrics that can be applied to all construction supply chains to provide a comparative assessment of the effectiveness of supply chains
- There is a need to provide structure and incentives to support supply chain management activities.

The theme of supply chain management in infrastructure projects is continued in the following chapter by Mills et al.

1.2.2   Collaboration and Integration

Anglian Water @one Alliance: A New Approach to Supply Chain Management is presented by Grant Mills, Dale Evans and Chris Candlish. In this chapter, they contend that within the construction management literature, supply chain management is frequently seen as a function performed by the client or contractor. They suggest a relational approach to supply chain management that engages and aligns the business goals of partners is more important. This approach should set both long- and short-term expectations for delivery, collaboratively manage supply chain risk, measure success, and provide both incentives and penalties for nondelivery when viewed against the client's business outcomes. The chapter draws on the experience and reactions from clients, contractors and suppliers to a long-established delivery model to provide an alternative strategic perspective on alliance supply chain management.

The case study provides evidence of the implementation of supply chain integration such as early supplier engagement, collaborative planning, and production. Mills et al. look at the evaluation of alliance supply chain management value and rated it as very important. In the Anglian Water case study, a project supplier work cluster and involvement map showed how alliance supply chain management was applied to allow all participants to work together to plan, design, and deliver integrated solutions. Mills et al. have put forward a new approach to alliance supply chain management developed while working alongside an Advanced Supply Chain Management Group at Anglian Water. The chapter promotes the use of alliancing as a vehicle for effective supply chain management.

1.2.3   Lesson Learned and Findings from Tier 1 Contractors

Understanding Supply Chain Management from a Main Contractor's Perspective by Emmanuel Manu and Andrew Knight begins with the proposition that there has been an emphasis on supply chain initiatives introduced by large construction clients and insufficient interest in the management of supply chains by Tier 1 contractors.

The chapter draws on evidence from a study undertaken with a large UK main contractor. Manu and Knight report on practices such as:

- The audit of supply chain actors' performance measurement across projects
- The use of collaborative IT systems
- Engagement in continuous performance measurement
- Motivation and incentivisation of the supply chain
- Prioritisation and promotion of long-term supply chain relationships with best performers.

The findings of the research were that where a main contractor achieves collaborative long-term relationships with a select group of Tier 2 contractors that can enhance competitive advantage, the benefits include:

- Support for the Tier 2 contractor in establishing higher levels of certainty about future workload
- The Tier 2 contractors provided support for Tier 1 during the bid stage for projects through pricing, specification development, and design contributions and planning
- The supply chain management practices seemed to promote integration of the supply chain and achieve mutual competitive advantage
- Visible supply chain management champions within individual supply chain firms were valuable in promoting the values of supply chain management across the firm and the supply chain.

The chapter ends with a plea for more extensive use of ethical practices in construction supply chains – supply chain management needs openness and trust to work effectively and opportunistic or exploitative behaviours will stifle the development of collaborative supply chain relationships.

1.2.4   Lean Practices in The Netherlands

Manu and Knight's chapter links to the chapter by Rafaella Broft, *Lean Supply Chain Management in Construction: Implementation at the Lower Tiers of the Construction Supply Chain*. Broft is interested in Tier 1 contractors and their role in achieving outstanding performance through management of the supply chain, with particular emphasis in this chapter on lean principles. She tackles the evolution of lean through its origins in the Japanese automotive sector and in particular the Toyota Production System. The project focus is seen as overemphasised and a limiting factor in the development and evolution of supply chain management in construction and, perhaps as a result, it is argued that supply chain management in construction lacks maturity. There is some discussion about the existence of fragmentation in construction. Interviews with practitioners revealed a lack of openness by Tier 1 contractors in their dealings with the lower tiers.

Broft turns her attention to making a case for lean principles in supply chain management. She argues that lean challenges the supply chain to fundamentally rethink value from the customers perspective. The value stream is identified as supporting and underpinning the entire processes of product or service delivery. Perhaps this is emphasising the point that value is created in the most unlikely places. A project mindset tends to focus upon the upper tiers of the supply chain, whereas much highly specialised knowledge and expertise is located lower in the chain. Leveraging the value located lower in the supply chain can very often provide opportunities for increased value for the clients and end-users. Finally, Broft questions whether construction is really much different from manufacturing; perhaps she is distinguishing 'manufacturing' (which increasingly incorporates manufacture to order) and 'repetitive manufacturing' where there is no element of manufacture to order.

1.2.5   Knowledge Transfer in Supply Chains

Towards the end of Part B, Hedley Smyth and Meri Duryan look at *Knowledge Transfer in Supply Chains.* No discussion about supply chain management would be complete without a case study dealing with knowledge sharing and management. The chapter starts with a reference back to the important but often forgotten work of The Tavistock Institute (1966) on the interdependence and uncertainty that are so much a part of construction supply chains. Although the original research by The Tavistock Institute (1966) presented interdependence as one of the challenges that construction faces, Smyth and Duryan suggest that interdependence might in fact be regarded as a benefit to construction; perhaps they are suggesting that interdependence creates dense linkages between supply chain actors and those linkages are the means of knowledge capture and management?

The case study material was drawn from an extensive research project conducted while one of the authors was working with one of the UKs largest rail infrastructure providers. In summary, the findings are:

- The lessons learned process is not routinely undertaken by organisations. Organisations and their supply chains rely on individuals to take the initiative or simply to retain useful lessons learned for application in subsequent projects
- IT platforms for knowledge management systems are not ubiquitous and a culture of learning is needed
- Even where some governance is present associated with knowledge capture and transfer, there was quite frequently a disconnect between knowledge captured and its transfer usefully to other supply chains
- Most fundamentally in the context of this book on supply chain management, Smyth and Duryan found that where knowledge management was successfully implemented it was done so with a *project* focus, rather than a *supply chain* focus. This inhibits the value of knowledge management and perhaps fosters an emphasis on *operational and technical issues* rather than *long-term strategic issues*
- A number of barriers to effective knowledge management in the supply chain were detected, both internal and external.

Smyth and Duryan develop a number of ideas about how construction supply chains can most usefully exploit knowledge management. There is also a discussion about which actor should initiate and take ownership of knowledge management systems.

1.2.6   The Role of Trust in Managing Supply Chains

Finally, Part B is brought to a close with Jing Xu's contribution on *Understanding Trust in Construction Supply Chain Relationships.* Supply chain relationships are procured through contract but essentially function through collaboration; and collaboration rests upon the establishment and maintenance of trust. Xu argues that trust sustains institutional, social, and organisational life. The case study comprises a major UK construction company operating in the sector of groundworks and piling. The project with which the case study was associated had been completed on site at the time of the study. The project had involved disputes resulting in a claim of loss and expense being directed at the client by the piling contractors. The claim amounted to 25% of the contract sum for the project. The outcome for the project was regarded by all concerned as a failure, despite having used two-stage tendering to encourage early contractor involvement and the existence of a strategy for supply chain integration. The findings of the case study-based research were, in summary, as follows:

- Trust in supply chains is emergent but needs to be fostered and promoted
- Self-reinforcing cycles of trust and collaboration are useful in relation to establishing efficiency, equity, reciprocity, and 'bounded solidarity'
- Through the process of creating a trusting environment, supply chain actors created higher levels of security, were more informative, flexible, and cohesive
- Better experiences in relation to trust tended to enable better performance overall.

1.3 Summary

The aim of this introductory chapter was to present an overview of the book's contents, to rationalise the content chosen and to highlight some of the areas of integration between the subjects covered. I have set out in brief some of the key arguments made in each of the chapters.

The focus of the book is not on explaining in detail the first principles of supply chain management, as this material is covered in the first book of the series (Pryke 2009), but instead to build upon the theoretical material in the first book and to present the latest case study findings aimed at improving the practice of supply chain management in construction. Part A of the book includes those chapters which are predominantly, but not exclusively, theoretical and Part B includes chapters which are primarily, but again not exclusively, case-study based.

References

Pryke, S. (ed.) (2009). *Construction Supply Chain Management: Concepts and Case Studies*, vol. 3. Wiley.

Tavistock Institute of Human Relations (1966). *Interdependence and Uncertainty: A Study of the Building Industry: Digest of a Report from the Tavistock Institute to the Building Industry Communication [s] Research Project*. Tavistock.

Part I

Chapters that Principally, but not Exclusively, Deal with Concepts and the Development of Theory

2

The Digital Supply Chain: Mobilising Supply Chain Management Philosophy to Reconceptualise Digital Technologies and Building Information Modelling (BIM)

Eleni Papadonikolaki

2.1 Introduction

In construction, the transition from analogue to digital processes, that is *digitisation*, is increasingly gaining traction and affecting the whole sector. The construction or architecture, engineering and construction (AEC) industry is a very diverse sector consisting of numerous multidisciplinary construction firms and professional service providers that are rarely integrated with each other (Vrijhoef 2011). Their various specialisations are diverse and range from managing real estate portfolios to manufacturing lighting fixtures. Typically, the AEC is geographically dispersed and its supply chain members are based in numerous different locations, working together to achieve common goals but predominantly with a project focus (Winch 2002). Currently, the discussion is shifting from the AEC to the built environment, an environment shaped by the involvement of more stakeholders than the AEC professionals, including policy makers, asset owners, users and so forth (Papadonikolaki 2016), although the two terms are often used interchangeably.

In the past, the geographically distributed character of the AEC was facilitated by sophisticated applications of information and communications technology (ICT). Until recently, a domain-specific version of computer-aided design (CAD), computer aided architectural design (CAAD) software, was the standard tool of computerisation in architecture and its use not only portrayed the contemporary architectural process but also increased the performance of AEC significantly by supporting automated, semiautomatic, and standardised processes (Aouad et al. 2012). More than a decade ago, the term building information modelling (BIM) was introduced, which is currently at the forefront of digitisation in the built environment. BIM is a technology-driven approach that includes integrated software solutions for AEC. It is an integrative technology with 'parametric intelligence' for the AEC (Eastman et al. 2008). As it generates, collects, represents, and manages building project information, it supports various information flows across the supply chain. Because BIM operates as a digital platform (Morgan and Papadonikolaki 2018) and allows other technologies to connect with it, it is a key technology for the digital supply chain.

Successful Construction Supply Chain Management: Concepts and Case Studies, Second Edition.
Edited by Stephen Pryke.
© 2020 John Wiley & Sons Ltd. Published 2020 by John Wiley & Sons Ltd.

Advancements in the implementation of digital technology in the built environment, through better computer infrastructure, associated with increasing computing power, mobile devices, for example hand-held devices or headsets, and various pervasive technologies, have taken the world closer to 'Industry 4.0'. The fourth industrial revolution or industry 4.0 is an industry supported by digital technologies and automation that leverages the power of cyber-physical systems, the internet of things (IoT), cloud, and cognitive computing (Lasi et al. 2014). Other sectors, for example manufacturing, are doing better than construction in leveraging the 'digital thread' – a connected flow of data from design to production. Etymologically, the term 'digital' refers to using or storing data or information and it has come to represent the key enabler of industry 4.0, although it is a relatively outdated concept, used to signify the move from the analogue era. Nevertheless, as information is meaning-laden data (Ackoff 1989) it is of key importance to industry 4.0. To this end, various digital technologies shape the *digitisation* in construction, which in turn allow for *digitalisation* of business and project processes, towards the eventual *digital transformation* of the industry. *Digitisation* refers to the transfer of information from analogue to digital, whereas, *digitalisation* refers to the process of changing business to digital business (Gartner 2016; Ross 2017). According to the Institute of Civil Engineers (ICE 2017, p. 2) digital transformation is:

> the application of digital technologies to all aspects of human life. [In this report] it applies to the wholesale changes in how our industry designs, builds, operates, maintains and decommissions assets. It also refers to the transformation of how we value data, and the impacts upon processes and systems, and ultimately decision making.

The manufacturing sector, which is more homogenous and simplistic than the AEC, benefitted immensely from advancements in ICT and operations research (OR) that led to the rationalisation of quality, logistics, business organisation, and partnerships. Supply chain research emerged after this period. Initially, a supply chain was represented by a set of flows: a downstream flow of material, an upstream flow of transactions, and a bidirectional flow of information (Christopher 1992). Supply chain management emerged as a philosophy that theorises and suggests activities for the regulation of these flows. Later, a supply chain was considered actually to be a network and not a linear chain per se (Pryke 2009), given that the multiple organisations that form this network generate different and multiple information streams simultaneously (Christopher 2005). Thus, a supply chain could be considered as a 'supply–demand network', or a complex and distributed network of organisations (Christopher 2011).

The intention here is to present an alternative view of digital technologies and digitalisation in the built environment and how digital technologies coupled with supply chain management philosophy can tackle the innate complexities of the AEC industry. On the one hand, supply chain management is approached as an integrative concept that addresses processual and relational issues in construction. On the other hand, digital technologies and particularly BIM are seen as information-driven means for structuring and regulating the supply chain information flows. The chapter follows a problematisation methodology (Alvesson and Sandberg 2011) using a narrative literature review to 'provide interpretation and critique' (Greenhalgh et al. 2018) and deepen the knowledge and understanding of the need for a *digital supply chain*.

2.2 The Nature of Construction

2.2.1 Addressing Existing Complexity and Fragmentation in Construction

Whereas the intense digitisation imposes an emerging complexity in construction, the sector is already considered an extremely complex industry from various perspectives and for different reasons from various scholars (Azambuja and O'Brien 2009; London 2009; Vrijhoef and London 2009). Complexity relates to the existence of numerous or infinite components and interrelations that comprise a system (Mitchell 2009). Winch (1998) identified high interconnectedness, unpredictability, and high user involvement in the innovation process as traits of a complex product system, such as the AEC industry. Complexity is multifaceted and relates to various aspects of the industry. Complexity in the AEC mainly concerns:

1. technical product complexity, due to the inherently complex design and construction processes (Gidado 1996; Dulaimi et al. 2002),
2. operational or processual complexity, from rigidities that develop along the various operations (Gidado 1996) and
3. organisational or relational complexity, which relates to the vast amount of involved multidisciplinary organisations (Nam and Tatum 1992).

Christopher (2011) further added network, range, customer, supplier, and information complexities to the above. AEC has additionally been susceptible to contextual complexities, due to external uncertainties, for example from the natural environment (Winch 2002). Simultaneously, given that other industries are also inherently complex but well-performing, complexity might not be the sole reason for the underperforming and weak image of the AEC industry (Papadonikolaki 2016).

Fragmentation is another metric, complementary to complexity, which refers not only to the number of actors but also to corrupted, broken, loose, nonexistent interrelations or linkages among them. The AEC industry is seen as fragmented into its constituent parts, for example numerous small and medium enterprises (SMEs) (London 2009) and this fragmentation creates additional complexity. Fragmentation relates mostly to the structure of the industry, and particularly to the high number of work specialisations involved, the low interdependency among them, and the local character of most projects (Azambuja and O'Brien 2009). Howard et al. (1989) described this phenomenon as a lack of 'horizontal integration' among the various project specialists or disintegration of organisational structures. De Bruijn and Heuvelhof (2008) noted that the rising professionalism automatically means more fragmentation for the organisations and their interorganisational relationships. Dubois and Gadde (2002) pointed out that the single-project focus of the AEC industry is responsible for this complex image that hinders organisational learning and innovation.

2.2.2 Advancements from Other Industries Applicable to Construction

It may be valuable to examine how other industries have historically dealt with similar challenges to manage technical, processual, and relational complexities. Dubois and Gadde (2002) highlighted that 'management techniques that improve performance in other industries are not readily transferable to this *(construction)* context'. A glimpse at the impact of specific advancements from manufacturing, retail, commerce, aerospace, and automotive industries might shed light on how such advancements deal with complexity and fragmentation in construction. Some of these advancements were precursors or present inherent affinities with supply chain management and digitisation.

From a processual aspect, the AEC industry is usually classified by reference to efficiency and effectiveness. Efficiency is a combination of reduced costs and smaller cycle times. In operations management (OM), the cycle time of a process consists of capacity, utilisation, and variability (Cachon and Terwiesch 2009). This variability is of major importance, being 'the primary factor that influences the parametrisation of a process', according to Reiner (2005, p. 433). OM research was developed after the 1950s to cover the need for scientific approaches in business research. The development of OM coincided with the positivistic philosophy of science, according to which 'the phenomenon under study can be isolated from the context' (Meredith et al. 1989, p. 306). Currently, the multiple operations involved in construction could be controlled by quantitative analysis, for example, scheduling, inventory analysis, and cost estimation, straight from the toolbox of OM research.

Other advancements for controlling the processual complexities in manufacturing that additionally apply to AEC are 'lean' and 'agile' methods. Lean construction refers to the 'adaptation of the underlying concepts and principles of the Toyota Production System to construction'(Sacks et al. 2009a). The fundamental focus of lean construction is to reduce waste, increase customer value, and attain continuous improvement (Koskela 1992). The lean approach requires, again, low variability of product to perform as well as in manufacturing, but it has good affinity with conceptualising collaborative work structures as a production system design (Ballard et al. 2001). Agility refers to 'using market knowledge and a virtual corporation to exploit profitable opportunities in a volatile market place' (Naylor et al. 1999, p. 778) to develop flexibility and responsiveness in the construction process.

The advent of ICT in delivery systems of the construction industry echoes the innovative systems used in the procurement of goods in the retail and commerce sectors. With the support of ICT, and particularly of data and object modelling techniques, electronic business (e-business) processes are employed by the organisations not only internally but also externally in collaboration with their partners, so as to regulate both technical and processual complexities. Among the improvements from e-procurement are savings in the acquisition process, alignment of all the suppliers, faster, continuous, and easier acquisition of goods, the integration with enterprise resource planning (ERP)-systems and elimination of rework, smart-tagging such as through radiofrequency identification (RFID) and near field communication (NFC), and accurate inventory predictions.

Another direction for the AEC industry is a focus on products rather than projects, similar to the aerospace and automobile industries. This relates to industrialisation by mechanisation and automation and could lead to further integrated products delivering efficiency and effectiveness; integration of the expertise of various disciplines in one confined and consistent 'building product'. These concepts have been successfully adopted in the AEC industry through platform strategies for product modularisation (usually via prefabricated components), interchangeability, standardisation, and distributed production systems (Halman et al. 2008). These approaches imply a decentralised rather than centralised design and engineering control as in the aerospace and automotive

Table 2.1 Complexities in the AEC and the extent to which these are tackled by advancements in other industries.

	Processual complexity	Technical (or product-related) complexity	Relational complexity and fragmentation
Operations management	x	—	—
Lean and agile approaches	x	—	x
e-Business	x	—	x
Modular design	x	x	—

Source: Adapted from Papadonikolaki (2016).

industries (Vrijhoef 2011) but do not necessarily address the structural fragmentation in AEC into numerous SMEs.

These concepts and practices in manufacturing, retail, commerce, aerospace, and automotive industries, might prove potentially useful to address the complexities of the AEC industry. Table 2.1 illustrates complexity and fragmentation areas that these approaches could improve. However, these approaches do not address the inherent fragmentation and complexity of construction, because they focus more on processual and technical complexities, rather than relational.

Based on the above, this chapter aims to reveal the extent to which the combination of supply chain management philosophy and digital technologies address processual, technical and relational complexities, and fragmentation in AEC.

2.2.3   Potential Synergies Between Supply Chain Management and Digitisation

Given that supply chain management philosophy is an ill-defined concept (London and Kenley 2001; Chen and Paulraj 2004), the concept of 'partnership' can facilitate the understanding of supply chain management in practice. Supply chain partnerships are constellations of firms, consisting of 'dyadic' partnering relations across multiple tiers, which attempt to manage the supply chain by adopting practices that focus, among others, on enhanced collaboration (Lambert et al. 1998). Digital technologies could be seen as long-term and strategic resources for the supply chain partnership and supply chain management. Such supply chain partnerships would not be disbanded and reinvented in each project, as the virtual corporation described by Davidow and Malone (1992) is. Instead, ideally the partners would carry experience, knowledge, and communication channels from project to project, without necessarily constructing a repetitive project. After all, supply chain collaboration is meaningful when intentional, strategic, and/or long-term, but less meaningful when project-based and/or short-term (Mentzer et al. 2001).

Digital technologies require not only digital literacy but also a managerial shift as BIM is accompanied by 'workflow transition difficulties' (Chien et al. 2014). Supply chain management and digital technologies are linked only conceptually and not substantially. Presently, there exist reports on how digital technologies, namely BIM, could facilitate virtual team collaboration (Becerik-Gerber et al. 2012) and team integration through BIM and integrated project delivery (IPD) (London and Singh 2013). However, in all these approaches, there is no emphasis on strategic supply chain relations – the relations in long-term partnerships. Often, digital technologies are approached by a small subgroup of the project team and used in an ad-hoc manner, partially, and incompletely. This is why considering digital through the lens of supply chain management philosophy as a digital supply chain is promising way forward to address the relational challenges in the built environment.

Both supply chain management and digital technologies are integrative in nature. BIM and digitisation are systemic innovations, and thus by their very nature interorganisational (Lindgren 2016), in the same way as the supply chain management philosophy. *This chapter argues that not only are supply chain management and digital technologies innately interdependent historically, but also their combination will facilitate moving towards the industry 4.0 paradigm in construction.* The aim of this chapter is to examine whether the supply chain management philosophy and digital technologies are compatible in theory by shedding light on their shared roots, and to justify their combination in order to work towards an integrated construction industry. The main question that this chapter addresses is: *How does the evolution of the supply chain management and BIM concepts allow us to understand the digital supply chain in the context of the construction industry?*

2.3 Origins and Development of Supply Chain Thinking in AEC

2.3.1 The Emergence of Supply Chain Thinking from Operations and Logistics Research

Supply chains emerged in manufacturing as cybernetics and developing information capabilities in the 1950s provided a fertile ground for regulating and optimising the physical distribution of goods from raw materials to end products. Cybernetics was closely tied to organisational research, statistics, pattern recognition, information theory, and control theory in an effort to coordinate, regulate, and control (Ashby 1956). From rationalising the processes of distribution and (later of) production, inevitably quality was improved, and this led to more strategic considerations for the procurement of products (Porter 1985; Christopher 1992). In construction, these ideas followed a few years later, with emphasis on materials and cost management.

In reality, supply chains work more as networks, rather than linear chains, as a supply chain is a network of multiple business and relationships (Lambert et al. 1998). Christopher (2011) advocated that the word 'chain' should be replaced by 'network', given that the suppliers and the customers are accordingly connected to multiple other suppliers within this network. Accordingly, he proposed that it should be termed 'demand chain management' rather than supply chain management since the supply chain should be driven by the market and not the suppliers (Christopher 2011). The acknowledgement of the ill-defined grounds of supply chain thinking also comes from the acceptance that supply chain thinking is 'part of an eclectic and developing hybridised field' (London and Kenley 2001, p. 778), despite being under study for more than 30 years (Chen and Paulraj 2004). The body of knowledge on supply chain management includes: (i) strategic purchasing, (ii) logistics integration and (iii) supply network coordination (Chen and Paulraj 2004).

The relevance of supply chain management to logistics is clearly articulated in manufacturing. London and Kenley (2001, p. 137) hypothesised that future research on the supply chain in the construction industry would move from 'the rhetoric that it is a management tool to improve the performance of the industry', to 'discussions of advantages of different types of networks, cluster or chain'. Leuschner et al. (2013) differentiated between the operational and relational view of supply chain integration in the supply chain management research field. Supply chain thinking was already a concept in transition, drifting between processual and relational aspects, by the time it was adopted in construction. Thus, supply chain research could be considered an 'umbrella' term that greatly depends on the epistemology of the researcher, the entity under study (material, actor, or information), and emphasis on either operations or relations.

2.3.2 The Adaptation of Supply Chain Management Concepts in AEC

Lambert et al. (1998) commented that constructing a definition of supply chain management is easier than implementing it in real life. This chapter is aligned with Pryke's (2009) intention to demonstrate that supply chain management in construction is much more than a trend and could potentially contribute to added value for the client and the other stakeholders in the built environment, beyond mere financial gains. In construction, supply chain management has been mainly considered as the management of the three flows: information, material, and cash (Arbulu 2009; Vaidyanathan 2009). Other scholars further simplified these flows to material and information flows (Cutting-Decelle et al. 2009), as cash flow could be potentially seen as part of the information flow. Others extended the set of flows to include material, labour, and equipment (Cox and Ireland 2002). In construction, supply chain management has multiple applications. Vrijhoef and Koskela (2000) defined four roles for supply chain management in construction:

1. Improving the interface between site activities and the supply chain.
2. Improving the composition and expertise of the supply chain.
3. Transferring activities from the construction site to the supply chain, through industrialisation and off-site production.
4. Integrated management of both the construction site and the supply chain.

In a sense-making effort, Green et al. (2005, p. 588) concluded that in contrast to aerospace, where supply chain management is an 'imperative of global competition', in construction, supply chain management is only applied for operational improvement with 'little evidence of strategic perspective'. However, Gosling et al. (2015) performed a longitudinal study of the impact of supply chain management on performance metrics and concluded that the higher – or more strategic – the level of partnership is, the more consistent the performance indicators are. Therefore, the usefulness of supply chain management closely depends on the strategic goals of the various supply chain networks that adopt it.

The concept of integration in construction supply chains is again appropriated from manufacturing supply chain research (Dulaimi et al. 2002). Integration in the supply chain context is attributed to a 'higher level' of management and to the combination of shared information flows or joint activities and operations (Vrijhoef 2011). Supply chain integration has been a preeminent concept in supply chain management research. The terms integrated supply chain or integration have already been used to describe a mature form of supply chain management (Lockamy III and McCormack 2004), who categorised supply chain management into (i) extended, (ii) integrated, (iii) linked, (iv) defined, and (v) ad hoc, in decreasing order of maturity. Eriksson (2015) used a more descriptive standpoint to analyse supply chain integration in four dimensions: (i) strength, for example through a combination of selection and incentives parameters, (ii) scope, (iii) duration, and (iv) depth, for example how many tiers are involved in the integration.

2.4 Pragmatic Impact of Supply Chain Thinking in Construction

2.4.1 Supply Chain Thinking Schools

In supply chain management and supply chain integration in construction, both qualitative aspects (for example, strategy and collaboration) and quantitative goals (for example, performance-related or input/output (I/O) approach) are involved. These 'qualitative' and 'quantitative' notions indicate that there exist two antithetical yet complementary fields pertinent to supply chain research. These antithetical notions resonate with the schools of thought in project management: one more task-oriented and the other more relational (Andersen 2016). Söderlund (2004) argued that project management has one root in engineering and the other in social sciences. These could be further related to the shift that took place after the 1950s from the positivist to the nonpositivist paradigm. London and Kenley (2001) identified two supply chain thinking research paradigms, one focusing on the I/O (operational) and one on relations. These two approaches should be seen as complementary, rather than conflicting.

Apart from the philosophical paradigm, the institutional setting contributes to varying interpretations of construction supply chain management. Dubois and Gadde (2002, p. 629) noted that various construction researchers, depending on their particular theoretical foundations and background, 'prescribe either more competition, or more cooperation, to increase the performance of the underperforming industry as a whole'. Arguably, the concept of supply chain management might also present various nuances, depending on market characteristics.

Supply chain thinking was first transplanted from the manufacturing sector to the UK construction market in the 1990s. Initially, the Latham (1994) Report 'Constructing the Team' criticised the industry for being adversarial, ineffective, fragmented, and low value for money for the client, and proposed partnering to increase teamwork and collaboration. Later, the Egan (1998) Report 'Rethinking Construction' proposed: (i) committed leadership, (ii) client-driven construction, (iii) process and team integration, (iv) quality-driven operations and (v) people-focused construction and was very keen on the use of ICT and digital to improve performance. Scholars of supply chain thinking in the UK construction industry are quite critical regarding its value and applicability as they saw that supply chain management hinders competition and implies unilateral control from one focal firm (Cox and Ireland 2002; Briscoe and Dainty 2005; Green et al. 2005; Fernie and Tennant 2013). Criticism of supply chain management in extant literature includes the repeated association of supply chain management with lack of human agency and relational aspects (Green et al. 2005; Fernie and Thorpe 2007).

The USA has seen a partial democratisation of the supply chain management concept, applied mostly through engineering, procurement, and construction (EPC) projects. Australia has adopted supply chain thinking with a focus on interfirm relationships, in the form of alliances with a tighter contractual mechanism to share risks and rewards. In the Netherlands, the inherent 'risk aversion' of the industry (Dorée 2004) is fertile ground for integrated, informal, and long-term partnerships (Vrijhoef 2011; Papadonikolaki 2016). In Scandinavia, supply chain thinking has been adopted as a both an I/O approach and simultaneously focusing on interfirm relations, with an emphasis on the procurement aspect of construction activities.

2.4.2 Supply Chain Concepts and Varying Interpretations

Usually, supply chain management is associated with the existence of contractual (but fundamentally relational) arrangements among the involved firms, for example partnering. Fernie and Tennant (2013) in their effort to establish a definition of supply chain management, based on grounded theory, observed that often supply chain management and partnering philosophy are used interchangeably by practitioners, hoping that client-led partnering through framework agreements has no distinction from supply chain management. Other scholars refer to supply chain partnerships (Briscoe et al. 2001) and supply chain partnering. Partnering entails either project-specific or long-term goals in temporary or permanent networks of firms respectively (Dubois and Gadde 2002; Gadde and Dubois 2010). Long-term partnering is strategic and could form a prerequisite for supply chain management, given that interfirm relations might require a definition before being managed.

Typically, contracts are considered prerequisites for defining a supply chain partnership through the use of supply chain contracts or framework agreements. Therefore, supply chain framework agreements are identifiers of a supply chain partnership, which could demonstrate an explicit agreement among the supply chain actors. The term 'supply chain partnership' denotes multiple 'dyadic' relations, usually stemming from the contractor towards other actors or even multiparty contracts. These different actors may span beyond the AEC, across the built environment, and also involve owners or asset managers. Figure 2.1 illustrates schematically different types of supply chain partnerships.

2.5 Origins and Development of Digitisation in the Built Environment

2.5.1 Development of Digital Capabilities in the Built Environment

With the advent of technology, solutions from ICT started receiving traction in managing construction projects. ICT approaches offer informational, analytical, and decision-making tools for assisting the involved actors across the AEC lifecycle to manage project complexity. Typically, a project is a nexus for processing information (Winch 2005) and managing information is an inherent aspect of project management (Turner 2006). Among the first areas of ICT applied to construction were the graphical and visualisation capabilities from CAAD, which started in the 1950s and supported

(a)
Typical 'dyadic' partnering
relations between two companies

(b)
Partnering relations extending
across different tiers

(c)
'Star' partnering relations from a focal actor
across multiple partners and multiple tiers

(d)
'Network' partnering relations with actors related
through a multiparty framework (dashed lines)

Figure 2.1 Different types of supply chain partnerships: (a) dyadic, (b) across tiers, (c) multiple dyadic across tiers from a focal actor, and (d) multiparty partnerships. Source: Adapted from Papadonikolaki (2016).

visualisation, communication, and process and product modelling that coordinate different types of construction information (Aouad et al. 2012). BIM is a new type of ICT, whose modelling capabilities combine benefits from both CAAD and process modelling (Aouad et al. 2012) with built-in features of generating and managing building information, for example three-dimensional (3D) design, visualisation, automated drawings and codes generation, and quantity take-off.

Construction ICT has focused either on design or management capabilities (Forbes and Ahmed 2010). For example, past solutions that used ICT for management in AEC focused primarily on exchanging information to manage invoices, quantities, and construction crews. For communication, most organisations used extranets (Ajam et al. 2010), ERP, and online project databases for information management. Through these massive information systems, actors exchange and share various project documents, such as planning in the form of electronic document management (EDM), orders and invoices in the form of electronic data interchange (EDI), and less often building information, either as printed documents, from CAAD applications or as object-oriented models, for example BIM. However, none of these technologies have been globally accepted by all the various construction disciplines (Samuelson and Björk 2013; Demian and Walters 2014; Samuelson and Björk 2014).

BIM offers capabilities for combining both design and management capabilities and thus is inclusive and intuitive. It not only offers information artefacts, such as CAAD and quantity take-off, but also contributes to information management, as the information can be shared across the supply chain using common standards on online platforms, called common data environment (CDE). Because of these features, the various artefacts of BIM make it more accessible as a technology across the AEC supply chain and all built environment stakeholders. At the same time, BIM could be described as a digital platform, because it allows other technologies to connect with it, for example virtual reality, augmented reality and IoT. These digital capabilities of BIM put it at the centre of digital transformation in the built environment, as it has applicability and linkages across the whole lifecycle.

2.5.2 From Building Product Models to Building Information Modelling (BIM)

Contrary to popular belief, BIM is not a newly found technological innovation, but naturally evolved from long-standing efforts led by industry consortia to structure building information. Thinking in products rather than processes was a paradigm shift for the AEC. As with the efforts to rationalise and manage the processual aspects of construction, inspired by logistics in manufacturing, similar efforts took place in the area of product-related research. From the 1970s, one of the most predominant lines of thought was dealing with the problem of structuring information to represent knowledge about facts and artefacts (Eastman 1999). This shift followed similar changes in manufacturing and aerospace. With initiative from the USA Air Force, product definitions were developed around the mid-1980s to support 'the direct and complete exchange or sharing of a product model amongst computer applications, without human intervention' (Dado et al. 2010, p. 105). These developments coincided with standardisation approaches, aimed at replacing the existing product definitions that were based on graphics and two-dimensional (2D) geometry and were prone to misconceptions.

The advancements in product modelling from other industries joined the long-standing debate on the computerisation of AEC (Eastman 1999). The product model is an integrated representation of information and data about an artefact over its product life cycle (Dado et al. 2010). In the AEC industry, the term building product model (BPM) is used to denote the information about a building component embedded in a product model (Eastman 1999). Nederveen et al. (2010) differentiated BPM from CAAD through the potential of the former to be stored explicitly in a formal, computer-interpretable manner, without being susceptible to human interpretation. The origins of BIM are found in the object-oriented building product modelling approaches started in the 1990s (Eastman et al. 2008). A massive uptake of BIM appeared in the last decade, although it was considered an unfulfilled prophecy for more than three decades. Since the 1990s various data standards were developed and subsequently discontinued to support BIM.

2.5.3 Importance of Standards in a Digital Built Environment

A standard in ICT is a set of solutions that aims to satisfy and balance the needs and requirements from a diverse group of actors in a seamless manner for electronic communication within and between computers (Laakso and Kiviniemi 2012). The standards are expected to be used 'during a certain period, by a substantial number of the parties for whom they are meant' (Vries 2005, p. 15). The efforts for standardisation of building information could be considered a form of 'horizontal standard', focusing on achieving compatibility among an array of building product entities (Vries 2005).

The human-derived need to trust the accuracy and conciseness of the exchanged information and the machine-based necessity for interoperability – that is the ability of the various computer systems to exchange data and use information consistently – motivated the efforts for standardisation. These were organised by industrial consortia and cross-organisational bodies and began in the 1970s (Björk and Laakso 2010). So far, the Industry Foundation Classes (IFC) standard is the most long-lived (Björk and Laakso 2010). The IFC standard was specified and developed by the international nonprofit consortium International Alliance for Interoperability (IAI), now named BuildingSMART, which focuses on standardising processes, workflows, and procedures for BIM (BuildingSMART 2014). The IFC standard is a common language used for transferring information among various BIM applications and has so far undergone various revisions, since 1994, to represent a neutral and open data schema for BIM.

In practice, the IFC standard still presents drawbacks. The disadvantages stem from the inherent properties and development of the standard, but also from the actual process of building information exchange, because it is not a native model to various commercial BIM applications. Subsequently, loss of information often takes place when the actors exchange IFC files that are converted from the proprietary formats. From a supply chain thinking perspective, IFC standards promise quite consistent information among actors, despite these losses, and highlights the importance of developing and maintaining relevant standards across the AEC supply chain and the built environment.

Other standards include Construction Operations Building information exchange (COBie), a data schema introduced to illustrate how information from the IFC standard could be provided from the designers to the facility management (FM) (East 2007). COBie presents building information in a nongraphical and accessible human-readable manner and aims to be used during the later stages of the projects for operation or maintenance. Another category of standards pertinent to BIM could be considered the collaboration standards that aim at describing the function of BIM implementation, a type of vertical standard (Vries 2005). For example, Constructive Objects and the INtegration of Systems (COINS), which is also affiliated with BuildingSMART, is an effort initiated in the Netherlands that contains both exchanging data formats, for example IFC, descriptions of collaboration methods, and principles of information management (Dado et al. 2010). Table 2.2 illustrates key milestones in the evolution of the BIM concept.

2.5.4 Pluralism of Digital Artefacts and BIM Maturity Assessment Methods

BIM is currently at the forefront of construction digitalisation. Apart from digital representation of buildings, it relates to artefacts that affect the processes that technologies are adopted and implemented through. BIM is a *'multifunctional set of instrumentalities for specific purposes'* (Miettinen and Paavola 2014) and affects various actors across the built environment, while policies, processes, and technologies interact to generate a digital building (Succar et al. 2012). BIM is a set of existing and new digital technologies for generating, controlling, and managing building information (Papadonikolaki and Oel 2016). Whereas currently BIM is synonymous with various commercial software solutions for the built environment, it is essentially an 'umbrella' term, similarly to the supply chain management philosophy. This umbrella term denotes (apart from its merely commercial instances, for example software applications such as Autodesk Revit, Bentley Microstation, Graphisoft Archicad and many more) the process of developing

Table 2.2 Key studies and milestones in the evolution of the concept of digitisation in the AEC.

Year	Milestone	Source
1975	Development of building design system	Eastman (1999)
1989	Basic structure of a proposed **building product model**	Björk (1989)
1992	Introduction of term **building information model**	van Nederveen and Tolman (1992)
1994	International Alliance for Interoperability (IAI) was founded	Bazjanac and Crawley (1997)
1995	Start of Industry Foundation Classes (IFC) initiatives	Bazjanac and Crawley (1997)
1999	Building product models book was published	Eastman (1999)
2004	Introduction of term **Virtual Design and Construction** (VDC)	Kam and Fischer (2004)
2005	IAI was renamed BuildingSMART	Buildingsmart.org
2007	National BIM Standards (NBIMS) in the USA was founded	Nationalbimstandard.org
2008	**Building Information Modelling** Handbook was published	Eastman et al. (2008)
2009	Introduction of **Building information Management** term	Becerik-Gerber and Kensek (2009)
2011	The UK BIM strategy was announced	GCCG (2011)
2015	The **Digital Built Britain** strategic plan was published	HMG (2015)

Source: Adapted from Papadonikolaki (2018)

and exchanging information models. Apart from BIM-related software and standards, BIM as a digital platform links to other digital technologies and artefacts used throughout the built environment lifecycle.

Various digital artefacts such as the CDE, an online platform to exchange files, BIM-specific contracts, BIM execution plans (BEP, a plan that defines BIM-related roles and team interactions), and so forth, affect how digital technologies are used and increase the complexity of innovation adoption and implementation. Typically, BIM assessment tools, developed by scholars and professionals, quantify maturity of digital and BIM-related work across projects, organisations, teams, or individuals (Azzouz et al. 2016). For example, an industry-developed tool measures digital and BIM maturity (ARUP 2015) through 11 criteria formed by the existence of various artefacts, such as the deployment of: BIM design data review, BIM innovation champions, CDE, BEP, document/model referencing and version control, knowledge-sharing practices, open standard deliverables, virtual design reviews, BIM-related contract, employers information requirements, and project procurement route. Thus, apart from software and standards, digital in the built environment includes documentation, online platforms, and remote or physical processes for collaboration, for example co-working and co-location sessions.

2.6 Pragmatic Impact of Digitisation and BIM

2.6.1 BIM and the Enterprise: Bottom-Up Adoption

The above review on how information flows self-organise in the era of digital technologies and BIM suggests that these digital innovations are still in flux. Some practical real-world cases of BIM implementation were reported as early as 2007 (Ballesty 2007; Sacks et al. 2009b). However, industry still struggles to adopt these digital technologies in projects. Market reports such as those from McGraw-Hill Construction (2009, 2012, 2014) underline the preliminary positive outcomes of adopting BIM but typically focus on perceptions of practitioners and mainly contractors in the USA. Because initially the adoption of digital innovations rests with individuals and their organisations, individual resistance to change and organisational challenges also have an impact.

Due to the variety of digital artefacts that professionals in the AEC industry need to master in order to embrace digital technologies, ad hoc and firm-centred approaches on BIM prevail (Papadonikolaki et al. 2016). This situation does not necessarily promote supply chain thinking. Organisations struggle to align and reconcile their business strategies, internal resources, and external institutional pressures to adopt digital technologies (Morgan 2017). Due to the relevance of digital technologies and BIM to information management in projects, these organisational challenges are further magnified by the need to align with other organisations in the supply chain. Adopting and implementing these digital technologies activate decision-making at both intra- and interorganisational levels (Papadonikolaki and Wamelink 2017). Digital technologies in the built environment are inevitably subject to many critics, pragmatists, and conservatives. Rogers' (1962) diffusion theory on how innovation travels through various communication channels over time is still very topical and applicable to digital technologies. Therefore, looking at digital technologies from a supply chain perspective might support communication, collaboration, and interorganisational learning of digital innovations.

2.6.2 BIM and the Institutional Setting: Top-Down Diffusion

Eastman (1999, p. 30) described that in the AEC industry, the 'bottom-up' innovation diffusion had been the norm:

> innovations have been adopted that are both organisational – such as the recent development of fast-track scheduling – and also technical – such as the use of finite element modelling and the electronic storing and transmission of drawings. However, all the changes have been incremental, adopted first in pilot cases, then slowly absorbed into the wider practices of the industry as a whole.

However, the opposite of the afore-described bottom-up innovation strategy (initiated at the enterprise level), is the top-down strategy. Numerous noteworthy publications by various government bodies, industry associations, communities of practice, and research institutions aim to increase awareness and support BIM adoption, in Australia, Europe, Asia, and North America (Kassem et al. 2015). Edirisinghe and London (2015, p. 156), analysed BIM standards and policy initiatives globally and concluded that there is an 'influential link between national policy initiatives and the adoption data' from practitioners. The USA has had the most advanced level of top-down BIM regulation and adoption, as mandatory BIM use in government projects has been in place since 2007. In the UK, a mandatory mature level of BIM implementation (Level 2) in public procurement started in 2016.

The top-down strategies for BIM adoption in the UK are particularly interesting. After all, as discussed around the pragmatic impact of the supply chain management concept, the UK traditionally had an interventionist policy in construction, highlighted through the Egan and Latham reports. Until 2014, BIM engagement among more than half of contractors in the UK was low (Edirisinghe and London 2015). In 2008, Mark Bew and Mervyn Richards developed a BIM maturity diagram and defined four maturity levels, where CAAD-based projects are at Level 0 and highly integrated and interoperable buildings data are considered as Level 3 (GCCG 2011). According to the UK BIM maturity levels, Level 1 concerns a primary 2D or 3D design. Level 2 BIM refers to a common environment where multidisciplinary data would be shared using COBie. Level 3 BIM is rebranded as a 'Digital Built Britain' vision and is a 'work in progress' focusing beyond the AEC on the whole built environment lifecycle. It refers to the integrated and interoperable version of BIM, and refers to the use of common dictionaries, IFC standards, and common processes, with the ultimate aim of facilitating lifecycle management via BIM and IoT (HMG 2015). Recently, the Cambridge Centre for Digital Built Britain (CDBB) has recommended moving away from the Level 3 and Level 4 terms to 'integrate and operate' of "BIM data to service-delivery processes, within secure information landscapes and across federated digital twins" (CDBB, 2019).

2.6.3 Mismatch Between Top-Down and Bottom-Up Strategies

Undoubtedly, as explored in the previous two sub-sections, the evolution of BIM has encouraged a broader discussion about digitisation uptake in the built environment across countries. However, challenges for BIM adoption remain from both bottom-up and top-down strategies. On the one hand, market reports such as those from McGraw-Hill Construction (2009, 2012, 2014) can capture only one fragment of the diffusion of BIM, and particularly from large contractors and not SMEs. These reports are probably not very relevant to Europe, given that, for example, the European construction sector is composed of around 99% of SMEs (UEAPME 2015). Dainty et al. (2017) highlighted the danger of a digital divide if SMEs are not included in digital diffusion efforts. On the other hand, the actual utilisation of BIM does not always evolve in practice as proclaimed, prescribed, and desired by the rigid top-down policy-making strategies. The mandatory downward implementation may disproportionately increase not only ruthless competition but also frustration.

Market coordination problems have been previously identified regarding the need to standardise file exchange formats (Laakso and Kiviniemi 2012). Despite efforts from industry consortia, these also need the catalytic intervention of policy-makers. Succar and Kassem (2015) point out that normative diffusion policies for BIM within a particular market might trigger 'through mimetic pressures' similar actions by other governmental and authoritative bodies in other countries. Apart from the bottom-up and top-down diffusion dynamics, the 'middle-out' level could potentially play a role in strengthening the adoption of digitisation (Succar and Kassem 2015). Succar and Kassem (2015) claim that large organisations and industry associations could influence (i) SMEs coercively (downwards), (ii) governmental bodies normatively (upwards) and (iii) the other large organisations mimetically (horizontally).

Supply chain partnerships and supply chain management philosophy could offer such an intermediate level between national policies and firms and could potentially play a role in increasing digitisation across the industry and facilitating digital transformation. Supply chain partnerships typically engage in increased communication, co-location practices that support collaboration, inter-firm learning and inclination for co-developing solutions (Papadonikolaki et al. 2017; Papadonikolaki and Wamelink 2017). Supply chain management and supply chain partnerships could likewise play a role in assisting the popularisation of digitisation in the AEC, digitalisation in projects and digital transformation in the built environment. Likewise, Winch (1998) stressed that although the bottom-up approach involves generating new ideas and patterns through problem-solving, the efficiency of this learning process and its applicability in projects greatly depends on firm capacity.

To this end, digitisation in the built environment calls for a relational approach. The coordination of firms through supply chain management and long-term partnering can provide firms with additional resources from the virtual corporation to successfully master the digital technologies. Similarly, digital capabilities can form value propositions and competitive advantage of construction supply chains. Dubois and Gadde (2002) suggested that inter-firm cooperation and reciprocal adjustments within the involved firms would eventually foster learning and innovative thinking in the AEC. After all, in manufacturing, commerce, and automotive industries, the 'competition shifts from a company orientation to a supply chain orientation' (Vonderembse et al. 2006, p. 224). Thus, there is scope for increasing digitisation in the built environment through the digital technologies supply chain.

2.7 Synthesis of Digital Technologies Construction Supply Chain

2.7.1 Potential and Outlook of Digital Technologies to Support Supply Chains

Digital technologies and BIM, as a digital platform, could support the material, or building product flows during the AEC lifecycle. During design, the object-oriented logic behind BIM could not only automatically produce bills of quantities but also dynamically enrich them with model properties (Eastman et al. 2008). Therefore, digital technologies could support the material flows, even before they are procured and placed on site. This could transfer site activities from the construction site to the production facilities of the supply chain (off-site production), as described by Vrijhoef and Koskela (2000). Subsequently, site preparation can start before the actual delivery of materials. During construction, the object-oriented capabilities of BIM could be used to support various site activities, such as transportation of materials, their distribution on site and site layout optimisation. Simultaneously, coupling various digital technologies such as BIM and tracking through barcodes, RFID, and global positioning system (GPS) could monitor the supply chains to enhance the visibility during the material delivery (Irizarry et al. 2013). Therefore, digitisation is characterised by connected technologies and needs technical interoperability.

Digitisation facilitates collaboration via CDEs and improves the traditional data management. Due to persistent challenges of interoperability, digital technologies, and BIM in particular, require special efforts from the various multidisciplinary actors to ascertain that the exchanged data are accurate, consistent, and usable from their collaborating partners (Owen et al. 2010). Besides interoperability, others advocate that the complexity of the BIM-based information flows has not been merely an ICT issue but 'implies a richer interweaving of more than technology' including processes, culture and values, and management of contractual issues among the stakeholders (Grilo and Jardim-Goncalves 2010, p. 529), thus a relational challenge. The emerging patterns of interactions among the project team suggest a self-organising network of information flows (Hickethier et al. 2013). Apart from the AEC professionals, clients and property developers have been playing a dominant role in the demand for including BIM in project delivery (Porwal and Hewage 2013) to receive both physical and digital assets. Procurement routes not only affect the relations among team members but also the success of digital delivery. Therefore, apart from technical interoperability, the organisational or business interoperability (Kubicek et al. 2011) in the supply chain is a key determinant of adopting and sustaining digital innovations.

Because digital technologies continuously develop and there exists no comprehensive software or platform to capture all needs and functionalities for the multidisciplinary AEC actors (Eastman et al. 2008), digital technologies also constantly evolve. BIM-based collaboration is hardly real-time and could become concurrent only in homogeneous environments, for example with native files shared on the network, on a remote database, or on a BIM server (Cerovsek 2011). Berlo et al. (2012) challenged the centrality of BIM-based collaboration and proposed workarounds through the federation of 'reference models' that offer acceptable BIM interoperability among actors. Therefore, the information flows of the involved actors develop in noncentral, potentially asymmetric, and asynchronous configurations (for an illustration of these networks see Papadonikolaki 2016). Hence, to master digital technologies, the supply chain needs to develop, invest in, and master ways to work together regardless of the software and hardware capabilities and collaborate in breaking the silos (Papadonikolaki et al. 2016). These asymmetrical, imbalanced, and asynchronous digital collaboration patterns that emerge are omnipresent and allow for a loose coupling of the various involved stakeholders (Papadonikolaki 2018). Whereas digital and BIM support collaboration via digital artefacts, they generate additional needs for coordination and collaboration because of their continuous evolution. Collaboration among actors is challenging even through digital technologies and this will become even more challenging as emerging technologies such as IoT, artificial intelligence (AI) and blockchain technologies enter the built environment digitisation space.

2.7.2 Co-Evolution of Supply Chain Management and Digital in AEC

From as early as the 1990s, the supply chains were described more as systems or networks rather than linear configurations (Christopher 1992; Lambert et al. 1998). Mentzer et al. (2001) suggested that supply chain management as a management philosophy is closely affiliated to system thinking. A system contains essential parts or subsystems with innate behaviours and properties that constitute a functional whole (Ackoff and Gharajedaghi 2003). Thus, focusing on supply chain systems, and particularly on both the pluralism of AEC actors and built environment stakeholders and their collaborative relations, is essential for understanding and utilising digital technologies. Based on this, this chapter conceptualised the AEC industry as a set of:

- processes
- products
- relations

In this spirit, this chapter focused on reviewing all these processual, product-related, and relational complexities of the built environment and how these are addressed by supply chain management philosophy and digitisation. This chapter has attempted a chronological review of the concepts of supply chain management and digitisation, to explore the areas of their potential compatibility and alignment. The concept of supply chain management matured from a processual view of managing the complexities of the AEC industry into a relational concept. Likewise, digital technologies and particularly BIM evolved from a product-related view of managing the complexities of the built environment into a relational concept. Therefore, it is concluded that the relational view of supply chain management can help coordinate and leverage the unprecedented volume of digital technologies that digitise the AEC. Figure 2.2 illustrates some advancements and milestones during the co-evolution of the supply chain management and digital constructs towards a relational perspective.

Both supply chain management philosophy and digitisation in the built environment evolved from processual and product-related and tangible concepts respectively to relational constructs that affect various actors across the institutional settings of the built environment. Surprisingly, digital technologies have been linked not only to the coordination of technological artefacts, but also to complex socio-technical processes to align actors and information (Liu et al. 2016; Papadonikolaki 2016). Digital technologies and BIM move beyond technological determinism towards a relational shift in the AEC. Undoubtedly, relational supply chain management is needed to sustain digital innovation in the built environment. This chapter drew upon theory to discuss not only the compatibility, relevance, and topicality of supply chain management and digitisation concepts in the built environment, but also outline implications for policy makers and industrial leaders who wish to take digitisation in the built environment forward.

2.8 Conclusion

This chapter has looked at various challenges in the management of AEC industry and particularly focused on the issues of processual, technical, and relational complexities, as well as the increasing fragmentation. In this narrative literature review, the concepts of supply chain management and digitisation in the built environment were analysed chronologically. First, after reviewing the origins and development of supply chain management philosophy and practices, it was concluded that relational, structured, strategic, and long-term partnering could form an opportunity for managing both processual and relational complexities. Whereas supply chain management was originally introduced as a concept synonymous with the positivist paradigm, this chapter presented

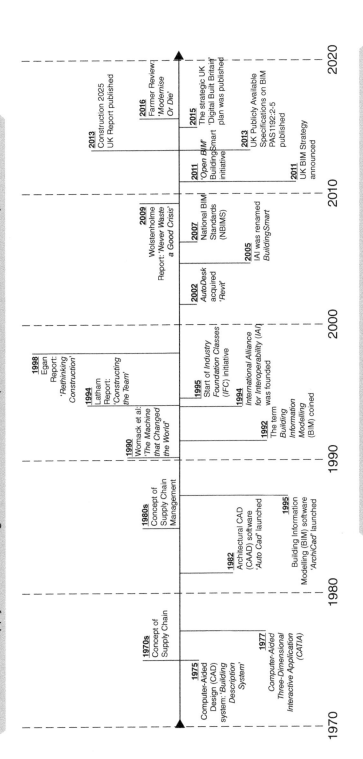

Figure 2.2 Co-evolution of supply chain management philosophy and digitisation in the built environment from processual and product to relational. Source: Original.

how understanding the concept of supply chain management from a relational stand-point supports digital transformation in the built environment. Whereas the concept of supply chain management emerged from processual thinking rooted in cybernetics, it has acquired a pragmatic relational outlook. Second, after reviewing the evolution of digitisation in the built environment, it is suggested that digital technologies could manage both technical and product-related complexities in the AEC industry. Therefore, the supply chain management philosophy is presented as a means to address the phenomenon of digital innovation and leverage from digitisation by creating digital supply chains.

Taking the analysis of the aforementioned complexities further, through their pragmatic relevance in the industry, the chapter advocated that supply chain management philosophy and digital technologies could manage not only the processual and product-related complexities in the built environment, but its relational complexities. The alignment of supply chain management and digitisation was explored through the theoretical lens of the conceptual framework around processes, technologies, and relations, which was defined after the chronological review of supply chain management and digital constructs. A combined routine of digital supply chain management could significantly promote supply chain integration and reduce industry fragmentation, and additionally offer a sustained attitude towards digital technologies. Likewise, adopting and implementing digital technologies could be greatly enriched from supply chain management philosophy in a middle-out diffusion strategy. This chapter highlighted that digital transformation in the built environment through digitisation needs adaptation to relational thinking. Creating digital supply chains can thus facilitate digital technologies transformation in the built environment through emphasis on long-term relations.

References

Ackoff, R.L. (1989). From data to wisdom. *Journal of Applied Systems Analysis* 16: 3–9.

Ackoff, R.L. and Gharajedaghi, J., 2003. *On the Mismatch between Systems and their Models* [online]. http://www.acasa.upenn.edu/System_MismatchesA.pdf

Ajam, M., Alshawi, M., and Mezher, T. (2010). Augmented process model for e-tendering: towards integrating object models with document management systems. *Automation in Construction* 19: 762–778.

Alvesson, M. and Sandberg, J. (2011). Generating research questions through problematization. *Academy of Management Review* 36: 247–271.

Andersen, E.S. (2016). Do project managers have different perspectives on project management? *International Journal of Project Management* 34: 58–65.

Aouad, G., Wu, S., Lee, A., and Onyenobi, T. (2012). *Computer Aided Design Guide for Architecture, Engineering and Construction*. London, UK: Routledge.

Arbulu, R. (2009). Application of integrated materials management strategies. In: *Construction Supply Chain Management Handbook* (eds. W.J. O'Brien, C.T. Formoso, R. Vrijhoef and K.A. London), 71–79. Boca Raton, FL: CRC Press.

Arup, 2015. *New BIM Maturity Measure model launches* [online]. UKMEA Press Office. Available from: https://www.arup.com/news-and-events/news/new-bim-maturity-measure-model-launches [Accessed 22 July 2019].

Ashby, W.R. (1956). *An introduction to cybernetics London.* UK: Chapman & Hall Ltd.

Azambuja, M. and O'Brien, W.J. (2009). Construction supply chain modeling: issues and perspectives. In: *Construction Supply Chain Management Handbook* (eds. W.J.O.'. Brien, C.T. Formoso, R. Vrijhoef and K.A. London). Boca Raton, FL, USA: CRC Press 259–268.

Azzouz, A., Shepherd, P. and Copping, A., 2016. Emergence of Building Information Modelling Assessment Methods. Paper presented at Integrated Design at 50: Building our Future, Bath UK

Badi, S.M., Li, M., and Pryke, S.D. (2016). The influence of communication network centrality on individual popularity: a case study of a Chinese construction project. In: *The proceedings of the annual conference of the Royal Institution of Chartered Surveyors* (Vol. 2016, September). Royal Institution of Chartered Surveyors (RICS).

Ballard, G., Koskela, L., Howell, G. et al. (2001). Production system design in construction. *Proceedings of the 9th Annual Conference of the International Group for Lean Construction.* Kent Ridge Crescent, Singapore.

Ballesty, S. (2007). *Adopting BIM for Facilities Management: Solutions for Managing the Sydney Opera House.* I.N.PIcon.Net Pty Ltd. ISBN: 978-0-9775282-2-6.

Bazjanac, V. and Crawley, D.B. (1997). *The Implementation of Industry Foundation Classes in Simulation Tools for the Building Industry.* Berkeley, CA: Lawrence Berkeley National Laboratory.

Becerik-Gerber, B. and Kensek, K. (2009). Building information modeling in architecture, engineering, and construction: emerging research directions and trends. *Journal of Professional Issues in Engineering Education and Practice* 136: 139–147.

Becerik-Gerber, B., Ku, K.H., and Jazizadeh, F. (2012). BIM-enabled virtual and collaborative construction engineering and management. *Journal of Professional Issues in Engineering Education and Practice* 138: 234–245.

Berlo, L.V., Beetz, J., Bos, P. et al. (2012). Collaborative engineering with IFC: new insights and technology. *9th European Conference on Product and Process Modelling (ECPPM 2012).* Reykjavik, Iceland.

Björk, B.-C. (1989). Basic structure of a proposed building product model. *Computer-Aided Design* 21: 71–78.

Björk, B.-C. and Laakso, M. (2010). CAD standardisation in the construction industry – —a process view. *Automation in Construction* 19: 398–406.

Briscoe, G. and Dainty, A. (2005). Construction supply chain integration: an elusive goal? *Supply Chain Management: An International Journal* 10: 319–326.

Briscoe, G., Dainty, A., and Millett, S. (2001). Construction supply chain partnerships: skills, knowledge and attitudinal requirements. *European Journal of Purchasing and Supply Management* 7: 243–255.

Buildingsmart, 2014. *BuildingSMART, International home of OpenBIM: Core Purpose* [online]. http://www.buildingsmart.org/about/vision-mission/core-purpose

Cachon, G. and Terwiesch, C. (2009). *Matching Supply With Demand: An Introduction to Operations Management*, 3e. New York, USA: McGraw-Hill.

CDBB, 2019. *BIM* [online]. Centre for Digital Built Britain. Available from: https://www.cdbb.cam.ac.uk/BIM [Accessed 14 October 2019].

Cerovsek, T. (2011). A review and outlook for a 'Building Information Model'(BIM): a multi-standpoint framework for technological development. *Advanced Engineering Informatics* 25: 224–244.

Chen, I.J. and Paulraj, A. (2004). Understanding supply chain management: critical research and a theoretical framework. *International Journal of Production Research* 42: 131–163.

Chien, K.-F., Wu, Z.-H., and Huang, S.-C. (2014). Identifying and assessing critical risk factors for BIM projects: empirical study. *Automation in Construction* 45: 1–15.

Christopher, M. (1992). *Logistics and Supply Chain Management: Strategies for Reducing Cost and Improving Services*, 2e. London, UK: Financial Times Professional Ltd.

Christopher, M. (2005). *Logistics and Supply Chain Management: Creating Value-Adding Networks*, 3e. New York, USA: Financial Times Prentice Hall.

Christopher, M. (2011). *Logistics and Supply Chain Management*, 4e. Dorset, UK: Financial Times Prentice Hall.

Cox, A. and Ireland, P. (2002). Managing construction supply chains: the common sense approach. *Engineering Construction and Architectural Management* 9: 409–418.

Cutting-Decelle, A.F., Young, R.I., Das, B.P. et al. (2009). Standards-based approaches to interoperability in supply chain management: overview and case study using the ISO 18629 PSL standard. In: *Construction Supply Chain Management Handbook* (eds. W.J. O'Brien, C.T. Formoso, R. Vrijhoef and K.A. London), 2766–2778. Boca Raton, , USA: CRC Press.

Dado, E., Beheshti, R., and Van De Ruitenbeek, M. (2010). Product modelling in the building and construction industry: a history and perspectives. In: *Handbook of Research on Building Information Modelling and Construction Informatics: Concepts and Technologies* (eds. J. Underwood and U. Isikdag), 104–137. Hershey, PA: IGI Global Publishing.

Dainty, A., Leiringer, R., Fernie, S., and Harty, C. (2017). BIM and the small construction firm: a critical perspective. *Building Research and Information*: 1–14.

Davidow, W.H. and Malone, M.S. (1992). *The Virtual Corporation: Structuring and Revitalising the Corporation for the 21st century*. New York: Harperbusiness.

De Bruijn, J. and Ten Heuvelhof, E. (2008). *Management in Networks: On Multi-Actor Decision Making. Abingdon*. UK: Routledge.

De Vries, H.J. (2005). IT standards typology. In: *Advanced Topics in Information Technology Standards and Standardization Research* (ed. K. Jakobs), 11–36. Hershey, PA, USA: IGI Global.

Demian, P. and Walters, D. (2014). The advantages of information management through building information modelling. *Construction Management and Economics* 32: 1153–1165.

Dorée, A.G. (2004). Collusion in the Dutch construction industry: an industrial organization perspective. *Building Research and Information* 32: 146–156.

Dubois, A. and Gadde, L.-E. (2002). The construction industry as a loosely coupled system: implications for productivity and innovation. *Construction Management and Economics* 20: 621–631.

Dulaimi, M.F., Ling, F.Y.Y., Ofori, G., and De Silva, N. (2002). Enhancing integration and innovation in construction. *Building Research and Information* 30: 237–247.

East, E.W. (2007). *Construction Operations Building Information Exchange (Cobie): Requirements Definition and Pilot Implementation Standard*. Washington, D.C: US Army Corps of Engineers.

Eastman, C. (1999). *Building Product Models: Computer Environments, Supporting Design and Construction*. Boca Raton, Florida, USA: CRC Press.

Eastman, C., Teicholz, P., Sacks, R., and Liston, K. (2008). *BIM Handbook: A Guide to Building Information Modeling for Owners, Managers, Designers, Engineers, and Contractors*, 2e. Hoboken, New Jersey, USA: Wiley.

Edirisinghe, R. and London, K. (2015). Comparative analysis of international and national level BIM standardization efforts and BIM adoption. In: *Proceedings of the 32nd CIB W78 Conference* (eds. J. Beetz, L. Van Berlo, T. Hartmann and R. Amor), 149–158. Eindhoven, The Netherlands.

Egan, J. (1998). *Rethinking Construction: Report of the Construction Task Force*. London, UK: HMSO.

Eriksson, P.E. (2015). Partnering in engineering projects: four dimensions of supply chain integration. *Journal of Purchasing and Supply Management* 21: 38–50.

Fernie, S. and Tennant, S. (2013). The non-adoption of supply chain management. *Construction Management and Economics* 31: 1038–1058.

Fernie, S. and Thorpe, A. (2007). Exploring change in construction: supply chain management. *Engineering, Construction and Architectural Management* 14: 319–333.

Forbes, L.H. and Ahmed, S.M. (2010). Information and communication technology/building information modeling. In: *Modern Construction: Lean Project Delivery and Integrated Practices*, 2032–2228. Boca Raton, FL: CRC Press.

Gadde, L.-E. and Dubois, A. (2010). Partnering in the construction industry — problems and opportunities. *Journal of Purchasing and Supply Management* 16: 254–263.

Gartner. 2016. Gartner IT Glossary [Online]. Available: http://www.gartner.com/itglossary/digitalization [Accessed 15/12 2016]

GCCG, 2011. *Government Construction Client Group: BIM Working Party Strategy Paper*.

Gidado, K.I. (1996). Project complexity: the focal point of construction production planning. *Construction Management and Economics* 14: 213–225.

Gosling, J., Naim, M., Towill, D. et al. (2015). Supplier development initiatives and their impact on the consistency of project performance. *Construction Management and Economics* 33: 1–14.

Green, S.D., Fernie, S., and Weller, S. (2005). Making sense of supply chain management: a comparative study of aerospace and construction. *Construction Management and Economics* 23: 579–593.

Greenhalgh, T., Thorne, S., and Malterud, K. (2018). Time to challenge the spurious hierarchy of systematic over narrative reviews? *European Journal of Clinical Investigation* 48: e12931.

Grilo, A. and Jardim-Goncalves, R. (2010). Value proposition on interoperability of BIM and collaborative working environments. *Automation in Construction* 19: 522–530.

Halman, J.I.M., Voordijk, J.T., and Reymen, I.M.M.J. (2008). Modular approaches in Dutch house building: an exploratory survey. *Housing Studies* 23: 781–799.

Hickethier, G., Tommelein, I.D., and Lostuvali, B. (2013). Social Network Analysis of Information Flow in an IPD-Project Design organization. *Proceedings of the International Group for Lean Construction*. Fortaleza, Brazil.

HM Government, 2015. *Digital Built Britain, Level 3 BIM Strategic Plan* [online]. HM Government. Available from: https://assets.publishing.service.gov.uk/government/uploads/system/uploads/attachment_data/file/410096/bis-15-155-digital-built-britain-level-3-strategy.pdf [Accessed 22 July 2019].

Howard, H., Levitt, R., Paulson, B. et al. (1989). Computer integration: reducing fragmentation in AEC industry. *Journal of Computing in Civil Engineering* 3: 18–32.

ICE, 2017. *State Of The Nation 2017: Digital Transformation* [online]. Institution of Civil Engineers. Available from: www.ice.org.uk/getattachment/news-and-insight/policy/state-of-the-nation-2017-digital-transformation/ICE-SoN-Report-Web-Updated.pdf.aspx [Accessed 22 July 2019].

Irizarry, J., Karan, E.P., and Jalaei, F. (2013). Integrating BIM and GIS to improve the visual monitoring of construction supply chain management. *Automation in Construction* 31: 241–254.

Kam, C. and Fischer, M. (2004). Capitalizing on early project decision-making opportunities to improve facility design, construction, and life-cycle performance—POP, PM4D, and decision dashboard approaches. *Automation in Construction* 13: 53–65.

Kassem, M., Succar, B., and Dawood, N. (2015). Building information modeling: analyzing noteworthy publications of eight countries using a knowledge content taxonomy. In: *Building Information Modeling: Applications and Practices in the AEC Industry* (eds. R. Issa and S. Olbina), 329–371. Reston, VA: ASCE Press.

Koskela, L. (1992). *Application of the new production philosophy to construction*. CA: Stanford University.

Kubicek, H., Cimander, R., and Scholl, H.J. (2011). *Layers of interoperability. Organizational Interoperability in E-Government*, 85–96. Springer.

Laakso, M. and Kiviniemi, A. (2012). The IFC standard: A review of history, development and standardization. *Journal of Information Technology in Construction* 17: 134–161.

Lambert, D.M., Cooper, M.C., and Pagh, J.D. (1998). Supply chain management: implementation issues and research opportunities. *International Journal of Logistics Management* 9: 1–20.

Lasi, H., Fettke, P., Kemper, H.-G. et al. (2014). Industry 4.0. *Business and Information Systems Engineering* 6: 239–242.

Latham, M. (1994). *Constructing the Team*. London, UK: HMSO.

Leuschner, R., Rogers, D.S., and Charvet, F.F. (2013). A meta-analysis of supply chain integration and firm performance. *Journal of Supply Chain Management* 49: 34–57.

Lindgren, J. (2016). Diffusing systemic innovations: influencing factors, approaches and further research. *Architectural Engineering and Design Management* 12: 19–28.

Liu, Y., Van Nederveen, S., and Hertogh, M. (2016). Understanding effects of BIM on collaborative design and construction: An empirical study in China. *International Journal of Project Management* 35: 686–698.

Lockamy Iii, A. and Mccormack, K. (2004). The development of a supply chain management process maturity model using the concepts of business process orientation. *Supply Chain Management: An International Journal* 9: 272–278.

London, K. (2009). Industrial Organization Object-Oriented Project Model of the Facade Supply Chain Cluster. In: *Construction Supply Chain Management Handbook* (eds. W.J. O'Brien, C.T. Formoso, R. Vrijhoef and K.A. London). Boca Raton, Florida, USA: CRC Press.

London, K. and Kenley, R. (2001). An industrial organization economic supply chain approach for the construction industry: a review. *Construction Management and Economics* 19: 777–788.

London, K. and Singh, V. (2013). Integrated construction supply chain design and delivery solutions. *Architectural Engineering and Design Management* 9: 135–157.

McGraw Hill (2009). The business value of BIM: getting building information modeling to the bottom line. In: *Smart Market Report* (ed. H.M. Bernstein). McGraw Hill Construction.

McGraw Hill (2012). The business value of BIM for construction in North America: multi-year trend analysis and user ratings (2007–2012). In: *Smart Market Report* (ed. H.M. Bernstein). McGraw Hill Construction.

McGraw Hill (2014). The business value of BIM for construction in major global markets: how contractors around the world are driving Innovation with building information modeling. In: *Smart Market Report* (ed. H.M. Bernstein). McGraw Hill Construction.

Mentzer, J.T., Dewitt, W., Keebler, J.S. et al. (2001). Defining supply chain management. *Journal of Business logistics* 22: 1–25.

Meredith, J.R., Raturi, A., Amoako-Gyampah, K., and Kaplan, B. (1989). Alternative research paradigms in operations. *Journal of operations management* 8: 297–326.

Miettinen, R. and Paavola, S. (2014). Beyond the BIM utopia: approaches to the development and implementation of building information modeling. *Automation in Construction* 43: 84–91.

Mitchell, M. (2009). *Complexity: A guided tour*. Oxford University Press.

Morgan, B., 2017. Organizing for technology in practice: implementing Building Information Modeling in a design firm. PhD. University College, London.

Morgan, B. and Papadonikolaki, E. (2018). Organizing for digitization: a balancing act. In: *Academy of Management Specialized Conference, Big Data and Managing in a Digital Economy*. Guildford, UK: Academy of Management.

Nam, C. and Tatum, C. (1992). Noncontractual Methods of Integration on Construction Projects. *Journal of Construction Engineering and Management* 118: 385–398.

Naylor, J.B., Naim, M.M., and Berry, D. (1999). Leagility: integrating the lean and agile manufacturing paradigms in the total supply chain. *International Journal of Production Economics* 62: 107–118.

Nederveen, S., Van Beheshti, R., and De Ridder, H. (2010). Supplier-driven integrated design. *Architectural Engineering and Design Management* 6: 241–253.

Owen, R., Amor, R., Palmer, M. et al. (2010). Challenges for integrated design and delivery solutions. *Architectural Engineering and Design Management* 6: 232–240.

Papadonikolaki, E. (2016). *Alignment of Partnering with Construction IT: Exploration and Synthesis of Network Strategies to Integrate BIM-enabled Supply Chains*. Delft: Architecture and the Built Environment Series.

Papadonikolaki, E. (2018). Loosely coupled systems of innovation: aligning BIM adoption with implementation in Dutch construction. *Journal of Management in Engineering* 34: 05018009.

Papadonikolaki, E. and Oel, C.V. (2016). The Actors' perceptions and expectations of their roles in BIM-based collaboration. In: *Proceedings of the 32nd Annual Association of Researchers in Construction Management Conference (ARCOM 2016)* (eds. P. Chan and C.J. Neilson), 93–102. Manchester, UK: Association of Researchers in Construction Management.

Papadonikolaki, E. and Wamelink, H. (2017). Inter- and intra-organizational conditions for supply chain integration with BIM. *Building Research and Information* 45: 1649–1664.

Papadonikolaki, E., Vrijhoef, R., and Wamelink, H. (2016). The interdependences of BIM and supply chain partnering: empirical explorations. *Architectural Engineering and Design Management* 12: 476–494.

Papadonikolaki, E., Verbraeck, A., and Wamelink, H. (2017). Formal and informal relations within BIM-enabled supply chain partnerships. *Construction Management and Economics* 35: 531–552.

Porter, M.E. (1985). *Competitive Advantage: Creating and Sustaining Superior Performance*. New York, USA: Free Press.

Porwal, A. and Hewage, K.N. (2013). Building information modeling (BIM) partnering framework for public construction projects. *Automation in Construction* 31: 204–214.

Pryke, S. (2009). *Construction Supply Chain Management (Innovation in the Built Environment)* West Sussex. Chichester, UK: Wiley-Blackwell.

Pryke, S., Badi, S., and Bygballe, L. (2017). Editorial for the special issue on social networks in construction. *Construction Management and Economics* 35: 445–454.

Reiner, G. (2005). Supply chain management research methodology using quantitative models based on empirical data. In: *Research Methodologies in Supply Chain Management* (eds. H. Kotzab, S. Seuring, M. Müller and G. Reiner), 431–444. Heidelberg, Germany: Physica-Verlag.

Rogers, E.M. (1962). *Diffusion of Innovations*, 1e. New York: Free Press of Glencoe.

Ross, J. (2017). Don't confuse digital with digitization. In: *MIT Sloan*. Retrieved from https://sloanreview.mit.edu/article/dont-confuse-digial-with-digitization/.

Sacks, R., Dave, B.A., Koskela, L., and Owen, R. (2009a). Analysis Framework for the Interaction Between Lean Construction and Building Information Modelling. In: *17th Annual Conference of the International Group for Lean Construction*, 221–234. Taipei, Taiwan: National Pingtung University of Science and Technology, LCI Taiwan and LCI Asia.

Sacks, R., Koskela, L., Dave, B.A., and Owen, R. (2009b). The interaction of lean and building information modeling in construction. *Journal of Construction Engineering and Management* 136: 968–980.

Samuelson, O. and Björk, B.-C. (2013). Adoption processes for EDM, EDI and BIM technologies in the construction industry. *Journal of Civil Engineering and Management* 19: S172–S187.

Samuelson, O. and Björk, B.-C. (2014). A longitudinal study of the adoption of IT technology in the Swedish building sector. *Automation in Construction* 37: 182–190.

Söderlund, J. (2004). On the broadening scope of the research on projects: a review and a model for analysis. *International Journal of Project Management* 22: 655–667.

Succar, B. and Kassem, M. (2015). Macro-BIM adoption: conceptual structures. *Automation in Construction* 57: 64–79.

Succar, B., Sher, W., and Williams, A. (2012). Measuring BIM performance: five metrics. *Architectural Engineering and Design Management* 8: 120–142.

Turner, R.J. (2006). Towards a theory of project management: the nature of the project. *International Journal of Project Management* 24: 1–3.

UEAPME, 2015. *European Association of Craft, Small and Medium-Sized Enterprise: Construction* [online]. European Association of Craft, Small and Medium-Sized Enterprise. Available from: http://www.ueapme.com/spip.php?rubrique17 [Accessed Access Date 2015].

Vaidyanathan, K. (2009). Overview of IT applications in the construction supply chain. In: *Construction Supply Chain Management Handbook* (eds. W.J. O'Brien, C.T. Formoso, R. Vrijhoef and K.A. London). Boca Raton, FL, USA: CRC Press.

Van Nederveen, G. and Tolman, F. (1992). Modelling multiple views on buildings. *Automation in Construction* 1: 215–224.

Vonderembse, M.A., Uppal, M., Huang, S.H., and Dismukes, J.P. (2006). Designing supply chains: towards theory development. *International Journal of Production Economics* 100: 223–238.

Vrijhoef, R. (2011). *Supply Chain Integration in the Building Industry: The Emergence of Integrated and Repetitive Strategies in a Fragmented and Project-Driven Industry.* Amsterdam, The Netherlands: IOS Press.

Vrijhoef, R. and Koskela, L. (2000). The four roles of supply chain management in construction. *European Journal of Purchasing and Supply Management* 6: 169–178.

Vrijhoef, R. and London, K. (2009). A review of organizational approaches to construction supply chain. In: *Construction Supply Chain Management Handbook* (eds. W.J. O'Brien, C.T. Formoso, R. Vrijhoef and K.A. London). Boca Raton, FL: CRC Press.

Winch, G. (1998). Zephyrs of creative destruction: understanding the management of innovation in construction. *Building Research and Information* 26: 268–279.

Winch, G.M. (2002). *Managing Construction Projects*, 1e. Oxford, UK: Blackwell Science.

Winch, G.M. (2005). Rethinking project management: project organizations as information processing systems. In: *Innovations: Project Management Research 2004* (eds. D.P. Slevin, D.I. Cleland and J.K. Pinto), 41–55. Newton Square, PA: Project Management Institute.

3

At the Interface: When Social Network Analysis and Supply Chain Management Meet

Huda Almadhoob

3.1 Introduction

The supply chain management discipline has increasingly focused on understanding the underlying supply chain relationships beyond the traditional buyer–supplier dyads, spanning a much larger scope and providing greater insight into the explanation of project functionality (Wichmann and Kaufmann 2016; Borgatti and Li 2009). Such focus stems from the recognised inherent complexity and uncertainty of the construction environment that arises from the need to manage adversarial relationships between large numbers of actors, with multiple interests and objectives, as well as the existence of many different interrelated risks and constraints that are usually subject to change over time (Flyvbjerg 2009; Winch 2002).

Research has shown that traditional static and deterministic approaches in managing supply networks are ineffective in dealing with uncertain and turbulent environments such as the construction industry (Marchi et al. 2014; Pryke 2017). An interest, therefore, has been developed in the past few years to understand the dynamics of project supply networks from a complexity theory perspective (Marchi et al. 2014). It is suggested that complexity theory can offer a deeper understanding of managing supply chains and their behaviours (Marchi et al. 2014).

The effective organisation, functionality, and adaptability of supply chains from such a perspective asserts that supply chains should be studied as complex adaptive systems that have an inherent tendency to self-organise (Gell-Mann 1994; Stacey 1996). This will not only help to derive explanatory insights but will also assist in navigating the turbulent waters of complex projects and the construction environment in general. A complex adaptive systems perspective should lead to a paradigm shift in the unit of analysis from the extensively studied economic attributes and organisational effectiveness of projects towards the examination of the relational and social dimensions that constitute project actualities. This drive towards the people-intensive nature of the construction industry inevitably shifts the emphasis towards the micro/individual interpersonal communicative relationships embedded within self-organising networks. Such an understanding entails moving away from the traditional top-down hierarchical and contractually prescribed management approaches which are usually confined by

Successful Construction Supply Chain Management: Concepts and Case Studies, Second Edition.
Edited by Stephen Pryke.
© 2020 John Wiley & Sons Ltd. Published 2020 by John Wiley & Sons Ltd.

'what is permissible' and 'what is not'. Focusing on the network of relationships involved in the design and delivery of projects also requires the adoption of new analytical tools (e.g. Pryke 2017; Choi et al. 2015). It is suggested that application of network science and associated social network analysis can offer great potential in analysing different levels of complexity in supply chains and their management (Wichmann and Kaufmann 2016).

The objective of this chapter is to reconceptualise project supply chains as complex adaptive systems and thus provide a theoretical and methodological basis that helps in understanding the complexity of supply chains in a construction environment. Fundamentally, the concepts of complex adaptive systems and social network analysis will be blended together to focus on the power of informal self-organised networks as counterpart to the formal contractual governance structures and their role in achieving successful project delivery in a complex transient project environment. Exploration of these paradigms generates several conclusions, managerial implications, and proposals for future research.

3.2   Reconceptualising Supply Chains

Although there is no universal definition for supply chain management (Edkins 2009), the definition given by Christopher (1992) is commonly used in literature where it is defined as 'the management of upstream and downstream relationships with suppliers and customers to deliver superior customer value at less cost to the supply chain as a whole'. This definition implies a construction project can be viewed as 'a network of relationships' (Pryke and Smyth 2006) and supply chains are created by linking series/networks of individuals and organisations together, all the way down from clients to suppliers in a manner that creates value in excess of the additional costs associated with maintaining the supply chain itself (Edkins 2009).

Increasingly, contemporary management scholars and the supply chain management discipline have focused on understanding the underlying supply chain relationships beyond the traditional buyer–supplier dyads to view construction projects as networks of social relationships[1] by adopting network science (Loosemore 1998; Borgatti and Li 2009; Pryke 2012). These have been increasingly termed 'supply networks' (Pathak et al. 2007; Nair et al. 2009). Such a network perspective enriches the traditional view of supply chains that have long been examined as 'linear structures' with sets of sequential, vertically organised transactions representing successive stages of value creation (Mabert and Venkataramanan 1998 cited in Nair et al. 2009), by moving closer to the realistic relational behaviours between the involved parties in the supply networks (Nair et al. 2009). Nevertheless, such studies in the construction industry remain rare and the majority of them have examined the relationships that are formally prescribed by hierarchical organisational structure and/or contractual obligations (Pryke 2012; Ruan et al. 2013). Thus, they fall short in understanding the actualities of delivering projects beyond the buyer–supplier contractual relationships.

Figure 3.1 illustrates the concept of the supply chain as a complex network of relationships between individuals and/or organisations. For clarity purposes, a social network 'consists of a finite set or sets of actors and the relation or relations defined on

---

1 Considering both 'hard' (e.g. money and material flow) and 'soft' (e.g. information exchange, friendship, advice) types of relationships (Borgatti and Li 2009; Pryke 2012).

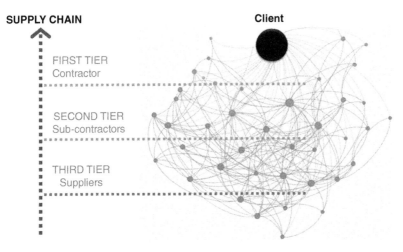

SUPPLY CHAIN   **Client**

FIRST TIER
Contractor

SECOND TIER
Sub-contractors

THIRD TIER
Suppliers

Figure 3.1 Supply chains as networks of individuals and organisations. Source: Adapted from Pryke (2009), p. 2.

them' (Wasserman and Faust 1994, p. 20). The actors are represented by nodes (i.e. circles) while the lines represent the relationships between these actors. This illustrates, in abstract terms, the complex multiscale interactions and interdependencies among different entities, processes, and resources comprising the different tiers of project supply networks. In Figure 3.1 the verticality of the project supply chain in its traditional perspective is abstractly juxtaposing with the nonlinear interactions in supply networks.

The figure aims to demonstrate a snapshot of the actualities of construction projects, where transitory networks of iterative flows can be observed. Examples are knowledge transfer, information exchange, instructions, advice and financial, and contractual relationships (Pryke 2012, 2017). These are activated as needed to respond to specific issues as encountered. Self-organising, nonlinear procurement relates the complex interplay between network structure and its functions which relate to the supply networks' dynamic environment (Surana et al. 2005). The next section will elaborate on this subject.

3.3 Supply Networks as Complex Adaptive Systems

The involvement of complex multiscale interactions within the dynamic environment (for example, changing organisational and market trends) of supply networks indicates that problems of coordination and decision making will loom large. This stresses the need for adaptive and flexible collective behaviour in the supply network to achieve 'client delight'.

A network perspective within a turbulent and dynamic environment can be rooted to complexity theory, which is concerned with 'the study of the dynamics of complex adaptive systems which are nonlinear, have self-organising attributes and emergent properties' (McMillan 2006, p. 25). That is, supply networks need to be recognised and analysed not only as 'systems' but as a complex adaptive systems that are 'emerging, self-organizing, dynamic and evolving' (Choi et al. 2001) exhibiting a highly structured collective behaviour based on local interactions without any centralised control (Stacey 2003).

Complex adaptive systems are composed of autonomously acting agents/actors, co-evolving through their interactions and adapt continuously to the changing environments by changing hierarchies, subgroups, or the whole network (Holland 2002; Wycisk et al. 2008; Stacey 2003). This is a novel approach to study supply networks and understand and manage their organisational and functional emergent dynamics (Choi et al. 2001; Surana et al. 2005).

An extensive review of complex adaptive systems relevant literature suggests that these self-organising systems have the following common characteristics (Auyang 1999; Cooke-Davies et al. 2008; Goldstein 1994; Comfort 1994; Heylighen 2011; von Hayek, 1980 cited in Hülsmann et al. 2007; Kauffman 1993; Lawhead 2015; Molleman, 1998 cited in Mahmud 2009; Prokopenko 2013; Stacey 2003; Haken, 1978 cited in Heylighen et al. 2006; Ulrich and Probst 1984):

1. The process of self-organising is driven intrinsically and collectively, as a joint action by the system's constituents to self-regulate through feedback loops, and does not involve any external interventions;
2. It occurs only when the system operates at the edge of chaos;[2]
3. It requires the internal components to adhere to the system's abstract values and principles which are embedded in generally accepted norms, cultural aspects, traditions, and customs (for example, higher level of trust, honesty, openness, mutual understanding, communication, etc). These are developed over time and as the network evolves;
4. It generates coordinated activity in order to achieve a shared goal and be resistant to any damage;
5. Coordinated activity happens spontaneously (i.e. natural action without any preplanning, controlling, and/or design);
6. Its consequences are unpredictable as each system is capable of exercising choice differently and behaves in a unique way;
7. It leads to the emergence of new structures, patterns, and/or forms of behaviour from randomness but at a higher level of order (i.e. complexity).

Interestingly, Wycisk et al. (2008) have described the parallels between the complex adaptive systems elements, properties and behaviours, and the supply networks. These three dimensions are illustrated in the Figure 3.2 and explained below with the aim of helping to understand the interplay between the different levels and thus the dynamics of complex adaptive systems and supply networks (Wycisk et al. 2008; Marchi et al. 2014).

Elements of Supply Networks

- Actors/agents are firms or individuals, such as suppliers, manufacturers, distributors, retailers, clients, and other firms constituting the entire supply network.
- Actors interact in an environment to exchange material resources, finance, information, and knowledge.

---

2 This is a transition phase at the crossroads of either falling into true chaos, which means full system disintegration, or reshaping the system by an inner 'anti-chaos' force and pulling it back towards order (Stacey 2003).

Figure 3.2 The three dimensions of complex adaptive systems. Source: Adapted from Marchi et al. (2014), Figure 1, p. 445, based on Wycisk et al. (2008).

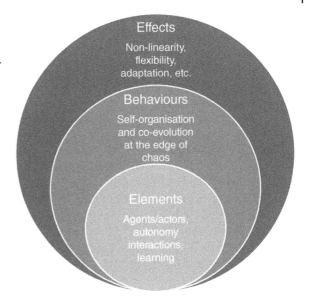

- Actors retain their relative autonomy to operate, for example, in different sectors and markets. Their actions are self-initiated and empowered by delegation and decentralisation of decision making.
- Learning results from interactions and exchange of information, knowledge, and capabilities and thus may present a competitive advantage for the entire supply network.

Behaviours of Supply Networks

- Self-organisation allows the system to reshape itself into a new order. This process of system reshaping arises as a result of an agent's autonomous interactions, that is, a bottom-up driven change (Gell-Mann 1994; Stacey 1996).
- Co-evolution is the result of agents interacting and learning over time, amplified by positive feedback loops (Stacey 1996); for example, contractors implementing and cultivating on subcontractors' knowledge to reduce cost, improve efficiency, etc. As a result, subcontractors innovate and simultaneously adapt to market needs.
- The optimal system behaviour occurs at the 'edge of chaos' – the place where the competing goals of adaptability and efficiency are balanced (Carroll and Burton 2000). This leads to maximum or optimum output (e.g. efficiency, effectiveness, adaptability, and robustness).

Effects of Supply Networks

- Through adaptation processes, the network can interact with its dynamic environmental demands by modifying the network structure (Choi et al. 2001; Comfort 1994), that is, by reshaping the network relational patterns/actors and adding or excluding relations. An example is subcontractors connecting with new suppliers or changing their capabilities (for instance, implementing new technologies) to respond to technological changes (Marchi et al. 2014).

- An understanding of the adaptive capacity of a network to cope with the uniqueness of issues/risks faced in the project sets the scene for unpacking a network's flexibility and resilience; this flexibility relates to the network's ability to maintain topology and functionality under perturbations without leading to catastrophic failure (Marchi et al. 2014; Stacey 1996).

Further description details and examples of these dimensions have been provided by Wycisk et al. (2008) and Marchi et al. (2014), but this chapter only highlights those that are most relevant.

Complexity theory suggests that any well-fragmented coordinated activity with an inherent diversity, such as large construction projects and their supply networks (Tavistock 1966), has a natural evolutionary capacity that is sufficient enough to trigger the self-organising process (Stacey 2003). This can happen at any given point in time and regardless of the phase of project life cycle (Stacey 2003; Pryke et al. 2015). This inevitability is underpinned by the fact that construction projects involve several parties with a multiplicity of authority power but each with different interests and values (Wild 2002). That is to say, the existence of such a large number of conflicting forces and consensus problems (mainly relating to resource allocation and methodologies used to achieve the desired results) will loom large over time and eventually push the system towards the edge of chaos. Thereafter, survival and success of the projects will be highly reliant on the ability to manage unpredictable and nonlinear interactions at the edge of chaos (Bertelsen 2003; Geraldi 2008). An interesting dichotomy exhibited by complex adaptive systems is the coexistence of paradoxical forces at the edge of chaos, for example competition and cooperation (Surana et al. 2005), which is an inherent risk in construction project supply chains in which disputes, agreements, and alliances change the network dynamics, adaptability, and co-evolution process (Marchi et al. 2014).

Dealing with competition as manifested by the rise of opportunistic behaviour, complexity theory postulates that, at the edge of chaos, this adverse phenomenon is counterbalanced by another form of cooperative relationships emerging as the project progresses (Anvuur and Kumaraswamy 2008; Bertelsen 2003). These emergent relationships are classified by some as 'informal', self-organising, and not necessarily closely related to contractual relationships, as illustrated in Figure 3.3 below (Anvuur and Kumaraswamy 2008; Bertelsen 2003; Hopper 1990 cited in Dainty et al. 2007; Cross et al. 2002, but cf. Pryke 2017). These self-organised networks aim to increase collaboration, coordinated behaviour, and goal alignment in the temporary project team and also help to improve problem solving, communication, fast track the processes, and expedite decision making, regardless of any financial incentives (Anvuur and Kumaraswamy 2008; Bertelsen 2003; Coleman 1999). This creates an environment that allows networks to be flexible, resilient, and function effectively, responding to any internal and external permutations in the project environment (Heylighen 2013; Cross et al. 2002).

Therefore, embracing the complex adaptive systems perspective to study supply chains and their management should move our thinking away from the traditional Newtonian paradigm that proposes that the leaders' or management's role in organisations is limited to maintaining equilibrium or stability (Foerster 1984; Mahmud 2009). In a complex adaptive systems paradigm, managers are considered the catalysts and cultivators of a self-organising process (Foerster 1984; Mahmud 2009; Stacey 1996, 2010). Their role is mainly about enabling and facilitating the process as the network evolves and adapts continuously to the changing circumstances of the project and

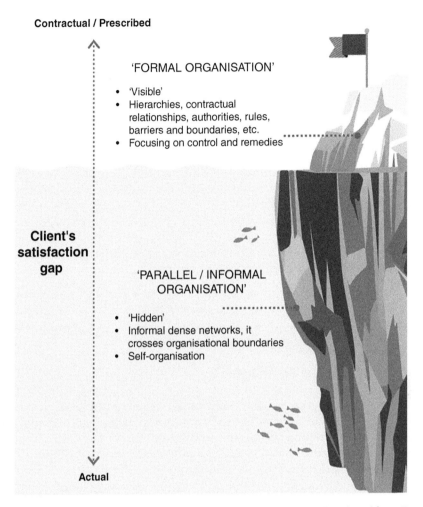

Figure 3.3 Organisation structure as patterns of interaction. Source: Developed from: Emmitt and Gorse 2009, p. 37.

needs of the involved actors (Heylighen 2013). Thus, supply network managers should be aware of these processes and appropriately balance how much they control and how much they allow to emerge (Choi et al. 2001). The latter calls for a redefinition of actors' roles in supply networks and thus the need to develop more appropriate governance structures (Pryke 2017). In this respect, Pryke (2017) calls for a move away from a focus on the functions associated with professional roles like architect, quantity surveyor, and main contractor, towards a focus on each actor's network role. Such redefinition will need to be reflected in reforms in procurement and management strategies in the construction industry (for example, cluster leader, design manager, etc.) (Pryke 2017).

Having said that, the complex adaptive system's perspective has recently been embraced by supply chain management scholars as it is believed to address the challenges of complexity in supply chain management (e.g. Choi et al. 2001; Pathak et al. 2007; Wycisk et al. 2008; Nilsson and Gammelgaard 2012), while social network analysis provides the means by which such complex and informal interactions that comprise projects and supply chains can be understood (Hollenbeck and Jamieson 2015 cited in Wichmann and Kaufmann 2016). In a similar vein, an interest in informal and emergent features of construction industry organisation has grown rapidly over the past two decades, particularly in the UK (Rooke et al. 2009). Although such interest has not been formalised and linked theoretically and methodologically to the complex adaptive systems perspective, such emerging and self-organising features have attracted the attention of Pryke (2017), among others. Pryke (2017) makes a number of points relevant to the discussion here:

- Conceptualising the projects and their associated supply chains as networks enables analysis of what otherwise might be described in quite abstract terms and possibly therefore ill-defined.
- Networks roles are distinct from delegated professional roles; network roles are important and need managing over time.
- Transience is not reflected in contractual relationships and affects behaviour on the ground regardless of those prevailing contractual relationships.

Having laid the theoretical foundation of conceptualising supply chains as networks and related the management of these networks to complexity theory, it might be appropriate to define social network analysis.

3.4 What Is Social Network Analysis?

Social network analysis is a contemporary application of network science to the modelling and analysis of social systems. It is the 'product of an unlikely collaboration between mathematicians, anthropologists and sociologists' (Pryke 2012, p. 77). It offers an alternative powerful formal language to investigate complex systems, by understanding how structure defines the overall social system. Its central axiom is not focused on the actors and their attributes and properties but instead on how they are interconnected (Wasserman and Faust 1994; Borgatti et al. 2014).

Over the past two decades, network science has established itself as a central theoretical framework across a variety of disciplines, relating to individuals, business units, and partnership networks (Wichmann and Kaufmann 2016). An increased interest in the systematic adoption of this paradigm in supply chain management research has been witnessed in the last decade. This is evidenced by the growth rate of the publications referencing social network analysis in supply chain management research as illustrated in Figure 3.4. However, this paradigm is still considered fairly new to supply chain management research with little application (Wichmann and Kaufmann 2016).

Detailed description of social network analysis analytical measures is beyond the limits of this chapter. Those interested in the mathematics behind the measures and a comprehensive review of social network analysis data collection procedures, methods, and application can refer to standard textbooks (e.g. Wasserman and Faust 1994; Borgatti et al. 2013; Prell 2012). Those interested in the application of social network analysis to project and supply chain networks might like to refer to Pryke (2012, 2017).

This section, however, aims to demonstrate why social network analysis is considered a paradigm shift, by highlighting the key features that distinguish social network

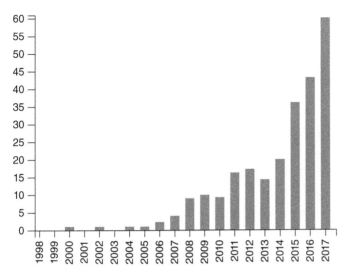

Figure 3.4 Number of publications indexed in Web of Science Core Collection with 'Social Network Analysis' and 'Supply Chain Management' in the topic in the period '1990–2017'. Source: Original – extracted from Web of Science.

analysis from other existing (linear) management and analytical methods (Wasserman and Faust 1994; Pryke 2012; Scott 2017; Nakoinz and Knitter 2016). Social networks have the following useful features:

1. *A topological space.* By appreciating the connectivity, social network analysis suggests a new kind of space where our traditional linear conception of space, which is based on Euclidean geometry,[3] is being stretched and distorted. The space created by connectivity is called a topology, which is of a nonlinear nature and can be used to abstract any inherent connectivity between any set of agents, regardless of their detailed form. This implies an ability to have different sets of global rules for different sets of agents.

2. *Flow.* Connectivity in social network analysis implies that along every connection there is a flow of something – if there is no flow, then there is no network. In communication networks, information flows; in financial networks, assets flow, and so on.

3. *Social actors/agents.* These are the different forces acting within the typological space where an exchange of flow happens between them, representing the connections. These can be 'discrete individual, corporate, or collective social units' (Wasserman and Faust 1994, p. 17). Establishment of the connections between them is based upon the perceived return which is expected to be higher than the investment of time, energy, interest, social capital, or any other resource that is of value.

4. *Micro/macro analysis.* Social network analysis has the capacity to analyse a network, by either taking a micro level bottom-up perspective or a more global perspective. In the former, the focus is on the actors and why and how they make connections,

---

3 Euclidean geometry: 'Relating to or denoting the system of geometry based on the work of Euclid and corresponding to the geometry of ordinary experience' (Trumble, B. and Pearsall, J. (Eds.). (1996) *The Oxford English Reference Dictionary.* Oxford University Press). In a nutshell, this is the study of flat space. (The high school geometry we all know and love!)

whereas in the latter, the focus is on the overall network and the environmental context to see how this shapes the system of connections.

5. *Environment.* Although network science and social network analysis are quite abstract representations, this does not mean researchers will lose sight of the fact that these networks exist within some context. Actually, they still help in contextualising and understanding the nature of the whole network and the kind of environmental forces that it is under.

After broadly defining social network analysis and its key distinguishing features, the next section will provide the rationale for adopting a network approach in understanding, analysing, and visualising construction supply networks and their management.

3.5 Rationale for a Network Approach

Most of the significantly important reviews of the UK construction industry and project procurement conducted during the second half of the twentieth century – such as Simon (1944), Emmerson and Emmerson (1962), Banwell (1964), Wood (1975), the British Property Federation Report (1983), Latham (1994), and Egan (1998) – have remarkably similar findings. However many of the issues discussed and specific recommendations made in these reports are still outstanding (Pryke 2012). This is mainly because these reports have highlighted the shortfalls of the construction industry's systems but failed to implement their recommendations effectively to achieve tangible change within the industry. With hindsight, implementation effectiveness is highly related to the analysis and presentation approaches used (Pryke 2012, 2017).

Clearly, the limitations of the previously used analytical approaches[4] highlight the need for a more rigorous, effective, and systematic approach that is able to provide the appropriate language and terminologies (Pryke 2004). It is argued that such a radical approach entails the turning of our thinking from the artificial boundaries that are imposed by traditional hierarchies and procurement terms (such as project phases, contracts and subcontracts, organisational charts, etc.) towards the idea of viewing value delivered to clients and stakeholders as a product generated by networks of relationships that span organisational and project-related boundaries (Pryke 2012).

The increased interest in informal and emergent features of construction industry organisation entailed extending the analysis and research scope beyond the usual economic attributes of organisational effectiveness (strategy, culture, and operations) and their associated legal obligations, towards focusing on the social dimensions at the individual levels (Ruan et al. 2013). Such a paradigm shift is attributed to the fact that the individual levels set the foundation for such informal and emergent features in the industry (Rooke et al. 2009). This is further underpinned by the fact that the construction industry is essentially a multidisciplinary, people-intensive arena which exhibits a high reliance on face-to-face interactions and interpersonal communications in order to coordinate and facilitate daily activities, problem solving and decision making (Hastings 1998; Pietroforte 1997; Middleton 1996). Moreover, increasingly it is being recognised that successful integration and waste minimisation of the entire

---

4 The review of the previously used analytical approaches and the critique of their limitations are beyond the scope of this chapter. However, this subject was extensively covered in Pryke (2012, Chapter 2).

supply chain depends heavily on information availability to supply network actors in an accurate and timely manner (Surana et al. 2005; Dainty and Brooke 2004). The latter have progressively highlighted the critical role of information technology in supply chain management, with its capability of setting up dynamic information exchange networks, to address the coordination and integration dilemma (Surana et al. 2005).

In this respect, several scholars have highlighted the benefits of implementing building information modelling (BIM) to supply chain management, ranging from design coordination, monitoring cost and time implications, to design changes and logistics (e.g. Irizarry et al. 2013; Cooper et al. 1997). Such discussions stimulated the research for complementary and integrated approaches, for example the geographic information system (GIS)–BIM integrated model (Irizarry et al. 2013) and BIM–social network analysis (Pryke 2017), to successfully design and deliver projects in dynamic and turbulent environments. Such information technology may have the potential to provide a centralised, real-time information flow to project actors while allowing decentralisation of decision making relating to the monitoring, integrating, and analysing of this information (Whyte and Levitt 2011 cited in Pryke 2017).

Therefore, to really understand the project supply network systems and their effectiveness, the associated social network analysis provides an opportunity to better analyse, both graphically (using sociograms) and mathematically in a more common and accessible language, the nature of the inter- and intraorganisational relationships and their function between individuals and firms seamlessly without any boundaries, and hence accurately describe success and failure by providing better focus for delivery of superior value that can achieve customer 'delight' (Pryke 2012).

According to Pryke (2012), social network analysis can overcome the following key problems confronting supply chain management in the construction industry:

- *Interdependence.* By relating networks to specific functions and hence reflecting the roles of the individual actors in the system, rather than what their job titles designate.
- *Nonlinear, complex iterative, and interactive processes.* Social network analysis is sophisticated enough to provide a more interactive means to deal with the complex, concurrent, and interdependent systems. For example, by taking a multilayered network perspective (i.e. the whole network constitutes other subnetworks, which are interdependent), such processes could be understood by analysing relationships of different types of ties among the same set of individuals (Dickison et al. 2016; Boccaletti et al. 2014).
- *The need for a single graphical representation.* By providing one systematic format to represent the different elements of systems and hence easily compare between them using the same language.
- *The need to make analysis at an appropriate level of detail.* By enabling analysis to be made at any selected level (individuals, groups of actors or whole network level) and hence presenting construction systems appropriately where decisions are processed on a concurrent and interdependent basis by groups of actors.
- *Quantification of differences.* By representing the differences between systems, supply chains and projects, and the actors' positions mathematically.
- *Nonhierarchical and nondyadic representation.* By providing a quantifiable measure for the actors' centrality that reflects role importance, prominence, or power within the project/organisation network rather than related to a dyadic contractual

relationship. Social network analysis gives the facility to cite more than two parties and hence represent more closely the team-based nature of construction systems.

- *Interfirm relationships versus interpersonal relationships.* Social network analysis has the ability to represent relationships between project actors in a comparable format to contractual relationships by gathering data from individuals and representing them as relationships between firms (by aggregation). Additionally, social network analysis is able to analyse and compare a variety of governance, procurement, and management forms in a simultaneous, uniform, and systematic manner that is readily understood by both academic and practitioner.
- *Explicit formal statements and measures.* Social network analysis provides very precise meanings to social terms, such as 'networking', 'webs of relationships', 'social position', etc., that might otherwise be defined only in metaphorical terms (Wasserman and Faust 1994). It also allocates mathematical definitions/formula to each of these terms.

The rationale presented here, and the benefits associated with adopting a network perspective in understanding and analysing the complex relationships comprising project supply networks, places great emphasis on studying the communication and information networks supporting these systems. Such an approach has been utilised to study a number of management phenomena, such as: project governance (e.g. Pryke 2004; Imperial 2005), risk management (e.g. Loosemore 1998; Pryke and Ouwerkerk 2003), learning and knowledge transfer (e.g. Reagans and McEvily 2003; Inkpen and Tsang 2005; Fritsch and Kauffeld-Monz 2010), collaboration (Imperial 2005), power and performance (Sparrowe et al. 2001), and supply chains management (e.g. Borgatti and Li 2009; Lee 2005).

Despite the increased momentum of adopting social network analysis to investigate the structure of supply chain networks, the application remains limited in the construction industry (Borgatti and Li 2009). Also, the comprehensive review of Wichmann and Kaufmann (2016) reveals that supply chain management scholars have not exploited the full potential of social network analysis and confined themselves to a handful of widely used measures. The next section will discuss the key challenges that may be behind such pitfalls in conducting social network analysis research.

3.6 Key Challenges in Conducting Social Network Analysis

Regardless of the wide consensus about the effectiveness of social network analysis in analysing complex systems (Ruan et al. 2013), there are, however, limitations and challenges inherent in collecting and analysing social network data in general (Halinen and Törnroos 2005; Kim et al. 2011 cited in Wichmann and Kaufmann 2016) and specifically in the construction industry. Wichmann and Kaufmann (2016) comprehensive review reveals that studies rarely address the limitations and analytical challenges that might occur when conducting network studies. These may include methodological limitations concerned mainly with data validation, limited sample size, and context and sampling bias (Wichmann and Kaufmann 2016). Thus, the paper calls for future social network analysis researchers to have a methodological rigour, conceding the limitations of their studies and how they might respond to them.

Therefore, in order to restore the balance of the discussion, some of these criticisms and limitations concerned with the construction industry can be summarised as follows (Pryke 2004, 2012):

- The special characteristics of the construction industry (e.g. multiplexity of networks, etc.) make it hard to clearly identify the boundaries between different networks. However, with social network analysis, precise classification of each network is essential, leading to a potentially large number of different relationships and volume of data; understanding the context for any study is important.
- Similar to most statistical analysis, social network analysis is a quantitative method used within an interpretive context (Loosemore 1998). Thus, it can only provide meaningful information when combined with qualitative data.
- Social network analysis is a complex method, creating a barrier to entry. It requires personal investment to establish the skills and knowledge to undertake successful research, particularly in terms of the application of social network analysis-specific software[5] for the analysis and visualisation of networks (e.g. UCINET, Gephi, etc.).
- Sampling is not appropriate with networks and a 100% response from all actors is preferred to achieve accuracy and thus meaningful results.[6]

3.7   Conclusions and Directions for Future Research

The novel contribution of this chapter lies in its analysis of supply networks from a new theoretical framework: complexity science. It highlights the need to understand and analyse supply chain networks as complex adaptive systems to effectively design and deliver construction projects. The main management emphasis, therefore, should shift to the informal and emergent networks that underlie supply chain networks and their associated projects (Pryke 2017).

As has been discussed earlier, adopting a complex adaptive systems perspective joined with social network analysis has the capability to respond to many challenges confronting supply chains and their management. Therefore, supply chain management scholars, particularly in the construction industry, should increasingly incorporate network science and associated social network analysis to understand the processes, systems, and functionality of project supply chains. Through our efforts to link the two approaches, we would like to identify new research directions and stimulate a growing interest and further substantial research in construction supply networks from a complex adaptive systems point of view.

Several proposals for future work arise from adopting such a paradigm. First, numerous phenomena, such as supply chain complexity, adaptability, resilience, and sustainability have been extensively studied through different theoretical frameworks.

---

5 Refer for example to Scott (2017) for a comprehensive overview of social network analysis and the different software programs.
6 Pryke (2017) notes this is a contentious point, and software exists to simulate missing data from large networks. He has challenged the idea that missing data can be simulated in the context of large complex projects and their associated supply chain networks, in the same way that one might do with a network for infectious diseases – this is because complex networks in construction are not homogeneous with a large number of specialised clusters dealing with project functions, such as design management and financial management for example.

However, they remain not fully understood. Such phenomena are inherently suited to investigation using social network analysis (Wichmann and Kaufmann 2016).

Second, from a methodological perspective Wichmann and Kaufmann (2016) and Ruan et al.'s (2013) comprehensive reviews call scholars to exploit the full potential of social network analysis. That is, greater insights about inherently complex interrelations can be unlocked by investigating a broader range of network measures; there is great potential to move beyond basic network analytical measures such as centrality (Borgatti et al. 2013).

Third, as social network analysis involves no assumptions about hierarchy, it provides the ideal method to understand the nature of how work is coordinated in the self-organised networks that we see in complex construction schemes (Pryke 2017). Therefore, applying social network analysis to supply chain management research provides opportunities in responding to recent calls to redefine project actor roles reflecting the reforms in procurement and management strategies prevailing in the construction industry (e.g. Pryke 2012, 2017). In the pursuit for more effective management approaches capable of supporting the integration and coordination of the supply chain, social network analysis, as an analytical method, has the capability to capture the hidden functionality of projects to a level of accuracy that cannot be obtained using the traditional scientific methods of analysis. For example, utilising centrality measures could highlight the dynamic nature of network roles that supports the functions performed collaboratively by the supply chain as they evolve through iterative information exchange relationships in an environment which is essentially complex and transient. Mapping these informal communication networks using social network analysis provide opportunities for managers to identify 'central connectors' (prominent actors, perhaps acting as bridges or brokers) and 'bottlenecks' (perhaps structural holes) in the information exchange process (Cross and Prusak 2002 cited in Wichmann and Kaufmann 2016). Such knowledge could enhance the effectiveness of these networks and possibly provide organisational costs savings (Pryke 2017).

Fourth, from a theoretical perspective it is essential to move beyond the simple 'single-layer' networks that have been studied under the 'traditional' network theory and investigate more complicated but more realistic frameworks (Dickison et al. 2016; Boccaletti et al. 2014). This means conceptualising project supply chain networks as multilayer networks, and thus what happens at a single level/layer of interaction affects the structure and function at other interconnected layers (e.g. Padgett and McLean 2006). Unfortunately, for so long these interactions have been deduced and have largely been studied through the 'traditional' single-layered perspective (Borgatti and Li 2009). This new paradigm in network science matches the concepts of complex adaptive systems well and is believed to be the next step towards a better and greater comprehension of any modern social system (Dickison et al. 2016), including project supply chains.

3.8 Managerial Implications

In the turbulent environment of the construction industry, the management of supply networks should move away from rigidity of structures and interactions imposed by the control efforts (Christopher and Holweg 2011). Supply networks should aim to dwell on the edge of chaos, and thus management efforts should focus on embracing complexity by stimulating interactions (Stacey 2010). For example, contractor and subcontractor relationships should move beyond the traditional buyer–supplier contractual relations, and include information and knowledge exchange with the aim for innovation, the development of new product/service, reducing cost, improving efficiency, etc. Therefore, the management role is mainly about enabling and facilitating the process as the network evolves and adapts continuously to the changing circumstances of the project and the needs of the involved actors (Heylighen 2013). And thus, supply network managers should be aware of these processes and appropriately balance how much they control and how much they let emerge (Choi et al. 2001)

Stimulating interactions should be encouraged beyond the supply network participants. Autonomous interactions beyond the network should be stimulated and supplier lock-in and hold-up situations should be avoided (Christopher and Holweg 2011; Marchi et al. 2014). Such stimulus for autonomy allows interactions with other networks and thus opportunities for flexibility, information, and knowledge exchange. Consequently, learning can be amplified (Marchi et al. 2014).

Firms and supply networks and their environments co-evolve in a nonlinear and dynamic manner when an interaction and autonomy of action is promoted. While balancing control and flexibility is of essence, the result of this stimulus is a network that is self-organised, and capable of learning and creating novelty (Marchi et al. 2014). Management, therefore, should identify the capabilities required to achieve positive outcomes and facilitate their continuous adaptation in the supply network (Marchi et al. 2014). Interdisciplinary clusters dealing with all tiers of the supply network might solve problems collectively, for example.

References

Anvuur, A. and Kumaraswamy, M. (2008). Better collaboration through cooperation. In: *Collaborative Relationships in Construction: Developing Frameworks and Networks* (eds. H. Smyth and S.D. Pryke). Oxford: Wiley Blackwell.

Auyang, S.Y. (1999). *Foundations of Complex-System Theories: In Economics, Evolutionary Biology, and Statistical Physics*. Cambridge University Press.

Banwell, S.H. (1964). *The Placing and Management of Contracts for Building and Civil Engineering Work: Report of the Committee [on the Placing and Management of Contracts for Building and Civil Engineering Work]*. HM Stationery Office.

Bertelsen, S. (2003). Construction as a complex system. In: *Proceedings of International Group for Lean Construction*, vol. 11. Blacksburg, Virginia.

Boccaletti, S., Bianconi, G., Criado, R. et al. (2014). The structure and dynamics of multilayer networks. *Physics Reports* 544 (1): 1–122.

Borgatti, S.P. and Li, X. (2009). On social network analysis in a supply chain context. *Journal of Supply Chain Management* 45 (2): 5–22.

Borgatti, S.P., Everett, M.G., and Johnson, J.C. (2013). *Analyzing Social Networks*. Sage.

Borgatti, S.P., Mehra, A., Labianca, G.J., and Brass, D.J. (eds.) (2014). *Contemporary Perspectives on Organizational Social Networks*, vol. 40. Emerald Group Publishing.

Carroll, T. and Burton, R.M. (2000). Organizations and complexity: searching for the edge of chaos. *Computational and Mathematical Organization Theory* 6 (4): 319–337.

Choi, T.Y., Dooley, K.J., and Rungtusanatham, M. (2001). Supply networks and complex adaptive systems: control versus emergence. *Journal of Operations Management* 19 (3): 351–366.

Choi, T.Y., Shao, B., and Shi, Z.M. (2015). Hidden suppliers can make or break your operations. *Harvard Business Review* 29: 1–5.

Christopher, M. (1992). *Logistics and Supply Chain Management – Strategies for Reducing Cost and Improving Services*. London: Financial Times Professional Ltd.

Christopher, M. and Holweg, M. (2011). "Supply Chain 2.0": managing supply chains in the era of turbulence. *International Journal of Physical Distribution & Logistics Management* 41 (1): 63–82.

Coleman, H. Jr., (1999). What enables self-organizing behavior in businesses. *Emergence* 1 (1): 33–48.

Comfort, L. (1994). Self-organization in complex systems. *Journal of Public Administration Research and Theory: J-PART* 4 (3): 393–410.

Cooke-Davies, T., Cicmil, S., Crawford, L., and Richardson, K. (2008). We're not in Kansas anymore, Toto: mapping the strange landscape of complexity theory, and its relationship to project management. *IEEE Engineering Management Review* 36 (2): 5–21.

Cooper, M.C., Lambert, D.M., and Pagh, J.D. (1997). Supply chain management: more than a new name for logistics. *The International Journal of Logistics Management* 8 (1): 1–14.

Cross, R., Borgatti, S., and Parker, A. (2002). Making invisible work visible: using social network analysis to support strategic collaboration. *California Management Review* 44 (2): 25–46.

Dainty, A.R. and Brooke, R.J. (2004). Towards improved construction waste minimisation: a need for improved supply chain integration? *Structural Survey* 22 (1): 20–29.

Dainty, A., Moore, D., and Murray, M. (2007). *Communication in Construction: Theory and Practice*. Routledge.

Dickison, M.E., Magnani, M., and Rossi, L. (2016). *Multilayer Social Networks*. Cambridge University Press.

Edkins, A. (2009). Risk management and the supply chain. In: *Construction Supply Chain Management: Concepts and Case Studies* (ed. S.D. Pryke). Wiley Blackwell.

Egan, J. (1998). *Rethinking Construction*. London: The Department of the Environment, Transport and the Regions.

Emmerson, H. and Emmerson, S.H.C. (1962). *Survey of Problems before the Construction Industries: Report Prepared for the Minister of Works*. HM Stationery Office.

Emmitt, S. and Gorse, C.A. (2009). *Construction Communication*. Wiley.

British Property Federation (1983). *Manual of the BPF System: The British Property Federation for Building Design and Construction*. British Property Federation.

Flyvbjerg, B. (2009). Survival of the unfit test: why the worst infrastructure gets built–and what we can do about it. *Oxford Review of Economic Policy* 25 (3): 344–367.

Foerster, v.H. (1984). Principles of self-organization – in a socio-managerial context. In: *Self-Organization and Management of Social Systems*, vol. 26 (eds. H. Ulrich and G.J.B. Probst), 2–29. Berlin, Heidelberg: Springer Berlin Heidelberg.

Fritsch, M. and Kauffeld-Monz, M. (2010). The impact of network structure on knowledge transfer: an application of social network analysis in the context of regional innovation networks. *The Annals of Regional Science* 44 (1): 21.

Gell-Mann, M. (1994). Complex adaptive systems. In: *Complexity: Metaphors, Models and Reality. Perseus Books* (eds. G. Cowen, D. Pines and D. Meltzer), 17–28.

Geraldi, J.G. (2008). The balance between order and chaos in multi-project firms: a conceptual model. *International Journal of Project Management* 26 (4): 348–356.

Goldstein, J. (1994). *The Unshackled Organization: Facing the Challenge of Unpredictability Through Spontaneous Reorganization*. Productivity Press.

Halinen, A. and Törnroos, J.Å. (2005). Using case methods in the study of contemporary business networks. *Journal of Business Research* 58 (9): 1285–1297.

Hastings, C. (1998). The virtual project team-part 3. *Project Manager Today* 10: 26–29.

Heylighen, F. (2011). Rationality, complexity and self-organization. *Emergence: Complexity and Organization* 13 (1–2): 133–145.

Heylighen, F. (2013). Self-organization in communicating groups: the emergence of coordination, shared references and collective intelligence. In: *Understanding Complex Systems*, 117–149.

Heylighen, F., Cilliers, P., & Gershenson, C. (2006). Complexity and Philosophy. arXiv preprint cs/0604072.

Holland, J.H. (2002). Complex adaptive systems and spontaneous emergence. In: *Complexity and Industrial Clusters*, 25–34. Physica-Verlag HD.

Hülsmann, M., Grapp, J., Li, Y., and Wycisk, C. (2007). Self-Organization in Management Science. In: *Understanding Autonomous Cooperation and Control in Logistics – The Impact on Management, Information and Communication and Information Flows* (eds. M. Hülsmann and K. Windt), 169–192. Berlin: Springer.

Imperial, M.T. (2005). Using collaboration as a governance strategy: lessons from six watershed management programs. *Administration & Society* 37 (3): 281–320.

Inkpen, A.C. and Tsang, E.W. (2005). Social capital, networks, and knowledge transfer. *Academy of Management Review* 30 (1): 146–165.

Irizarry, J., Karan, E.P., and Jalaei, F. (2013). Integrating BIM and GIS to improve the visual monitoring of construction supply chain management. *Automation in Construction* 31: 241–254.

Kauffman, S.A. (1993). *The Origins of Order: Self Organization and Selection in Evolution*. Oxford University Press.

Kim, Y., Choi, T.Y., Yan, T., and Dooley, K. (2011). Structural investigation of supply networks: a social network analysis approach. *Journal of Operations Management* 29 (3): 194–211.

Latham, M. (1994). *Constructing the Team*. London: The Stationery Office.

Lawhead, J. (2015). Self-Organization, Emergence, and Constraint in Complex Natural Systems, arXiv:1502.01476.

Lee, P.D. (2005). Measuring supply chain integration: a social network approach. *Supply Chain Forum: An International Journal* 6 (2): 58–67.

Loosemore, M. (1998). Social network analysis: using a quantitative tool within an interpretative context to explore the management of construction crises. *Engineering, Construction and Architectural Management* 5 (4): 315–326.

Mahmud, S. (2009). Framework for the role of self-organization in the handling of adaptive challenges. *Emergence: Complexity and Organization* 11 (2): 1–14.

Marchi, J.J., Erdmann, R.H., and Rodriguez, C.M.T. (2014). Understanding supply networks from complex adaptive systems. *BAR-Brazilian Administration Review* 11 (4): 441–454.

McMillan, E. (2006). *Complexity, Organizations and Change: An Essential Introduction*. Routledge.

Middleton, D. (1996). Talking work: Argument, common knowledge and improvisation in multidisciplinary teams. In: *Cognition and Communication at Work* (eds. Y. Engestrom and D. Middleton), 233–256. Cambridge University Press, Cambridge.

Nair, A., Narasimhan, R., and Choi, T.Y. (2009). Supply networks as a complex adaptive system: toward simulation-based theory building on evolutionary decision making. *Decision Sciences* 40 (4): 783–815.

Nakoinz, O. and Knitter, D. (2016). *Modelling Human Behaviour in Landscapes: Basic Concepts and Modelling Elements.* Springer.

Nilsson, F. and Gammelgaard, B. (2012). Moving beyond the systems approach in SCM and logistics research. *International Journal of Physical Distribution and Logistics Management* 42 (8/9): 764–783.

Padgett, J.F. and McLean, P.D. (2006). Organizational invention and elite transformation: the birth of partnership systems in Renaissance Florence. *American Journal of Sociology* 111 (5): 1463–1568.

Pathak, S.D., Day, J.M., Nair, A. et al. (2007). Complexity and adaptivity in supply networks: building supply network theory using a complex adaptive systems perspective. *Decision Sciences* 38 (4): 547–580.

Pietroforte, R. (1997). Communication and governance in the building process. *Construction Management and Economics* 15 (1): 71–82.

Prell, C. (2012). *Social Network Analysis: History, Theory and Methodology.* SAGE Publications.

Prokopenko, M., Boschetti, F., and Ryan, A. (2013). *Guided Self-Organization: Inception.* Springer Science and Business Media.

Pryke, S.D. (2004). Analysing construction project coalitions: exploring the application of social network analysis. *Construction Management and Economics* 22 (8): 787–797.

Pryke, S.D. (2009). *Construction Supply Chain Management: Concepts and Case Studies.* Oxford: Wiley Blackwell.

Pryke, S. (2012). *Social Network Analysis in Construction*, vol. 1. Supply Chain Management.

Pryke, S. (2017). *Managing Networks in Project-Based Organisations.* Wiley.

Pryke, S.D. and Ouwerkerk, E. (2003) Post-completion risk transfer audits: an analytical risk management tool using social network analysis. Proceedings of 2003 Construction and Building Research Conference of the RICS Research Foundation (COBRA 2003).

Pryke, S. and Smyth, H. (2006). *The Management of Complex Projects: A Relationship Approach.* Oxford: Blackwell.

Pryke, S., Badi, S. M., Soundararaj, B., Watson, E., and Addyman, S. (2015). Self-organising networks in complex projects. Proceedings of COBRA AUBEA 2015.

Reagans, R. and McEvily, B. (2003). Network structure and knowledge transfer: the effects of cohesion and range. *Administrative Science Quarterly* 48 (2): 240–267.

Rooke, J.A., Koskela, L., and Kagioglou, M. (2009). Informality in organization and research: a review and a proposal. *Construction Management and Economics* 27 (10): 913–922.

Ruan, X., Ochieng, D. E. G., Price, A. D., and Egbu, C. O. (2013) Time for a real shift to relations: appraisal of Social Network Analysis applications in the UK construction industry. *Australasian Journal of Construction Economics and Building*, 13, 92–105.

Scott, J. (2017). *Social Network Analysis.* Sage.

Simon, S.E. (1944). *The Placing and Management of Building Contracts: Report of the Central Council for Works and Buildings.* HM Stationery Office.

Sparrowe, R.T., Liden, R.C., Wayne, S.J., and Kraimer, M.L. (2001). Social networks and the performance of individuals and groups. *Academy of Management Journal* 44 (2): 316–325.

Stacey, R.D. (1996). *Complexity and Creativity in Organizations*. Berrett-Koehler Publishers.

Stacey, R. (2003). *Complex Responsive Processes in Organizations: Learning and Knowledge Creation*. Routledge.

Stacey, R.D. (2010). *Complexity and Organizational Reality: Uncertainty and the Need to Rethink Management after the Collapse of Investment Capitalism*. Routledge.

Surana, A., Kumara, S., Greaves, M., and Raghavan, U.N. (2005). Supply-chain networks: a complex adaptive systems perspective. *International Journal of Production Research* 43 (20): 4235–4265.

Tavistock Institute of Human Relations (1966). *Interdependence and Uncertainty: A Study of the Building Industry: Digest of a Report from the Tavistock Institute to the Building Industry Communication [s] Research Project*. Tavistock.

Ulrich, H. and Probst, G. (1984). *Self-Organization and Management of Social Systems*. Berlin: Springer-Verlag.

Wasserman, S. and Faust, K. (1994). *Social Network Analysis: Methods and Applications*, vol. 1, 116. Cambridge University Press.

Wichmann, B.K. and Kaufmann, L. (2016). Social network analysis in supply chain management research. *International Journal of Physical Distribution & Logistics Management* 46 (8): 740–762.

Wild, A. (2002). The unmanageability of construction and the theoretical psycho-social dynamics of projects. *Engineering Construction and Architectural Management* 9 (4): 345–351.

Winch, G. (2002). *Managing Construction Projects: An Information Processing Approach*. Wiley.

Wood Report (1975). *Building and Civil Engineering Economic Development Committee's Joint Working Party Studying Public Sector Purchasing, the Public Client and the Construction Industries*. London: National Economic Development Office (NEDO).

Wycisk, C., McKelvey, B., and Hülsmann, M. (2008). "Smart parts" supply networks as complex adaptive systems: analysis and implications. *International Journal of Physical Distribution & Logistics Management* 38 (2): 108–125.

4

Green Supply Chain Management in Construction: A Systematic Review

Niamh Murtagh and Sulafa Badi

4.1 Introduction

As a sector, construction imposes a heavy toll on the natural environment. The industry consumes enormous quantities of raw materials. It produces prodigious amounts of waste and is responsible for a significant proportion of global carbon emissions. The Fifth Intergovernmental Panel on Climate Change Assessment Report attributes a third of total final energy use globally to buildings. More significant still, the report finds that the built environment produces a third of black carbon emissions which are a form of carbon emissions with greater potential to speed up the melting of Arctic ice sheets (Lucon et al. 2014). Environmental sustainability in construction therefore is a pressing concern for the industry – and society – globally (Kibert et al. 2000). In response, sustainable or green supply chain management has much to offer. Practitioner and research interest in green supply chain management in construction is developing, complementing the burgeoning literature on green supply chain management in sectors such as the automotive and electronics industries. However, the challenges for the construction sector are in some ways different from other industries and a systematic review of the application of green supply chain management in our sector has not yet been published. This chapter aims to address the gap by providing a timely review of existing research on green supply chain management in construction. It complements the other chapters in this book in its focus on the environmental performance of supply chains in delivering the built environment. The objective is to collate, describe, and synthesise existing work in the area, to provide a comprehensive and accessible summary to postgraduate students, researchers, and interested practitioners.

The chapter is based on a systematic review of the literature. We first set the context for the review by summarising the evidence for the environmental impact of construction, and governmental and industry concerns and initiatives. After clarifying key definitions, we explain the methodology of systematic review. An analysis of the literature is then described. A synthesis of the findings from the literature is discussed and finally, practical implications and areas for future research are presented.

Successful Construction Supply Chain Management: Concepts and Case Studies, Second Edition.
Edited by Stephen Pryke.
© 2020 John Wiley & Sons Ltd. Published 2020 by John Wiley & Sons Ltd.

4.1.1   Environmental Impact of Construction

The built environment consumes high levels of energy (Lucon et al. 2014). It is also recognised to be heavily dependent on extraction of raw materials (European Commission 2008). Construction and demolition waste constitutes one of the highest volume waste streams in the EU, generating between a quarter and a third of all waste (European Commission 2015). With the global population now over seven billion, increasing urbanisation and the spreading built environment are contributing to rapid escalation of rates of loss of biodiversity. Biodiversity loss has been identified as having already passed the 'planetary boundary' threshold which brings potentially disastrous consequences for humanity (Rockström et al. 2009). The construction sector's high consumption of resources and energy, both in delivery of the built environment and in operation of buildings and infrastructure thereafter, its effect on biodiversity, and its generation of heavy, voluminous, and mixed waste, contribute to its particularly detrimental environmental impact. This impact is exacerbated by the long-term nature of the final product: today's buildings and infrastructure decrease future territory for many species, 'lock in' energy consumption over the next decades (Lucon et al. 2014), and represent a potential legacy of future waste. National and regional governments worldwide have recognised the challenges and are reacting with tighter legislation. Examples include energy efficiency in UK building regulations and waste reduction directives in the EU. There are also initiatives to lead by example, such as California's success in reducing the greenhouse gas emissions from government buildings by 40% between 2010 and 2016 (CA Governor 2016) and the Masdar City project in the United Arab Emirates which plans to develop a wholly clean energy-powered city.

The construction industry too has begun to address the challenges, with over 70 national Green Building Councils working to advise stakeholders of the built environment, offer leadership and co-ordination, and conduct campaigning programmes. For example, the UK Green Building Council has brought together business leaders from across the industry to lobby government on raising environmental standards and to aim for net-zero carbon standards required of all new buildings by 2030 (March 2018). In Australia, the Green Building Council contributed to a city dashboard, which will be used to monitor performance of 21 cities on sustainability criteria and liveability, amongst other factors. Singapore's Green Building Council runs a Green Schools Initiative which aims to educate students on the built environment and environmental sustainability, and to impart basic principles of green building design.

Environmental performance assessments such as BREEAM and LEED are increasingly applied to new developments. As of March 2018, over two million buildings across 77 countries were registered with BREEAM (the UK-led Building Research Establishment Environmental Assessment Method) and over two million square feet of building development are certified daily with LEED (Leadership in Energy and Environmental Design, a US-initiated globally recognised standard of achievement in sustainability). Innovations in construction materials are gaining wider acceptance. These include both product innovation, such as sustainable structural wood products like glulam, and process innovation, such as off-site prefabrication, which greatly reduces waste. Individual construction firms have begun to pursue innovative management practices. These have often taken a lead from the manufacturing industry, for example in applying lean operation and supply chain management (Green and May 2005; Fulford and Standing 2014). Major contractors, in particular, have championed supply chain management principles in construction (Karim et al. 2006; Eriksson et al. 2007) and these supply chain management practices have been implemented mostly at the client/main contractor interface.

The need for collaboration within and across sectors in the construction supply chain, and for systemic approaches, has been proposed as crucial for more rapid progress (Fischedick et al. 2014). The domain of green supply chain management offers such a systemic and collaborative approach. Green supply chain management is therefore likely to be critical to enabling more sustainable construction and reducing the sector's negative impact on the natural world of today and tomorrow.

The burgeoning literature on green supply chain management in different sectors over the past two decades has influenced the growing interest by practitioners and the research community in green supply chain management in construction. For example, industry and government collaboration produced an action plan for plasterboard aimed at reducing the life cycle impacts of the material (Dadhich et al. 2015). However, the challenges for the construction sector are in some ways different from other industries. A systematic review of the application of green supply chain management in construction has not yet been performed, to our knowledge. Addressing this gap, the chapter describes the status of research in the field of green supply chain management in construction. What has been done and where? What topics have been included? What insights have been uncovered and what are the practical implications?

4.1.2   Definition

Before describing the methodology for the research, it is important to clarify the primary term and distinguish it from similar concepts. Christopher (2011, p. 26) defines a 'supply chain' as 'the network of organizations that are involved, through upstream and downstream linkages, in the different processes and activities that produce value in the form of products and services in the hands of the ultimate consumer'. Building on this, and an understanding of supply chain management offered in earlier chapters, at this point we define green supply chain management as all initiatives aimed at reducing the environmental impact of the supply chain for the built environment. A more detailed definition will be outlined in Section 4.4.2 below. We distinguish green supply chain management from sustainable supply chain management: sustainable supply chain management is the broader concept which incorporates economic and social sustainability as well as environmental (Ahi and Searcy 2013; Pagell and Shevchenko 2014) and forms part of sustainable operations management (Walker et al. 2014). Here, however, the focus is *green* supply chain management – the impact of supply chains on the natural environment, in terms of energy consumption, use of materials, and generation of waste. Green supply chain management also differs from lean supply chain management: lean principles address the identification and managing out of waste with the objectives of increasing efficiency, lowering cost, and providing improved value to the customer (Banawi and Bilec 2014). Although lean and green overlap, the objectives of green supply chain management centre on environmental performance and target a broader range of criteria.

Having defined green supply chain management, we next describe the research methodology.

4.2   Research Methodology

In order to provide a comprehensive and rigorous review of the literature to date, we chose a method known as a systematic review of the literature. This approach offers a means to conduct a critical appraisal using a methodical and transparent process. The systematic review of literature method originated in the field of health as a method which provides a rigorous approach to the synthesis of empirical data. It is now applied increasingly widely in a broad variety of domains where a thorough and comprehensive review can contribute to knowledge (Briner and Denyer 2012; Caiado et al. 2017). The systematic review of literature differs from a nonsystematic or expert review in its explicit recording of all decisions pertinent to the review. Nonsystematic reviews may not make clear the extent and sources of their search or the criteria for inclusion or exclusion of material. The reader depends on the author's expertise in having selected the most important or highest quality papers. But the reader may then be left unsure if a study has been omitted for deliberate reasons or through oversight. There is uncertainty as to what constituted 'importance' or 'quality' in selecting the studies to review. The authors may have been prone to unconscious biases in selecting studies they were most familiar with or enjoyed reading. In contrast, the systematic review aims to specify its method in sufficient detail for replicability. The selection criteria are presented in detail, as are the bases for decision making. The systematic review of literature involves more than one author to allow cross-verification so that the reader can be assured that every attempt was made to avoid biased selection. Having mapped the territory in an explicit manner, the systematic review then critically examines the included studies. The findings are then brought together to provide a coherent account of the field.

The objective of our review here was to evaluate the status of the field of green supply chain management in construction. The research questions were broadly focused:

- What is the status of research on green supply chain management in construction?
- What are the key insights and practical recommendations?

The stages in the method followed those set out in Gough et al. (2012):

Stage 1 – Define eligibility criteria
Stage 2 – Define search terms
Stage 3 – Search, screen, and compile a list of included papers
Stage 4 – Code and critically evaluate included studies
Stage 5 – Formulate synthesis

4.2.1   Stage 1: Define Eligibility Criteria

Papers referencing sustainable, green, or environmental supply chains were included – a small number of papers focusing only on the social aspects of sustainable supply chains were excluded. In determining whether a paper was on green supply chain management, the guiding rule was that, for inclusion, the research or discussion had to consider the perspectives of different supply chain actors or to cross stakeholder boundaries. Where there was doubt about whether a paper qualified, it was included in the final dataset so that potentially valuable insights were not lost. Papers that addressed any aspect of construction were included, whether focused on a specific subsector such as road maintenance or construction more generally.

4.2.2   Stage 2: Define Search Terms

The search terms used were:

- Supply chain AND (green OR sustainable) AND construction in Subject
- Publication date up to August 2017
- In articles (papers)
- Language English

4.2.3   Stage 3: Search, Screen, and Compile List of Included Papers

In order to ensure high quality, only papers from peer-reviewed journals were selected – books, book chapters, conference papers, and articles in trade journals were excluded as peer review may be either absent or not consistently applied across these media. In order to ensure a comprehensive search for the current study, the search engine selected was Explore, a proprietary front-end populated with meta-data from over 500 sources. In addition, searches were conducted on SCOPUS, Web of Knowledge, and the ARCOM (Association of Researchers in Construction Management) abstract database directly, to ensure the widest possible range of sources. The search yielded 207 papers with 44 papers finally selected for detailed analysis.

4.2.4   Stage 4: Code and Critically Evaluate Included Studies

The coding structure is presented in Table 4.1. Data from each paper were entered into a table of codes. Analysis proceeded by writing a short overview by code, going back to papers where additional questions emerged.

Table 4.1 Coding structure.

1	**Geographic location**
2	**Paper focus** – construction industry generic or specific (e.g. road maintenance, residential)
3	**Stage** – planning, design, procurement, construction, operation, demolition, disposal
4	**Definition** of supply chain management/green supply chain management/sustainable supply chain management
5	**Study/paper aims**
7	**Method** – case study/mixed/quantitative/qualitative
8	**Tools and techniques** – e.g. decision-making, environmental regulation
9	**Stakeholders** – including responsibilities of stakeholders
10	**Main findings**
11	**Practical implications**

4.2.5   Stage 5: Formulate Synthesis

Working from the descriptions by code produced in the analysis stage, the most salient themes were drawn out. In addition, the authors applied a critical stance to the data to look for gaps. This stage in the process is in part subjective, depending on the themes judged to be important by the authors. Other researchers may have chosen other themes to highlight.

4.3   Analysis

In this section, we present a short description of the literature under each of the main codes, in order to characterise the field to date.

4.3.1   Research Interest over Time

No qualifying papers were found from before the year 2000. A small proportion (16%) were published up to and including 2011 and the field shows an increase in interest from 2012 and a more dramatic upturn in 2016 and 2017 (see Figure 4.1). With growth in interest in green supply chain management in general being mapped to the early 1990s (Zhu and Sarkis 2006), it would appear that construction and allied research fields have been slow to adopt the concept but that research interest in application of the concept is now firmly underway.

4.3.2   Source Journals

The 44 papers appeared in a surprisingly large variety of journals, 31 in total, with only four publications including multiple papers on the topic. *Journal of Cleaner Production*

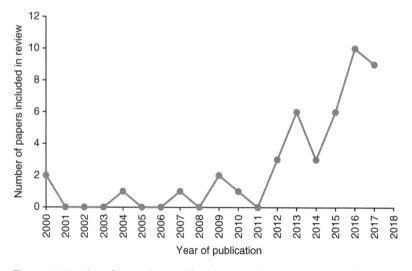

Figure 4.1 Number of papers by year of publication. Note: papers published up to 31 August 2018 were reviewed so the 2017 total is partial. Source: Original.

Figure 4.2 Geographic location of studies. Source: Original.

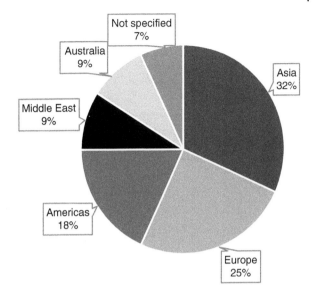

published 10 of the papers, *Sustainability* (Switzerland) published three, and *Journal of Construction Engineering and Management*, and *WIT Transactions on Ecology and Environment*, each published two. With papers published in disciplinary outlets ranging from Waste Management to Resource Policy to Building Research and Information, there is evidence of wide interest in green supply chain management in construction and related topics, with recognition of its importance across many disciplines. The range of publication outlets also speaks to the multifaceted nature of the green supply chain management construct.

4.3.3 Geographic Spread

The papers in the review provided generally good international representation, although only one study took place in South America and none of the papers discussed construction supply chains in sub-Saharan Africa (see Figure 4.2).

4.3.4 Methods

Table 4.2 presents the methods used in the papers reviewed.

A variety of methods has been used to examine different facets of green supply chain management in construction. Fifteen papers applied specialised quantitative methods, including material flow analysis, modelling, life cycle analysis, and organisational environmental footprint to examine particular questions within the green supply chain management process. Balasubramanian (2014) applied a robust quantitative method, unusual in this literature, to derive structural factors influencing an organisation's move to green supply chain management. The remainder of the studies used a range of methods: interviews only (5), survey only (6), mixed methods of interviews and surveys (5), and case studies (6). Across the methods, in most cases the number of cases is appropriate for detailed analysis, with up to 31 individual professionals being interviewed, and usable responses on surveys ranging between 39 and 455. The

Table 4.2 Methods used in review papers.

Method	Papers	Metrics
Interviews	Zuo et al. (2009)	n not given
	Rizzi et al. (2014)	$n = 28$, observations, secondary data
	Udawatta et al. (2015)	$n = 16$
	Bohari et al. (2017)	$n = 4$
	Balasubramanian and Shukla (2017b)	$n = 31$
Surveys	Zou and Couani (2012)	$n = 91$; response rate $= 36\%$
	Ketikidis et al. (2013)	$n = 60$; response rate $= 52\%$
	Adawiyah et al. (2015)	$n = 94$; response rate not given
	Seth et al. (2016)	$n = 347$; response rate $= 69\%$
	Balasubramanian and Shukla (2017a)	$n = 455$; response rate $= 19\%$
	Shen et al. (2017)	$n = 39$; response rate not given
Mixed	Dainty and Brooke (2004)	Interviews n not given; survey $n = 27$; response rate $= 60\%$
	Ruparathna and Hewage (2015)	Interviews $n = 9$; survey $n = 30$; response rate $= 10\%$
	Chileshe et al. (2016)	Interviews $n = 6$; survey $n = 49$; response rate $= 9\%$
	Salzer et al. (2016)	Interviews $n = 29$; field observations $n = 16$
	Wong et al. (2016)	Interviews $n = 7$; survey $n = 84$; response rate $= 70\%$
Case study	Da Rocha and Sattler (2009)	Cases $= 1$ reuse process, 1 project supply chain. Interviews $n = 25$; 5 buildings observed; documentation; plus 1 firm, interviews $n = 2$; observation of 4 meetings with $n = 5$
	Albino and Berardi (2012)	Cases: 3 construction projects. Interviews n not given; observations; report validation with practitioners
	Elbarkouky and Abdelazeem (2013)	Cases: 2 firms; interviews $n = 6$
	Arroyo et al. (2016)	Case: 1 global refurbishment project, 5 locations. Researcher embedded in firm, undertaking product analysis, running training and 1 decision-making and 1 post-decision meeting; report validation with practitioners
	Kim et al. (2016)	Case: 1 contractor firm. Survey of $n = 106$ first tier suppliers of the firm
	Woo et al. (2016)	Case: 1 contractor firm. Survey of $n = 103$ supplier–buyer pairs within the firm's supply chain

Table 4.2 (Continued)

Method	Papers	Metrics
Action research	Uttam and Roos (2015)	Case study on procurement on an infrastructure project; one author was a team member; meeting minutes and other project documentation, survey instrument to 3 shortlisted contractors, 6 interviews.
Life cycle analysis (LCA)	Blengini and Garbarino (2010)	Recycled aggregates
	Kucukvar et al. (2014)	Construction and demolition waste
	Dadhich et al. (2015)	Plasterboard
	Faleschini et al. (2016)	Recycled aggregates
	Kucukvar et al. (2016)	Construction and demolition waste
	Abdul Ghani et al. (2017)	US building stock
	Nasir et al. (2017)	Insulation
Other modelling	Hendrickson and Horvath (2000)	Economic input–output analysis-based life-cycle assessment (EIO-LCA) model using 1992 US economic data
	Sarkis et al. (2012)	Decision model for subcontractor selection, based on analytic hierarchy process (AHP) and analytic network process (ANP)
	Hsueh and Yan (2013)	Assessment model for contractor selection, based on AHP and fuzzy logic
	Zhou et al. (2013)	Mathematical model aiming to maximise the aggregate profits of normalised construction logistics
	Balasubramanian (2014)	Interpretative structural modelling analysis of green supply chain management enablers based on literature review and focus group $n = 12$
	Chen, R.-H. et al. (2015)	Assessment model for sustainability development indicators in minerals
	Chen, P.-C. et al. (2017)	Interactive flow (Sankey) diagram for material and waste flows
	Neppach et al. (2017)	Organisational environmental footprinting
Commentary, literature review or theory building	Irland (2007)	Commentary
	Aho (2013)	Commentary
	Ofori (2000)	Literature review
	Mahamadu et al. (2013)	Literature review
	Sertyesilisik (2016)	Literature review; extends Porter's diamond framework on competitive advantage
	Ahmadian et al. (2017)	Secondary data and illustrative case study; framework for materials selection

Source: Original

literature, therefore, is generally founded on a solid empirical base. The exceptions are a small number of studies with very low numbers of interviewees, and two papers in which data sources were not quantified, making it difficult to assess robustness. Only one study was conducted as action research although green supply chain management would appear to be a particularly appropriate domain in which academic research and practitioner knowledge could cross-fertilise.

4.3.5 Tools and Techniques

Specialised tools and techniques were proposed in nearly one-third of the papers. Life cycle analysis was applied primarily to specific materials: recycled aggregate, plasterboard, and construction and demolition waste (Blengini and Garbarino 2010; Dadhich et al. 2015; Kucukvar et al. 2016). In most cases, a hybrid form of life cycle analysis was used, with geographic information included in two cases to facilitate calculation of transportation impacts. Such papers presented a detailed analysis of a range of environmental impacts, contributing to knowledge on the particular material or resource in question. Taking a different focus, one paper (Abdul Ghani et al. 2017) applied a form of life cycle analysis followed by a secondary analysis to evaluate comparative contributions of greenhouse gases by different supplier industries to construction, concluding that ready-mix concrete, electricity supply, and lighting fixtures were the most environmentally intensive. A second study that moved away from an elemental focus examined the relative impacts by construction sector: industrial and commercial, residential, road infrastructure, and other infrastructure (Hendrickson and Horvath 2000). They concluded that the four major subsectors in US construction appear to use fewer resources and have lower rates of environmental emissions and waste than their share of the gross domestic product might suggest.

Beyond life cycle analysis, several of the papers focused on decision-making support, examining techniques including Choosing By Advantages (CBA) and Technique for Order of Preference by Similarity to Ideal Solution (TOPSIS). These are methods to systematically approach complex, multicriteria decisions and were applied here to selection of a contractor and project team (Sarkis et al. 2012; Hsueh and Yan 2013), as well as to sustainable development indicators (Chen et al. 2015). Although of potential benefit in the complex decision making required by environmental concerns, the papers tended to focus on an improved method but offered little in terms of consideration of practical issues in applying the method. In particular, the studies omitted practitioner feedback which could have shed light on the challenges of complex decision making within an everyday operational context. In contrast, through their detailed case study, Arroyo et al. (2016) found that time to collect data, time to train practitioners on a decision-making method, and the practical difficulties of bringing all stakeholders together limited the extent to which a method for complex decision making could be applied in practice.

Two further techniques were studied in the papers reviewed: as part of an interactive decision support tool, the use of Sankey diagrams to offer visual representation of the intricate life cycle impacts of a material (Chen et al. 2017), and organisational environmental footprinting (Neppach et al. 2017). While Sankey diagrams may facilitate decision making by clustering complicated information, organisational environmental footprinting was found to be of limited use in the construction sector. Although organisational environmental footprinting can bring potential benefits for a construction organisation in identifying 'hotspots' of particularly high environmental impact, Neppach et al. concluded that the method is more applicable to manufacturing firms. For construction companies, the continuously changing portfolio of projects makes it difficult to track performance as the mix changes year to year. Comparison between construction companies is also fraught as the diversity of services and offerings varies considerably. In addition, the upstream impacts from suppliers are difficult to assess. Neppach et al.'s (2017) study provided valuable insights into the challenges within construction and the (in)applicability of organisational environmental footprinting as an assessment method.

External frameworks (environmental certification, regulation) were discussed in three papers (da Rocha and Sattler 2009; Albino and Berardi 2012; Irland 2007) and internal organisational or management approaches, were the focus of a further eight. These papers considered the business model (Aho 2013), alliances and supplier integration (Dainty and Brooke 2004; Ofori 2000), critical success factors and performance measures (Seth et al. 2016), lean methods (Sertyesilisik 2016), risk management (Zou and Couani 2012), communications, and environmental capability (Kim et al. 2016; Woo et al. 2016). In addition, a few papers provided extensive lists of enablers in addition to the tools and techniques mentioned (e.g. Balasubramanian 2014; Seth et al. 2016; Wong et al. 2016), but it is somewhat difficult to distinguish conceptually between a technique applied to accomplishing greener supply chain management, and an enabling factor.

4.3.6 Stakeholders

Many organisations and individuals are involved in a construction supply chain. Using Freeman's definition of stakeholders as 'those groups and individuals who can affect or be affected' by a firm's activities (Freeman et al. 2010, p. 9), we examined the stakeholders identified in each paper, if any. The range of stakeholders referred to illustrated the complexity of the construction supply chain. Beyond the suppliers and logistics operators who would be expected to feature in supply chain management more generally, reference was made to developers and clients (10), construction professionals including architect/designers, engineers, project managers and specialist subcontractors (15), and principal contractors (4). Opinions varied on the level of commitment and motivation of different stakeholder roles. While most authors who commented on stakeholder responsibility considered the client/developer as an important driver, not all considered the general contractor to be motivated to green the supply chain. Wong et al. (2016) argued that clients and developers in Hong Kong are not engaged, as green procurement is not legally mandated, and developers do not see the investment benefit. Although some scholars argued that the design team has a major influence on the final product through design and materials selection (Albino and Berardi 2012; Arroyo et al. 2016; Sertyesilisik 2016), others viewed designers as having limited incentive to collaborate in green supply chains (Wong et al. 2016; Balasubramanian and Shukla 2017b). Wong et al. (2016) attributed this to an immature market, and to designers applying time and cost but not environmental criteria to their output. These authors suggested that financial incentives, such as tax relief, and more information on the environmental benefits offered by green design could act to incentivise stakeholders such as developers and designers.

In almost one third of the papers, however, stakeholders were not addressed and this absence represents a weakness in the literature reviewed, through omission of one of the primary foci of green supply chain management (Ahi and Searcy 2013). Further, as Balasubramanian and Shukla (2017a,b) argued, stakeholder perspectives differ and therefore the different standpoints, approaches, and objectives of each stakeholder should be examined. A further noticeable gap was in consideration of a wider group of stakeholders: dating back to the Brundtland definition (WCED 1987), sustainable development encompasses society today and future generations as stakeholders due to the impact of current consumption of natural resources on lifestyles and quality of life. None of the papers, however, considered in any detail the implications of the natural world or future generations as stakeholders, or discussed how the needs of these external stakeholders could be represented in construction projects.

4.3.7 Definitions of Green Supply Chain Management

Not all authors sought to define green supply chain management. Many of the papers with a complementary focus to green supply chain management defined ancillary concepts such as lean or sustainable construction (Ofori 2000; Sertyesilisik 2016), green manufacturing (Seth et al. 2016), and sustainable supply mix and sustainable materials management (Blengini and Garbarino 2010; Chen et al. 2015). Where definitions of green supply chain management were presented, most were operationally focused, specifying the constituent processes. Green procurement featured extensively, along with green manufacturing, green distribution, reverse logistics (Adawiyah et al. 2015), green purchasing, green production, green consumption and green recycling (Zhou et al. 2013), green transportation (Balasubramanian 2014), green design, packaging, and waste minimisation (Dadhich et al. 2015). These processes were rarely themselves defined, leaving the terms open to differing interpretations.

The objectives of green supply chain management were considered in some papers, and these were proposed to be: to enhance competitiveness (Woo et al. 2016), to add value for stakeholders (da Rocha and Sattler 2009), to improve environmental, economic, and operations performance (Balasubramanian 2014), to improve service, increase market share and sustainability of supply (Adawiyah et al. 2015), to increase operational efficiency, cut costs and minimise risks or for ethical reasons (Dadhich et al. 2015), and to reduce environmental impact (Balasubramanian and Shukla 2017b). Surprisingly few authors attempted to consider a more conceptual perspective although in a small number of papers, there was recognition of the holistic, end-to-end perspective (da Rocha and Sattler 2009; Balasubramanian and Shukla 2017a). A few authors placed emphasis on integration of processes between suppliers and clients (da Rocha and Sattler 2009; Zhou et al. 2013), and integration of green practices into business processes (Balasubramanian 2014; Balasubramanian and Shukla 2017b; Nasir et al. 2017) and into interorganisational supply chain management (Elbarkouky and Abdelazeem 2013; Ketikidis et al. 2013). Nasir et al. were the only authors in the papers reviewed to expand the traditional linear concept of the supply chain to incorporate circular economy principles (Nasir et al. 2017). As they note, such an approach yields environmental gains (Elbarkouky and Abdelazeem 2013; Ketikidis et al. 2013) and may become increasingly essential in addressing dwindling stocks of scarce resources. Our findings here echo those of Ahi and Searcy's (2013) sector-independent systematic review of green supply chain management definitions, in which they found limited consideration of principles of business sustainability such as a stakeholder focus and long-term perspective.

4.4 Discussion

Having described the selected papers by code, we now draw together the findings, looking at patterns, salient themes, and gaps.

4.4.1 Overview

The review of 44 papers showed rapidly increasing interest in topics associated with green supply chain management in construction from 2012, following somewhat later than a more general surge in research interest in green supply chain management in other industries (Zhu and Sarkis 2006). The spread of journals in which relevant studies have been published demonstrates wide interest and is a promising basis for a thriving research domain. Generally good geographic coverage has been achieved but this could be extended. Developing economies in sub-Saharan Africa and in South America (where only one study was based) were noticeable gaps. A mixture of methods have been used, including in-depth interviews, surveys, mixed methods, case studies, and modelling, which provides a generally robust empirical base, with a few exceptions. Only one action research study was identified, however, and this is an approach which could yield particularly valuable insights in future research (Seuring and Gold 2013).

Multiple drivers for firms pursuing green supply chain management were identified (Balasubramanian and Shukla 2017b), with evidence for a positive relationship between implementation of green supply chain management practices and economic performance (Woo et al. 2016; Ketikidis et al. 2013).

4.4.2 Definition

A point of note was the failure of many papers to define explicitly what they understood by green supply chain management. Across the set of papers, it is noteworthy that there is limited discussion of how green supply chain management may be understood as similar to but differentiated from supply chain management. A final striking omission in terms of definition is the concept of 'true sustainability' – the notion that the end goal (even if not wholly achievable) must be no adverse environmental impact, the potential for positive environmental contribution, and for indefinite continuance of the supply chain (assuming nonsupply chain factors remain favourable) (Pagell and Wu 2009). The definition of green supply chain management should also incorporate activities and desired outcomes or objectives (Stock and Boyer 2009). We offer a more comprehensive definition of green supply chain management for construction, drawing on the papers reviewed (see Box 4.1).

Box 4.1 Definition of Green Supply Chain Management for Construction

Green supply chain management for construction comprises the management of all activities in an organisation, operating in or serving the construction sector, related to minimising the environmental impact of all of its supply chains which contribute to its final products, with the aim of achieving zero net harm and the potential to operate indefinitely, given an available market.

As the purchase of products or services is the fundamental relationship along the supply chain, the *activities* comprise green purchasing, at a minimum, that is, the use of environmental criteria in selection of products and services.

Green supply chain management can also include:

1. Green design, which aims to limit impact on the natural world through product or service design, considering materials, production, and product in-use resource consumption and toxicity;
2. Green manufacturing, which seeks to reduce negative environmental impact in manufacturing processes, considering energy, water and toxicity and includes design, process and packaging;
3. Green transportation, which tries to reduce environmental impact caused by transportation of goods, considering fuel, volume and weight of goods, and number of journeys;
4. Waste management, which seeks to minimise resource usage and to reuse where possible;
5. Green operation, which minimises environmental damage when the product is in use, particularly through energy and water efficiency;
6. End-of-life management which maximises the reuse of materials at end-of-life for a product and includes reverse logistics.

The *objectives* of green supply chain management comprise improved environmental performance and improved business performance through greater efficiency, increased competitiveness, and increased value to stakeholders.

The *stakeholders* of the built environment are understood to include not only the client but also building users, workers in construction, local and distant communities, future generations, and the natural world.

Across all supply chain actors, green supply chain management requires *management* of the above activities to achieve the required objectives, that is, planning, control, measurement, monitoring, and evaluation.

The activities, objectives, and management of green supply chain management will vary by role of firm.

4.4.3   Nature of Construction

The focus of our systematic review was green supply chain management in construction and thus the papers were examined for how they engaged with the construction sector. Almost all of the papers began with an overview of the considerable negative environmental impact of construction but few went on to address the specific nature of the sector. Mention was made of fragmentation (Dainty and Brooke 2004; Aho 2013) and of complexity and multiple stakeholders (Arroyo et al. 2016; Neppach et al. 2017). Some consideration of the processes and structure of the construction regime were offered in a few of the papers (Sertyesilisik 2016; Wong et al. 2016) while a more extensive review of the unique character of the construction sector was discussed only in a small number of articles, as we now summarise. The project-based nature of construction, multiple dyadic relationships, high numbers of firms within supply chains, the frequently one-off nature of contracts, and the potentially adversarial relationships were considered by Balasubramanian and Shukla (2017a,b). Albino and Berardi (2012) also referred to the slowness of change in the industry, the reluctance of suppliers and contractors to invest in new approaches on the assumption that the benefit accrues only to the client, the uncertainty and instability inherent in the sector, and how the values embedded within contractual relationships frequently vary considerably from the ideals of successful supply chain management.

4.4.4 Stakeholder Roles

There was implicit recognition in the papers reviewed that the benefits – and costs and risks – of green supply chain management in construction may not fall equally. For example, Albino and Berardi (2012) found that willingness to adopt additional risk may be a factor in stronger integration between suppliers, designer, and contractors on green projects. Wong et al. (2016) specified different levels of incentives for clients, developers, and designers. Aho (2013) argued that the number of stakeholders in construction projects should be minimised through shorter supply chains with each member taking on greater responsibility and risk. However, he notes that this must come with greater reward but the current cost-plus model, in which each point in the supply chain charges for integrating the products and services of their own suppliers plus a margin, acts to constrain such change. Although Ofori (2000) described a generalised failure to accept responsibility for environmental improvement across the construction sector, most authors viewed the construction sector client, the firm closest to the end-product, as bearing primary responsibility (Ahmadian et al. 2017).

Within construction, critically, the role of the firm influences its primary green supply chain management activities, and the potential risks, costs, and benefits. For example, the primary impact of architectural and engineering consultancy firms may be in green design (although green design will in turn influence other processes); the primary impacts of the contractor managing the site may be in green transportation and waste management, etc. Thus green supply chain management differs with the organisation's role but few authors acknowledged this. Balasubramanian and Shukla (2017a) have led the way in their analysis by organisation type but further research specific to the role in the construction project team is needed. In particular, more in-depth research is needed on the different balance of potential costs, risks, and benefits for each level of the supply chain.

4.4.5 Practical Recommendations

A number of the papers reviewed provided observations or recommendations for practitioners. The importance of understanding the context in order to implement green supply chain management was noted, including geographic (Blengini and Garbarino 2010), and socioeconomic (da Rocha and Sattler 2009; Chileshe et al. 2016). A revised business model is needed that aligns value to the client and to society more generally with pricing (Aho 2013). Regulation and standards facilitate progress, signalling government commitment, and industry standards work as guidelines and benchmarks (Bohari et al. 2017). More green procurement standards, certification schemes from industry, publicly accessible databases of environmental data on products, and exemplars of success are all enablers of green supply chain management (Wong et al. 2016).

In terms of organisational management, although Seth et al. (2016) argued that previous work on green manufacturing had tended to ignore management aspects, the papers reviewed here addressed a number of aspects:

- The need to develop more integration and stronger relationships with suppliers, including alliances and greater trust, has been noted (Dainty and Brooke 2004; da Rocha and Sattler 2009), extending to co-makership, with early and extensive contact with the general contractor and design team (Albino and Berardi 2012).
- Organisations in the supply chain should share understanding of values and performance (Aho 2013) and measures of performance should be aligned with critical success factors (Seth et al. 2016).
- Alterations in contracts are needed, to implement green procurement: for example tenders for design, contractor selection, build, and maintenance should include environmental requirements, and such requirements should be used to evaluate tender responses (Bohari et al. 2017). Prequalification on environmental criteria can be a practical approach (Sertyesilisik 2016).
- Greater risk in green innovation should be covered by warranties (Albino and Berardi 2012).
- Adequate management resources and commitment to monitoring new and changed processes should be available (Dainty and Brooke 2004), and both operational and strategic management, together with a change in culture, are necessary (Sertyesilisik 2016).
- Expertise in sustainability is crucial (Chen et al. 2017) but can be bought for a project (Bohari et al. 2017).

4.5 Looking to the Future

Based on the analysis of the papers in the dataset, it is now possible to suggest areas which merit further work.

The literature showed a particular focus on applying detailed quantitative methods to the complexity of decision making and environmental criteria, with 15 of the 44 papers investigating quantitative techniques. Although much valuable work has been, and is being, undertaken on systematic methods for complex decision making (Malek et al. 2017; Fazil-Khalaf et al. 2017), feedback from practitioners demonstrated that organisations are not likely to have the time to deploy such methods in their everyday work, or to dedicate the time needed to develop expertise in project team members to apply such methods (Arroyo et al. 2016). More work is needed to bridge the detailed analyses with more pragmatic methods that can be rapidly applied in day-to-day decision making. We noted above that action research was conducted in only one of the papers. This approach could be particularly valuable in working alongside and gaining deep understanding of the pragmatic challenges faced by practitioners, as well as trialling insights from theory and previous research in practice. A useful contribution has pointed to the value of examining 'hotspots' (Dadhich et al. 2015) – the links in the supply chain with the greatest impact. The work of Abdul Ghani et al. (2017) usefully sought to identify the most intensive supplier industries to construction with respect to greenhouse gas emissions, and this represents a fruitful avenue for further examination in practice and research, as one possible approach for providing a rapid overview of the supply chain, and practical and quick methods to identify hotspots.

The literature has provided insights into factors which contribute to successful green supply chain management. Of further benefit to practitioners would be studies which can advise on where to start and how to prioritise, potentially following the approach of Bossink (2007), who proposed an eight-stage model of moving towards sustainable innovation in construction. Although organisational environmental footprinting can help bring attention to supply chain management, it has been found to be of limited use in construction due to the project-based nature of work in the sector (Neppach et al. 2017). Guidance for practitioners should consider real world complexity including the number of products a construction project or supplier may deal with, the volume of information potentially available, the availability and accessibility of product and supply chain information, the practicalities of maintaining currency of information, ownership of information and liability for errors (Zhao et al. 2017), and planning for further development of life cycle analyses. Guidance on improving communications with suppliers, integrating systems, and developing trust would be of practical use. Xu et al. (Chapter 15) offer insights that could be applied in green supply chain management research. Studies elsewhere have highlighted the role of cooperative procurement arrangements such as partnering, alliances, Private Finance Initiatives (PFI), and relational contracting in driving superior environmental performance in construction (Anderson et al. 2004; Eriksson and Westerberg 2010; Badi and Pryke 2015) and this knowledge could usefully be harnessed in future research on green supply chain management in construction.

Innovation plays a key role in green supply chain management (Gao et al. 2017) and, hence, greater understanding of innovation in construction green supply chain management is needed. In the construction industry, creativity and innovation is often seen to be underpinned by effective communication and collaboration among the diverse, multidisciplinary, and multiskilled project actors (Bresnen and Marshall 2000). This is particularly important in the development of sustainable buildings and in the introduction of innovative environmental technologies due to the need for effective expertise to handle the complexity of environmental optimisation (Rohracher 2001). The integration of design, construction, and operation disciplines is seen as pivotal in the delivery of higher environmental performance (Cole 2000) and innovation (Badi and Pryke 2015). With interorganisational optimisation at its core, further research is needed to elucidate the systematic, complex, collaborative, and complementary nature of construction green supply chain management innovation.

Further focus is needed on the unique character of the construction industry. As noted in the synthesis, some of the papers acknowledged attributes of the sector such as uncertainty and instability, adversarial contractual arrangements, and slowness to change and these factors may directly prevent progress on green supply chain management. Further, the construction sector remains dominated by goods dominant and project-focused logics which emphasise transactional value, short-term evaluation, minimisation of costs, and short-term profit (Smyth 2015). These are concepts which research on green supply chain management has yet to tackle.

The complexity of supply chains in construction speaks to the need for both detailed, subdomain specific, and sector generic research. For example, we do not know if the same principles or the same priorities in operational processes and procedures apply across all types of projects – are there differences, for example, in managing the supply chains for residential development where high numbers of units of similar design will be constructed, and in managing the supply chains for a mixed-use commercial development? Similarly, we do not yet know if the same approach should be taken for different components and materials – are the processes necessary for the supply chain for steel beams the same as those for the supply chains of window systems? Studies to add to knowledge are required which consider the holistic, end-to-end issues by size of focal organisation (major contractor, Tier 2 subcontractor, specialist subcontractor), by material (further work on timber and aggregate, new studies on window systems, roof systems, heating, ventilation and air conditioning, for example) and by type of project (in addition to the work on residential, studies on commercial, health, education, hospitality, and infrastructure projects).

In parallel with an end-to-end perspective, potentially owned by the client, there is a need to investigate further what green supply chain management means for different roles within the supply chain. Building on the work of Balasubramanian and Shukla (2017a), research is needed on the similarities and differences of managing a green supply in developers, principal and general contractors, Tier 2 contractors, specialist subcontractors, trades, different types of consultancy, and process (e.g. cement) and product-oriented (e.g. façade systems) suppliers. Such work needs to build on insights that few subcontractors are engaged with supply chain management practices despite widespread understanding of partnering in the industry (Mason 2007). A particular focus for guidance should be on the challenges for small-to-medium enterprises (SMEs) which comprise the majority of the construction sector (ONS 2017) but received little explicit attention in the literature. The critical role of industry bodies (Wong et al. 2016) in setting standards, collating data, and providing access to information, could be further explored.

Finally, there is a need to develop theory in this area. There was very limited application of theory in the papers reviewed and use of supply chain management theory was similarly rare. In one of the few exceptions, Balasubramanian and Shukla (2017a) proposed and tested a nine-construct structural model, in which they demonstrated the relationship between internal and external drivers and barriers to core and facilitating green practices. Further testing and development of this model, and of other theoretical frameworks, will help to develop what our review has found to be a young field.

4.6 Conclusion

In terms of research, much remains to be done but this is not a criticism of, and does not detract from, the valuable work presented in the papers we reviewed. The papers demonstrated a growing research domain and provided useful insights across many areas. In a domain as complex, and as relatively new, as green supply chain management, there remain many gaps in knowledge. Our objective in reviewing the literature to date was to encourage further work by suggesting areas for attention as elements of a future research agenda. We hope that future research on greening the construction supply chain will harness advances in supply chain management and management more generally in progressing the sector's response to one of society's most urgent problems.

References

Abdul Ghani, N.M.A.M., Egilmez, G., Kucukvar, M., and Bhutta, M.K. (2017). From green buildings to green supply chains: an integrated input-output life cycle assessment and optimization framework for carbon footprint reduction. *Management of Environmental Quality* 28 (4): 532–548.

Adawiyah, W.R., Pramuka, B.A., and Najmudin, J.D.P. (2015). Green supply chain management and its impact on construction sector small and medium enterprises (SMEs) performance: a case of Indonesia. *International Business Management* 9 (6): 1018–1024.

Ahi, P. and Searcy, C. (2013). A comparative literature analysis of definitions for green and sustainable supply chain management. *Journal of Cleaner Production* 52: 329–341.

Ahmadian, F.F.A., Rashidi, T.H., Akbarnezhad, A., and Waller, S.T. (2017). BIM-enabled sustainability assessment of material supply decisions. *Engineering, Construction and Architectural Management* 24 (4): 668–695.

Aho, I. (2013). Value-added business models: linking professionalism and delivery of sustainability. *Building Research and Information* 41 (1): 110–114.

Albino, V. and Berardi, U. (2012). Green buildings and organizational changes in Italian case studies. *Business Strategy and the Environment* 21 (6): 387–400.

Anderson, P., Cook, N., and Marceau, J. (2004). Dynamic innovation strategies and stable networks in the construction industry: implementing solar energy projects in the Sydney Olympic Village. *Journal of Business Research* 57: 351–360.

Arroyo, P., Tommelein, I.D., and Ballard, G. (2016). Selecting globally sustainable materials: a case study. *Journal of Construction Engineering and Management* 142 (2): 05015015.

Badi, S. and Pryke, S. (2015). Assessing the quality of collaboration towards the achievement of sustainable energy innovation in PFI school projects. *International Journal of Managing Projects in Business* 8 (3): 408–440.

Balasubramanian, S. (2014). A structural analysis of green supply chain management enablers in the UAE construction sector. *International Journal of Logistics Systems and Management* 19 (2): 131–150.

Balasubramanian, S. and Shukla, V. (2017a). Green supply chain management: an empirical investigation on the construction sector. *Supply Chain Management* 22 (1): 58–81.

Balasubramanian, S. and Shukla, V. (2017b). Green supply chain management: the case of the construction sector in the United Arab Emirates (UAE). *Production Planning and Control* 28 (14): 1116–1138.

Banawi, A. and Bilec, M.M. (2014). A framework to improve construction processes: integrating lean, green and six sigma. *International Journal of Construction Management* 14 (1): 45–55.

Blengini, G.A. and Garbarino, E. (2010). Resources and waste management in Turin (Italy): the role of recycled aggregates in the sustainable supply mix. *Journal of Cleaner Production* 18: 1021–1030.

Bohari, A., Skitmore, M., Xia, B., and Teo, M. (2017). Green oriented procurement for building projects: preliminary findings from Malaysia. *Journal of Cleaner Production* 148: 690–700.

Bossink, B.A.G. (2007). The interorganizational innovation processes of sustainable building: a Dutch case of joint building innovation in sustainability. *Building and Environment* 42 (12): 4086–4092.

Bresnen, M. and Marshall, N. (2000). Partnering in construction: a critical review of issues, problems and dilemmas. *Construction Management and Economics* 18 (2): 229–237.

Briner, R.B. and Denyer, D. (2012). Systematic review and evidence synthesis as a practice and scholarship tool. In: *The Oxford Handbook of Evidence-Based Management* (ed. D.M. Rousseau), 112–128. Oxford: Oxford University Press.

CA Governor (2016). State of California Green Buildings. www.green.ca.gov/Buildings (accessed 13 August 2019).

Caiado, R.G.G., Dias, R.D.F., Mattas, L.G. et al. (2017). Towards sustainable development through the perspective of eco-efficiency: a systematic literature review. *Journal of Cleaner Production* 165: 890–904.

Chen, R.-H., Lin, Y., and Tseng, M.-L. (2015). Multicriteria analysis of sustainable development indicators in the construction minerals industry in China. *Resources Policy* 46: 123–133.

Chen, P.-C., Liu, K.-H., and Ma, H.-W. (2017). Resource and waste-stream modeling and visualization as decision support tools for sustainable materials management. *Journal of Cleaner Production* 150: 16–25.

Chileshe, N., Rameezdeen, R., Hosseini, M.R. et al. (2016). Analysis of reverse logistics implementation practices by south Australian construction organisations. *International Journal of Operations and Production Management* 36 (3): 332–356.

Christopher, M. (2011). *Logistics and Supply Chain Management*, 4e. Harlow: Financial Times/Prentice Hall.

Cole, R. (2000). Building environmental assessment methods: assessing construction practices. *Construction Management and Economics* 18 (8): 949–957.

Dadhich, P., Genovese, A., Kumar, N., and Acquaye, A. (2015). Developing sustainable supply chains in the UK construction industry: a case study. *International Journal of Production Economics* 164: 271–284.

Dainty, A.R.J. and Brooke, R.J. (2004). Towards improved construction waste minimisation: a need for improved supply chain integration? *Structural Survey* 22 (1): 20–29.

Elbarkouky, M.M.G. and Abdelazeem, G. (2013). A green supply chain assessment for construction projects in developing countries. *WIT Transactions on Ecology and the Environment* 179: 1331–1341.

Eriksson, P.E. and Westerberg, M. (2010). Effects of cooperative procurement procedures on construction project performance: a conceptual framework. *International Journal of Project Management* 29 (2): 197–208.

Eriksson, P., Dickinson, M., and Khalfan, M. (2007). The influence of partnering and procurement on subcontractor involvement and innovation. *Facilities* 25 (5): 203–214.

European Commission (2015). *Construction and demolition waste (CDW)*. http://bit.ly/ 1ERulE1 (accessed 13 August 2019).

European Commission (2008). *Raw materials initiative – meeting our critical needs for growth and jobs in Europe*. COM(2008) 699. Brussels.

Faleschini, F., Zanini, M., Pellegrino, C., and Pasinato, S. (2016). Sustainable management and supply of natural and recycled aggregates in a medium-size integrated plant. *Waste Management* 49: 146–155.

Fazil-Khalaf, M., Mirzazadeh, A., and Pishvaes, M.S. (2017). A robust fuzzy stochastic programming model for the design of a reliable green closed loop supply chain network. *Human and Ecological Risk Assessment* 23 (8): 2119–2149.

Fischedick, M., Roy, J., Abdel-Aziz, A. et al. (2014). Industry. In: *Climate Change 2014: Mitigation of Climate Change. Contribution of Working Group III to the Fifth Assessment Report of the IPCC*. Cambridge: Intergovernmental Panel on Climate Change.

Freeman, R.S., Harrison, J.S., Wicks, A.C. et al. (2010). *Stakeholder Theory: The State of the Art*. New York: Cambridge University Press.

Fulford, R. and Standing, C. (2014). Construction industry productivity and the potential for collaborative practice. *International Journal of Project Management* 32 (2): 315–326.

Gao, D., Xu, Z., Ruan, Y.Z., and Lu, H. (2017). From a systematic literature review to integrated definition for sustainable supply chain innovation (SSCI). *Journal of Cleaner Production* 142 (4): 1518–1538.

Gough, D., Oliver, S., and Thomas, J. (2012). *An Introduction to Systematic Reviews*. London: Sage.

Green, S. and May, S. (2005). Lean construction: arenas of enactment, models of diffusion and the meaning of "leanness". *Building Research and Information* 33 (6): 498–511.

Hendrickson, C. and Horvath, A. (2000). Resource use and environmental emissions of US construction sectors. *Journal of Construction Engineering and Management* 126 (1): 38–44.

Hsueh, S.-L. and Yan, M.-R. (2013). A multimethodology contractor assessment model for facilitating green innovation: the view of energy and environmental protection. *The Scientific World Journal* 2013: 624340.

Irland, L.C. (2007). Developing markets for certified wood products: greening the supply chain for construction materials. *Journal of Industrial Ecology* 11 (1): 201–216.

Karim, K., Marosszeky, M., and Davis, S. (2006). Managing subcontractor supply chain for quality in construction. *Engineering, Construction and Architectural Management* 13 (1): 27–42.

Ketikidis, P.H., Hayes, O.P., Lazuras, L. et al. (2013). Environmental practices and performance and their relationships among Kosovo construction companies: a framework for analysis in transition economies. *International Journal of Services and Operations Management* 14 (1): 115–130.

Kibert, C.J., Sendzimir, J., and Guy, B. (2000). Construction ecology and metabolism: natural system analogues for a sustainable built environment. *Construction Management and Economics* 18 (8): 903–916.

Kim, M.G., Woo, C., Rho, J.J., and Chung, Y. (2016). Environmental capabilities of suppliers for green supply chain management in construction projects: a case study in Korea. *Sustainability (Switzerland)* 8 (1): 1–17.

Kucukvar, M., Egilmez, G., and Tatari, O. (2014). Evaluating environmental impacts of alternative construction waste management approaches using supply-chain-linked life-cycle analysis. *Waste Management and Research* 32 (6): 500–508.

Kucukvar, M., Egilmez, G., and Tatari, O. (2016). Life cycle assessment and optimization-based decision analysis of construction waste recycling for a LEED-certified university building. *Sustainability* 8 (1): 89.

Lucon, O., Ürge-Vorsatz, D., Zain Ahmed, A. et al. (2014). Buildings. In: *Climate Change 2014: Mitigation of Climate Change. Contribution of Working Group III to the Fifth Assessment Report of the Intergovernmental Panel on Climate Change* (eds. O. Edenhofer, R. Pichs-Madruga, Y. Sokona, et al.). Cambridge, United Kingdom and New York, NY, USA: Cambridge University Press.

Mahamadu, A.M., Mahdjoubi, L., and Booth, C.A. (2013). Challenges to digital collaborative exchange for sustainable project delivery through building information modelling technologies. *WIT Transactions on Ecology and the Environment* 179: 547–557.

Malek, A., Ebrahimnejad, S., and Tavakkoli-Moghaddam, R. (2017). An improved hybrid grey relational analysis approach to green resilient supply chain network assessment. *Sustainability* 9 (8): 1433–1461.

Mason, J. (2007). The views and experiences of specialist contractors on partnering in the UK. *Construction Management and Economics* 25 (5): 519–527.

Nasir, M.H.A., Genovese, A., Acquaye, A.A. et al. (2017). Comparing linear and circular supply chains: a case study from the construction industry. *International Journal of Production Economics* 183: 443–457.

Neppach, S., Nunes, K.R., and Schebek, L. (2017). Organizational environmental footprinting in German construction companies. *Journal of Cleaner Production* 142: 78–86.

Ofori, G. (2000). Greening the construction supply chain in Singapore. *European Journal of Purchasing and Supply Management* 6 (3): 195–206.

ONS (2017). *Construction Statistics Annual Tables.* Online:. Office of National Statistics.

Pagell, M. and Shevchenko, A. (2014). Why research in sustainable supply chain management should have no future. *Journal of Supply Chain Management* 50 (1): 44–55.

Pagell, M. and Wu, Z. (2009). Building a more complete theory of sustainable supply chain management using case studies of 10 exemplars. *Journal of Supply Chain Management* 45 (2): 37–56.

Rizzi, F., Frey, M., Testa, F., and Appolloni, A. (2014). Environmental value chain in green SME networks: the threat of the Abilene paradox. *Journal of Cleaner Production* 85: 265–275.

da Rocha, C.G. and Sattler, M.A. (2009). A discussion on the reuse of building components in Brazil: an analysis of major social, economical and legal factors. *Resources, Conservation and Recycling* 54 (2): 104–112.

Rockström, J., Steffen, W., Noone, K. et al. (2009). A safe operating space for humanity. *Nature* 461 (7263): 472–475.

Rohracher, H. (2001). Managing the technological transition to sustainable construction of buildings: a socio-technical perspective. *Technology Analysis and Strategic Management* 13 (1): 137–150.

Ruparathna, R. and Hewage, K. (2015). Sustainable procurement in the Canadian construction industry: current practices, drivers and opportunities. *Journal of Cleaner Production* 109: 305–314.

Salzer, C., Wallbaum, H., Lopez, L.F., and Kouyoumji, J.L. (2016). Sustainability of social housing in Asia: a holistic multi-perspective development process for bamboo-based construction in the Philippines. *Sustainability* 8 (2): 151.

Sarkis, J., Meade, L.M., and Presley, A.R. (2012). Incorporating sustainability into contractor evaluation and team formation in the built environment. *Journal of Cleaner Production* 31: 40–53.

Sertyesilisik, B. (2016). Embending [sic] sustainability dynamics in the lean construction supply chain management. *YBL Journal of Built Environment* 4 (1): 60–78.

Seth, D., Shrivastava, R.L., and Shrivastava, S. (2016). An empirical investigation of critical success factors and performance measures for green manufacturing in cement industry. *Journal of Manufacturing Technology Management* 27 (8): 1076–1101.

Seuring, S. and Gold, S. (2013). Sustainability management beyond corporate boundaries: from stakeholders to performance. *Journal of Cleaner Production* 56: 1–6.

Shen, L., Zhang, Z., and Zhang, X. (2017). Key factors affecting green procurement in real estate development: a China study. *Journal of Cleaner Production* 153: 372–383.

Smyth, H. (2015). *Market Management and Project Business Development*. London: Routledge.

Stock, J.R. and Boyer, S.L. (2009). Developing a consensus definition of supply chain management: a qualitative study. *International Journal of Physical Distribution and Logistics Management* 39 (8): 690–711.

Udawatta, N., Zuo, J., Chiveralls, K., and Zillante, G. (2015). Attitudinal and behavioural approaches to improving waste management on construction projects in Australia: benefits and limitations. *International Journal of Construction Management* 15 (2): 137–147.

Uttam, K. and Roos, C.L.L. (2015). Competitive dialogue procedure for sustainable public procurement. *Journal of Cleaner Production* 86: 403–416.

Walker, P.H., Seuring, P.S., Sarkis, P.J., and Klassen, P.R. (2014). Sustainable operations management: recent trends and future directions. *International Journal of Operations and Production Management* 34 (5).

WCED (1987). *Our Common Future*. Oxford: WCED.

Wong, J., Chan, J., and Wadu, M. (2016). Facilitating effective green procurement in construction projects: an empirical study of the enablers. *Journal of Cleaner Production* 135: 859–871.

Woo, C., Kim, M.G., Chung, Y., and Rho, J.J. (2016). Suppliers' communication capability and external green integration for green and financial performance in Korean construction industry. *Journal of Cleaner Production* 112: 483–493.

Zhao, X., Feng, Y., Pienaar, J., and O'Brien, D. (2017). Modelling paths of risks associated with BIM implementation in architectural, engineering and construction projects. *Architectural Science Review* 60 (6): 472–482.

Zhou, P., Chen, D., and Wang, Q. (2013). Network design and operational modelling for construction green supply chain management. *International Journal of Industrial Engineering Computations* 4 (1): 13–28.

Zhu, Q. and Sarkis, J. (2006). An inter-sectoral comparison of green supply chain management in China: drivers and practices. *Journal of Cleaner Production* 14 (5): 472–486.

Zou, P.X.W. and Couani, P. (2012). Managing risks in green building supply chain. *Architectural Engineering and Design Management* 8 (2): 143–158.

Zuo, K., Potangaroa, R., Wilkinson, S., and Rotimi, J.O. (2009). A project management prospective in achieving a sustainable supply chain for timber procurement in Banda Aceh, Indonesia. *International Journal of Managing Projects in Business* 2 (3): 386–400.

5

Connecting the 'Demand Chain' with the 'Supply Chain': (Re)creating Organisational Routines in Life Cycle Transitions
Simon Addyman

Crucial in a number of people organizing their inter-activities with each other is their being able, as they act, to arouse in each other transitory understandings *of 'where' so far in their activities they have 'got to', and* action guiding anticipations *of 'where' or 'how' next they are likely 'to go on'. In other words, it is only in the course of their actions that they can* organize *their conduct of them, not before by planning them, nor after by criticizing them*

(Shotter 2008, p. 510).

5.1   Introduction

The above quotation draws our attention to both the '*ongoing*' and '*dialogical*' nature of organising from a 'process ontology' perspective (Rescher 1996; Tsoukas and Chia 2002; Hernes 2014; Lorino and Tricard 2012; Cunliffe et al. 2014). I define ongoing and dialogical in combination as being understood as a way of looking beyond any fixed structural or behavioural pattern of an organisation and towards how participants continuously engage between themselves and their organisational structure to achieve shared goals (Shotter 2008; Emirbayer and Mische 1998). From such a perspective of organising, perceived stable 'patterns of action' are understood as both an effortful and emergent accomplishment (Pentland and Reuter 1994; Feldman 2000) through engaging in the ongoing change and flux of organisational life, which is both situated in a particular context and in the flow of time (Langley et al. 2013).

In this chapter I suggest that such a perspective of organising raises questions about the efficacy of the deterministic, time delimited, staged life cycle model (Lundin and Söderholm 1995; Winter et al. 2006) that has become so prevalent in developing organisational and governance structures for the management of construction projects (Söderlund 2012; Morris 2013). It draws particular attention to the activity of '*transitioning*' across temporal boundaries that are created by delimiting time and its influence on the interdependent patterns of action between multiple project participants. Arguably,

Successful Construction Supply Chain Management: Concepts and Case Studies, Second Edition. Edited by Stephen Pryke.
© 2020 John Wiley & Sons Ltd. Published 2020 by John Wiley & Sons Ltd.

there would be no better place to observe interdependent patterns of action between multiple project participants than in a construction project supply chain.

Morris (2013), who placed the creation of value as the central tenet for the existence of any project, recognised that the life cycle model can have various forms but characterised it as having two dominant organisational stages – 'definition', led by the client organisation in what could be termed the 'demand chain', and 'delivery', led by the contractor in what is most commonly termed the 'supply chain' (Pryke 2009). In construction projects, specifically large engineering infrastructure projects (Miller et al. 2001), the temporal boundary between these two phases is traditionally separated by procurement and formal sanction to proceed, at which point the supply chain is mobilised. It is an important *transition* in the projects' life cycle that marks a shift, and a potential breakdown, in the ongoing structures of the organisation (Miller et al. 2001; Miller and Hobbs 2005; Jones and Lichtenstein 2008) and is the focus of this chapter.

Within the construction industry in the UK, over the last half a century and more, there have been a number of industry reports that have set out various problems associated with organising the construction process (Latham 1994; Egan 1998; Holti et al. 1999; Murray and Langford 2003; ICE 2017). It is suggested here that two of the most common 'organisational difficulties' highlighted in these reports are: firstly, that of reconciling the trade-off associated with long-term versus short-term relationships due to the often unique, one-off nature of temporary construction project organisations; secondly, the relationship this has with the resulting commercial arrangements between the parties, not least the client and the main contractor. Söderlund (2012) terms these as the 'problem of cooperation' and 'the problem of coordination', which although often theoretically explored separately, are in practice interrelated. In this chapter, I term these 'relational' and 'transactional' difficulties (respectively) and suggest that they influence interdependent patterns of action between those in the demand chain, and those in the supply chain.

To explore these difficulties theoretically and from a process ontology of organising I want to suggest that managing (not removing) these two organising difficulties, in the context of *transitioning* across a predetermined time boundary within the life cycle of a temporary organisation, may be relative to what I would like to term the *'temporal paradox'* of temporary organising. This paradox is created by the conflicting sense of *'newness'* in the creation of a temporary organisation (Lundin and Söderholm 1995) for each construction project, and often each life cycle stage, with the need for *'repetition'* to create organisational routines – *'recognisable patterns of interdependent action'* – that help build organisational capabilities (Feldman and Pentland 2003; Davies and Brady 2016). Relationships in construction project supply chains, particularly long-term ones, could provide a means of carrying organisational routines across temporal boundaries, it could be argued, and so potentially contribute to mitigating the conflicting forces of newness and repetition.

In the research project upon which I draw for this chapter, I explored 'how' organisational routines *'transition'* between life cycle stages with *'incomplete'* information and how the predefined time boundaries of life cycle stages influence their (re)creation. From the results of this research, I propose that potential disruptions in 'patterns of action' for managing the relational and transactional difficulties at life cycle stage boundaries as team members leave, and new team members join, has the potential to create inherently unstable demand and supply chain relationships, until they are given 'time' to become capable. This chapter draws upon a large infrastructure case study during its period of transition from definition (design) to delivery (construction) to explore this phenomenon.

In this chapter I will firstly explore the theoretical foundations of the temporal paradox that I have suggested exists within (temporary) construction project organising, that of the conflicting demands of 'newness' and 'repetition'. The second section will look at the issue of the life cycle model and how we might think about *transitioning* across life cycle stages with incomplete information. Finally, I will present my 'recursive process model of transitioning' as a way of helping to understand the process of *transitioning* with incomplete information in construction supply chains.

5.1.1   The Temporal Paradox in Temporary Organising

In 'a theory of the temporary organisation' presented by Lundin and Söderholm (1995) there is an underlying assumption that the temporary organisation does not exist in the *'ongoing present'* until it is *'created anew'* and given, ex ante, a time delimited life cycle. Lundin and Söderholm (1995) suggest that 'time' is a 'basic' concept of temporary organisations, and that temporary organisations are discreet organisational units that are created to control capital investment initiatives that fall outside of the normal operation of the parent organisation. Human subjects, referred to as 'agents', and non-human objects, referred to as 'artefacts', are temporarily brought together in 'relational' and 'transactional' arrangements to create this new organisational unit and they work 'interdependently' to process 'information', to reduce 'uncertainty', and produce outputs as they 'transition' through the predefined stages of the life cycle, before the organisation is finally terminated.

Conversely, in 'a new theory of organisational routines' by Feldman and Pentland (2003) there is an underlying assumption that the organisation *'already exists'* in the *'ongoing present'* and its capability is based on the ongoing (re)creation of routines. Time is an implicit concept within organisational routines as it is their repetition and so recognisability over time that creates their identity. 'Organisational routines' are a central feature of organisations and organisational research (Parmigiani and Howard-Grenville 2011). Organisations (and in our case we are applying to supply chains as temporary organisations) are said to become capable through the evolution of routines. Organisational routines are defined as *'repetitive, recognisable patterns of interdependent actions, carried out by multiple participants'* (Feldman and Pentland 2003, p. 95). Routines are generative in that, over time, they sustain both change and stability within organisations as efforts to regular patterns of action often result in the emergence of new patterns.

This 'newness' characteristic of temporary organisations suggests a lack of 'repeated interaction' and 'recognisable patterns of action'. Temporary organisations could therefore be characterised by 'organisational uncertainty' at the start of each life cycle stage, making them potentially unstable structures until organisational routines (patterns of action) are (re)created and levels of perceived uncertainty reduced. The ability to recreate routines is partly a function of the extent to which the supply chain is 'standing' or entirely 'one-off' and project-based. 'Organisational uncertainty' can be understood as both 'relational' and 'transactional' uncertainty (Söderlund 2012; Jones and Lichtenstein

2008) and these will be discussed further in the following section, which looks at the nature of the construction industry

5.2 The Construction Industry – Procurement and Relational Difficulties

The government is the largest client in construction in the UK and recognises specifically the role that the infrastructure sector plays in the long-term economic future of the country (IPA 2017a). Since 2010, the UK government has sought to firm up its approach and commitment to investment in infrastructure as demonstrated through its publication of the National Infrastructure and Construction Pipeline (IPA 2017c), which expects to see continued investment of over £600 billion over the next 10 years. Throughout the last century, the capability of government (and others) as a client, and the market as a supply chain, and the way that they engage together, has been the subject of a number of industry reports that have predominantly been led or commissioned by both government and industry (Murray and Langford 2003). The main exception to this is the Tavistock Institute report, which was produced outside of the construction industry's main participants. They identified that 'uncertainty' and 'interdependence' were the two key characteristics that dominated the construction process and that 'these twin aspects of interdependence and uncertainty have been interpreted in terms of communication and information flow' (Tavistock Institute 1966, p. 18). I will return to the Tavistock report later in the next section.

A review of the themes and drivers behind these reports (Langford and Murray 2003) shows a continuous focus on the relationship between the parties to the construction process, most notably between the client (demand) and designers and contractors (supply), and the resulting performance of the construction process. Langford and Murray (2003) highlight that a common feature that influences these relationships is the role of procurement and the contractual relationship between the parties, the connection between the demand chain and the supply chain. Towards the end of the last century, the Latham (1994) and Egan (1998) reports emphasised the relational characteristics between the parties, and the needed improvements in the construction process, respectively, and this work was built on by authors such as Holti et al. (1999) who espoused methodologies for supply chain management. These reports could be argued to have driven many of the recent developments in the industry, as highlighted by the Wolstenholme report (Wolstenholme et al. 2009).

It is here that we can start to explore theoretically the influence of the relational and transactional difficulties and how the concept of organisational routines may be a way of helping us reduce the uncertainties associated with these difficulties. It is beneficial to express this graphically, which I have done in Figure 5.1. The left-hand side I have built top down from Mintzberg (1979) who explains that 'Every human activity – from the making of pots to the placing of a man on the moon – gives rise to two fundamental and opposing requirements: the *division of labor* into various tasks to be performed and the *coordination* of these tasks to accomplish the activity' (Mintzberg 1979, p. 2).

I have mirrored this on the right hand side, drawing on the construction industry reports and the work of Söderlund (2012) with regards to what he terms as the organising problems of cooperation and coordination – 'The problem of cooperation originates

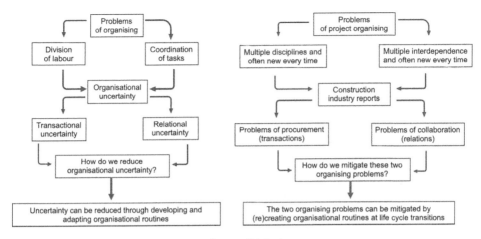

Figure 5.1 Problems of project organising. Source: Original.

from the fact that individuals and actors have conflicting goals and behave opportunistically, whereas the problem of coordination stems from the complexity of the task and the necessity to communicate and synchronise activities to achieve action efficiencies' (Söderlund 2012, p. 46). In this chapter I have termed these *'transactional'* and *'relational'* difficulties.

On the left of the graphic this results in organisational routines as a way of reducing what has been termed in the routines literature as 'pervasive' uncertainty. Pervasive uncertainty is that which cannot be removed by further information search but can potentially be reduced through the (re)creation of organisational routines (Becker and Knudsen 2005). The right-hand side is concerned with how these routines are (re)created as the project organisation transitions through the life cycle stages.

More recent publications espousing approaches not that dissimilar to that of Holti et al. (1999), such as the UK's Infrastructure Client Working Group (CWG) report into the performance of infrastructure projects entitled Project 13 (ICE 2017), have continued to highlight the transactional and relational challenges associated with infrastructure delivery. The resulting desire has been to design governance and procurement approaches that focus not just on the creation of value but are designed towards developing longer-term relationships. The orientation towards this approach is becoming more recognised by government through the work of the Infrastructure and Projects Authority who in their recent publication 'Transforming Infrastructure Performance', highlight their strategy of 'Procurement for Growth' and the development of smarter commercial relationships.

While I support the approach espoused within these reports, I argue that the temporal paradox of project organising strongly influences our understanding of the transactional and relational difficulties and that both of these difficulties will persist to some degree or another whichever form of contract or relational structure is designed and whether this is associated with short-term or long-term relationships. Our challenge as construction project managers is then not to suggest we have removed any one difficulty, but to minimise the resulting organisational uncertainty through (re)creating organisational routines and so minimising disruption as the organisation 'transitions' through the life cycle stages.

In the following section I build on this by understanding further the life cycle model, before looking at organisational routines and their relationship with organisational capabilities.

5.3 Temporary Organisations and the Project Life Cycle

Packendorff (1995) sets an agenda for theorising about temporary organising beyond the traditional planning and control theories in project management. Lundin and Söderholm (1995), influenced by Cyert and March (1963), presented a framework for the foundations of 'a theory of the temporary organisation' that along with Packendorff (1995) led to further research into such topics as understanding the importance of the wider context within which the project sits (Engwall 2003) and differentiating between 'functional' and 'temporary' organisations (Lundin and Steinthórsson 2003).

Lundin and Söderholm (1995) called for a new action-based theory developed from the inside out, where action takes primacy over decision making, a reversal of the approach from Cyert and March (1963) and which aimed at positioning the temporary organisation through four basic concepts of time, task, team, and transition that set the organisational boundary within which this action takes place. From a process ontology, Bakker et al. (2016) suggest that in temporary organising, the resulting patterns and outcomes from these actions are relative to the capability of project actors to reflect on and adapt their practices. In doing so they recognise the necessary incompleteness associated with the organising process … 'some conditions of actions will always remain unknown and unintended consequences may feed with or without recognition into conditions of the next sequence of actions' (2016, p. 3). And so, they suggest it is the relative dynamic nature of temporary organisations that influences their outcomes and that it is the use of 'rules, routines and resources to coordinate, enable and restrain the actions of actors, both inside and outside the focal entity' (2016, p. 3). I will look in further detail at project capabilities and organisational routines, and their impact on supply chains, in Section 5.4.

What theoretically binds temporary organisations together is what Söderlund (2012) explains as the 'project life cycle': 'One of the most salient features of projects is their organisational dynamics. Projects are born – they are created by man and they are designed to dissolve. The matter of birth and death of projects has accordingly been a core element of project management since the introduction of the project life cycle' (2012, p. 49). It is this life cycle that differentiates the project organisational form from other forms of organisation: '*All projects, no matter how complex or trivial, large or small, follow this development sequence*' (Morris 2013, p. 13 [emphasis in original]). Söderlund (2012) highlights how this feature of projects led to theories predominantly in the behavioural school, related to how projects function as temporary organisational forms (Gersick 1988; Goodman and Goodman 1976; Lundin and Söderholm 1995). Winter et al. (2006) propose that the life cycle model, while not rejected, has potentially constrained the understanding of projects, particularly from a 'process ontology' (Rescher 1996; Tsoukas and Chia 2002), by becoming accepted as the actual way that projects behave.

While this life cycle model of predefined dates can take on a number of forms, a governance structure that follows the traditional life cycle is expected to be able to monitor the progression of work through stage gates, where progress against predefined project or individual stage outputs can be assessed before gaining sanction to proceed to the next stage. Winch (2010) terms this as 'gating the process' and suggests that this is an area where project organisations have their own organisational routines for the measurement and control of the project.

In exploring this notion of 'gating the process', large engineering or mega projects have specific characteristics (Miller and Lessard 2001; Flyvbjerg 2014; Davies et al. 2017), but most notably it is the issue of their 'front end' that influences their ongoing capability (Morris and Hough 1987; Samset and Volden 2016) and more specifically the point of sanction to move from the front end into execution (Miller and Lessard 2001; Miller and Hobbs 2005). Morris (1997 and 2013), building on the earlier work, developed the term the 'Management of Projects' (Morris 1997) that led to work researching the definition of the 'front end' of projects (Edkins et al. 2013), and what it means for project governance and management (Samset and Volden 2016), specifically large engineering projects (Miller and Lessard 2001). Such work highlights the extent of influence that both internal and external stakeholders have, specifically their involvement in the statutory planning process in the front end of the project. In this chapter I have termed these stakeholders collectively as the 'demand chain' as a way of aiming to represent their importance and the interdependency between the client and their stakeholders with the 'supply chain', recognising that organisational routines are rooted in 'interdependent actions'.

Morris (2013) continued to highlight the challenges of this front end and in understanding the specific 'gate' when this 'definition' stage comes to an end and the project receives full sanction to enter into the 'delivery' phase – the time and space where the demand and supply chain connect. As Morris (2013) and Jones and Lichtenstein (2008) point out, this may be a number of different stages, depending on the type of project. nevertheless, in large or mega projects, this main transition from the definition stage to the delivery stage has been identified as an important step in the life cycle, as this quotation from Miller and Hobbs (2005) highlights: 'In most major projects, a time can be identified when most of, if not all, the pieces come into place, and when significant and irreversible commitments are made. This is typically the time when major contracts are signed, and financing is secured. This point marks the end of the strategic structuring phase and the beginning of the design and execution phase' (2005, p. 45).

It has been argued that the temporary organisation has developed organisational routines (patterns of action) during the front end of the project (Eriksson 2015), including the involvement of stakeholders (Edkins et al. 2013) and that these routines create project capabilities (Davies and Brady 2016) and inform organisational design (Eriksson and Kadefors 2017), creating 'perceptions' of organisational stability through the 'ongoing' action of processing information and reducing uncertainty. Yet as we seek to connect the demand chain with the supply chain as the project moves from development to delivery, it could also be argued that closing down a stage and starting a new stage in the organisation – 'transitioning' – while theoretically creating a measure of performance control, can momentarily disrupt the patterns of action through its dramatic spatial and temporal shift (Jones and Lichtenstein 2008; Miller and Hobbs 2005; Miller and Lessard 2001).

Returning to Winch (2010), in respect of 'gating the process', this 'transition' from stage to stage, and particularly the formal sanction to move from one stage to the next, is

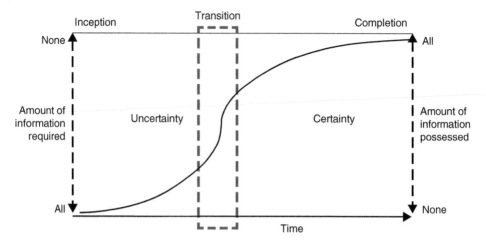

Figure 5.2 Transition, information, and uncertainty. Source: Adapted from Winch (2010).

governed by a predetermined set of outputs. It is around this that prescriptive and deterministic models of organisational and governance structures are designed and where traditionally there is an established stage gate assurance process, predominantly led by the production of a checklist of items (mainly physical documents) that have to be produced for the end of a stage (i.e. a design, a risk register, etc.) in order for the project to be granted approval to proceed, or not.

Winch (2010) made reference to the stage gates being points in the 'information flow' and indeed Winch, building on the work of Galbraith (1973) and the behavioural school of thinkers (March and Simon 1958; Cyert and March 1963), sees project organisations as information processing systems (Winch 2015). Figure 5.2 adapts the model by Winch (2010) and shows the potential levels of organisational uncertainty, relative to the availability of information at the stage of transition between definition and delivery.

The incompleteness of this information is something that Pryke (2017), who in developing his theoretical position on social networks in project-based organisations, draws our attention to in relation to the work of the Tavistock Institute (1966). In dealing with 'uncertainty' and 'interdependence' discussed earlier in this chapter, Pryke (2017), while focusing more specifically on the role of the delegation of work through contracts within the construction industry, highlights the notion of contractual 'incompleteness' as projects transition through the life cycle stages. While from a 'normative' project management governance perspective we can specify, through contracts or corporate project management procedures, the specific outputs at the end of a stage of the life cycle, what both Winch (2010) and Pryke (2017) argue, albeit from different theoretical positions, is that at each stage in the life cycle, there is a 'necessary' level of incompleteness in information. I argue here that it is the management of this level of incompleteness during the transition that influences the recognisability in the patterns of actions at the beginning of the next stage.

In this section I have focused on the nature of the temporary organisation and the staged life cycle model around which so much of the organisational and governance structures of construction projects are built. In seeking to understand further the temporal paradox and the relational and transactional difficulties of project organising

I focused on the specific transition between the definition and delivery stages of the life cycle and discussed how we can conceive this as being the time and space where the 'demand chain' and 'supply chain' become more formally connected together through the process of procurement and how from a project management perspective this is governed through the development of organisational routines and influenced by the incompleteness of information. In the following section I will explore further our understanding of organisational routines and their role in developing project capabilities.

5.4   Routines and the Capability of Projects

So far, I have highlighted the issue of patterns of action and how the life cycle model of projects, while valuable, potentially disrupts these patterns of action as they transition across temporal boundaries. In this section I look at the concept of organisational routines as a way of understanding these patterns and how this concept has been applied to the understanding of project organisations and their capability.

Organisations are said to become capable through the evolution of organisational routines (Nelson and Winter 1982), and organisational routines create and recreate patterns of action (Feldman and Pentland 2003). They have been suggested as a way of managing conflicting goals, coordinating tasks, and reducing uncertainty (Salvato and Rerup 2017; Jarzabkowski et al. 2012; Becker and Knudsen 2005; Dionysiou and Tsoukas 2013). Organisational routines are an integral part of organisations: 'To understand routines is to understand organizations. Routines are ubiquitous in organizations, and an integral part of organizations. One is hard put to identify an organization where no routines are present' (Becker 2008, p. 3). They have become a more common theme in organisational theory over the last 50 years (Cohen et al. 1996; Becker 2004; Parmigiani and Howard-Grenville 2011) with their understanding shifting from programmable, to evolutionary in nature (March and Simon 1958; Cyert and March 1963; Nelson and Winter 1982).

The early part of this century saw a number of studies that challenged the thinking that routines, while open to adaptation were generally static, stable, and unchanging entities (Cyert and March 1963), and building on Nelson and Winter (1982) started to break them open and understand their constituent parts, focusing on the role of agency in understanding the routine as a source of flexibility and change in organisations, and so being generative in nature (Feldman 2000; Feldman and Pentland 2003; Howard-Grenville 2005). Feldman's' (2000) notable work changed the view that it was solely exogenous change that caused routines (and therefore organisations) to adapt and brought forward a 'practice' oriented perspective (Feldman and Orlikowski 2011) that influenced the proposition of a process (becoming) ontology of organising (Tsoukas and Chia 2002) and recognised that organisational change may emerge over time from within the enactment of the routine itself, within a given situation.

Feldman and Pentland (2003) then develop this work further into 'a new theory of organizational routines'. Their work breaks open the routine, explaining their structural and characteristic make up, specifically in understanding three aspects that make up the routines internal structure, namely: the 'ostensive' aspect – cognitive representation of the expected action; the 'performative' aspect – actual performances of the action; and,

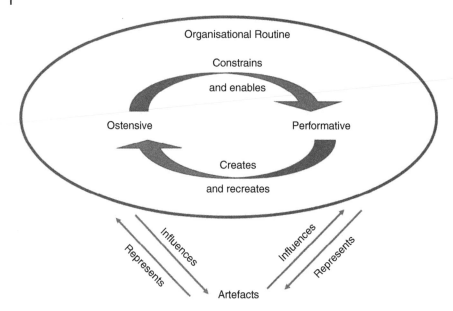

Figure 5.3 Organisational routines are generative systems. Source: Taken from Pentland and Feldman (2008a).

the 'artefact' aspect – document, process, or physical tool where knowledge is formally encoded. These three aspects of the routine enable the understanding of the generative mechanisms that influence stability and change in organisations, and so the (re)creation of organisational routines over time, within a given situation.

The ostensive, performative, and artefact aspects work in a recursive manner creating and recreating, enabling and constraining, and influencing and representing the patterns of action by multiple participants. This is expressed in graphical form in Figure 5.3. This recursive nature of routines allows us to explore how patterns of action may be (re)created at the boundaries of the projects' life cycle stages and it is within these patterns of action, resulting from enacting the routines, that I want to suggest we can find some knowledge to help understand the temporal paradox and transactional and relational difficulties associated with project organising.

Despite their role in organisational theory and their underpinning of a process ontology of organising (Tsoukas and Chia 2002; Feldman 2000), the opportunity that organisational routines offer to better understand the dynamic life cycle of projects has been predominantly limited to understanding the management and complexity of large projects (Stinchcombe and Heimer 1985; Eriksson 2015), learning across and between permanent and temporary organisations (Bresnen et al. 2005; Sydow et al. 2004; Manning 2008), and project-based organisations and organisational capability (Brady and Davies 2004; Davies and Hobday 2005; Davies and Brady 2016). As I highlighted earlier in section 5.3, for large engineering or mega projects, it has been argued that organisational routines in general are developed in the early stages of projects (Eriksson 2015), influence the way that they are designed and governed (Eriksson and Kadefors 2017), and are involved in developing the capability of the project organisation (Stinchcombe and Heimer 1985; Ahola and Davies 2012; Davies and Brady 2016).

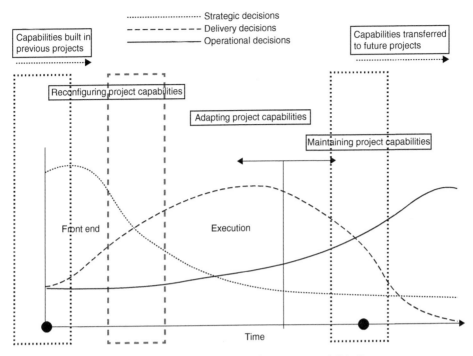

Figure 5.4 Transitioning capabilities. Source: Adapted from Zerjav et al. (2018).

Earlier work by Hobday, Davies, and Brady (Hobday 2000; Brady and Davies 2004; Davies and Hobday 2005) looked at the capability of project-based organisations and recognised the role of organisational routines (Nelson and Winter 1982) in developing organisational capability through the development and maturity of project management systems. Davies and Hobday (2005) provide a wider discussion on the paradox of project tasks as ranging from unique to repetitive as defined by Lundin and Söderholm (1995) and so the relationship between the need to 'exploit' existing routines and to 'explore' new routines to meet the demands of entering new markets or undertaking one-off 'vanguard' projects (March 1991; Brady and Davies 2004; Frederiksen and Davies 2008).

More recently, Davies and Brady (2016) have looked at the relationship between capabilities at the project level and those at the level of the firm through the lens of organisational routines and this work has been developed to look at how these routines are reconfigured, adapted and maintained from previous projects, through the whole life of the current project and on into future projects (Zerjav et al. 2018). This work can be drawn on to show that from a capabilities perspective, routines are (re)created in that transition from the front end to execution. This is presented in Figure 5.4 where the Zerjav et al. (2018) model is adapted to present three main transitions in a projects life cycle, with a rectangle in a dashed line being the transition under investigation from the empirical study I draw on in this chapter.

In this section I have drawn attention to the role of organisational routines in developing organisational capabilities, reducing uncertainty, the generative nature of their internal dynamics, and their adaptation through the life cycle of a project. In the following section, I present the recursive process model of transitioning.

5.5 A Recursive Process Model of Transitioning

In the research project upon which I draw for this chapter, I used the work of Pentland and Feldman (2008b) and identified six organisational routines used by the senior management team to manage the transition from design to construction, namely: organising, governing, contracting, designing, constructing, consenting. I then followed a practice approach (Sandberg and Tsoukas 2011) and the strategy of Van de Ven (2007) and van Maanen (1979) in identifying incidents and practical events within the routines. Using the work of Langley (1999), through temporal bracketing and visual mapping, I put the incidents I identified from the practical events of the routines into 'objective' chronological order, through the project's four-weekly, thirteen-period business rhythm and supported this with a composite narrative of the practical events (Jarzabkowski et al. 2014). To help understand the more 'subjective' temporal aspects, I drew on further literature on organisational and social transition (Abbott 2001; Gersick 1988), and used the concept of transition rituals (Söderlund and Borg 2017; van den Ende and van Marrewijk 2014) to assist in identifying the boundaries between the 'abstract' event sequences of change, which I identified over the year of the transition.

From this, I have suggested that the 'patterning of action' within and between the six routines as they transition across the life cycle stage boundary from design to construction can be presented as a five-stage 'recursive process model of transitioning' (Van de Ven 2007, p. 197). These five stages (Figure 5.5) present a more complex and socially entwined relationship between the demand chain and supply chain participants to the two stages (Sandberg and Tsoukas 2011), one that is not ordinarily presented in traditional life cycle models. I seek to present this by describing the activities I observed within each of the stages:

Abstract Event 5.1 Realising

Drawing attention and problematising an emergent breakdown in the purpose or task of the temporary organisation, or in the performance of an organisational routine.

On the 29 June 2015 we held a senior management team workshop at an off-site venue that sought to reflect on progress to date (the patterns of action we had developed together) and look ahead to the formal contractual transition from stage 1 to stage 2. At this workshop we realised that our current practices in collaborating together were not conducive to a successful transition and the type of organisation we needed for construction. This caused us to break from our current patterns, restructure our work, and make a start on what I later understood to be the (re)creation of our patterns of action. Our shared view of the values and objectives for the future remained intact, but we clearly needed to adapt organisational, contractual, and governance arrangements (the way we managed *relational* and *transactional* difficulties) if we wanted the mitigate the inherent uncertainties of moving from a design organisation to a construction organisation, specifically the large change in personnel.

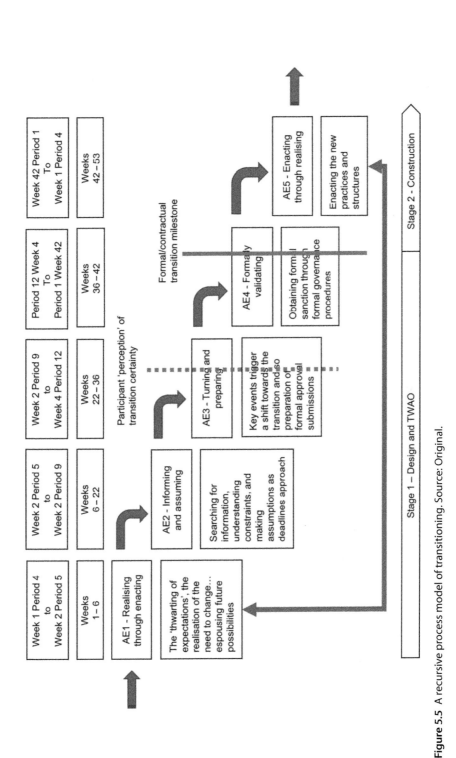

Week 1 Period 4 to Week 2 Period 5	Week 2 Period 5 to Week 2 Period 9	Week 2 Period 9 to Week 4 Period 12	Period 12 Week 4 To Period 1 Week 42	Week 42 Period 1 To Week 1 Period 4
Weeks 1 – 6	Weeks 6 – 22	Weeks 22 – 36	Weeks 36 – 42	Weeks 42 – 53

AE1 - Realising through enacting

The 'thwarting of expectations', the realisation of the need to change... espousing future possibilities

AE2 - Informing and assuming

Searching for information, understanding constraints, and making assumptions as deadlines approach

Participant 'perception' of transition certainty

AE3 - Turning and preparing

Key events trigger a shift towards the transition and so preparation of formal approval submissions

Formal/contractual transition milestone

AE4 - Formally validating

Obtaining formal sanction through formal governance procedures

AE5 - Enacting through realising

Enacting the new practices and structures

Stage 1 – Design and TWAO

Stage 2 - Construction

Figure 5.5 A recursive process model of transitioning. Source: Original.

Abstract Event 5.2 Informing and Assuming

Searching for 'information' transition routine goals. A recognition of limited time bringing closure to the information search.

This was a period of uncertainty characterised by high degrees of information search about the future. We realised that much of what we had to deal with was not risk, but uncertainty (Sanderson 2012) and so our approach shifted towards a process of scenario analysis, challenging schedule constraints and optioneering, to gain as much information as was available at that time. This stage drew to a close when we realised that time to the fixed transition date was running out fast and this led us into a process of making assumptions with the available information we had. We moved with what we termed 'necessarily incomplete' or 'sufficiently complete' information. With this suite of information and assumptions we made decisions that we believed would enable us to gain formal sanction to proceed into the next stage. These decisions then allowed us start understanding 'how' we needed to adapt our organisational, contractual, and governance arrangements to meet the fixed date for formal transition to stage 2 – construction. The flexibility given to us by a design and build relational contract and the management protocol facilitated this approach.

Abstract Event 5.3 Turning and Preparing

The period before formal validation, when single events or a group of actions provide pivotal 'turning points' towards the transition and the turn towards the final preparation of transition governance documents. It felt like informal approval.

During this period, based on the information and assumptions we had, we started to adapt and create new project governance arrangements and prepare documents for the formal assurance process, which commenced during this stage. These documents acted as the carriers of knowledge (the information, assumptions, and decisions we had made) through the transition and across the stage gate boundary. Most interestingly, a number of key events occurred in this phase, such as the formal granting of the Transport and Works Act Order (TWAO), the joining of senior construction staff, and signing off designs, that led us to have a 'felt' sense as a team that we were really transitioning (presented as the vertical dotted line in Figure 5.5). A sense that we were somehow not fully undertaking design, not yet fully in construction, nor had formal approval, but just a felt sense that we had the relational and transactional difficulties under control and a shared certainty regards formal approval to proceed being granted. This was emphasised by the fact that we had been granted approval by the project board to implement the new organisational arrangements of the senior management team set out in a revised management protocol.

Abstract Event 5.4 Validating

Gaining formal governance sanction at corporate level, issuing the S2WCN at project level, and agreeing the relational management protocol at project level.

This event was shorter in duration and characterised by formal sanction, with the Transport for London (TfL) board approving stage 2 and the project board approving issue of the Stage Two Works Commencement Notice (S2WCN) prior to 21 April 2016 (the solid vertical line in Figure 5.5).

Abstract Event 5.5 Enacting

Enacting the adapted practices in order to adjust to the change of temporal and spatial frame.

   This stage was characterised by regular breakdowns in our performance; our new patterns of action were struggling to take hold. The temporal and spatial shift from design to construction could not be avoided and it was clearly going to take time to embed the new practices that we had designed and embedded in new governance documents (the artefact aspect of the routine) and with our shared sense of what we were seeking to achieve (the ostensive aspect of the routine). It was clear that the adaptation of our governance documents was incomplete, in the sense that they could only become complete in our performance of them (the performative aspect of the routine). Importantly, it was our shared experiences and meanings of the breakdown in Abstract Event 5.1 and the subsequent work we had done together through Abstract Events 5.2–5.4, which enabled us to mitigate the extent of disruption in our new, emerging patterns of action. This relationship is presented with the arrow connecting abstract event stages 1 and 5 and so draws attention to the recursive nature of change in transition. Our relational difficulties were mitigated, and our goals and incentives remained aligned, as we had shared together in the breakdown of our patterns, the incompleteness of our information, our understanding of the uncertainties, and the ongoing planning and actions to deal with them. Most notably, it was the way we had flowed down stakeholder commitments from the 'front end' (the demand chain), through to the multiple subcontract arrangements to deliver the work (the supply chain).

   Figure 5.5 presents a graphic explanation of this model, showing the client's accounting periods and the data collection weeks within which each abstract event took place, with the end and beginning weeks overlapping. The arrows represent the movement into, through, and out of the sequence of abstract events and the recursive relationship between enacting and realising. The two boxes along the bottom present the two main life cycle stages. The dotted vertical line represents the timing of the 'felt' transition, while the solid vertical line presents the formal milestone date of 21 April 2016 when the project formally transitioned to stage two.

   In this section I have presented what I have termed a 'recursive process model of transitioning' that I identified as the project transitioned from front end definition (design and statutory planning) and into the execution stage (construction).

   In the following concluding section, I reflect on the process of managing transitions across life cycle stages and the relational and transactional difficulties, relative to the temporal paradox. I look at how to connect the demand chain with the supply chain for more capable construction project organisations.

5.6   Discussion

The 'ongoing' and 'dialogical' nature of our interdependent actions between the multiple participants of the 'demand' and 'supply' chains during the transition from design to construction both created and recreated the ostensive, performative, and artefact aspects of the six routines that I identified. I would like to propose here that having a predefined transition date was a strong influence on how our patterns of action were (re)created, as described in abstract event 2 for example, where due to the predefined date for transition, we perceived the need to bring our information search to a close with a degree of incompleteness. This is not to suggest that some form of patterning would not have taken place without this date (Jarzabkowski et al. 2012; Dionysiou and Tsoukas 2013) but that the five stages and the patterns of action we (re)created were as a result of this ex-ante defined date.

More specifically, I want to suggest that it is the shared sense of managing together the interdependencies (Tavistock 1966; Holti et al. 1999) in the patterns of action through the five abstract event stages of transition that was important. We were able to break away from old patterns as the design and statutory planning came to a close and team members departed, and adapt those old patterns towards the new members joining and the revised organisational and governance structures necessary for the construction stage. We were able to both maintain, and build on, our interpersonal and interorganisational relationships to mitigate potentially difficult contractual and relational conflicts that emerged as we completed the front-end definition stage of the project.

So while the ex-ante defined date was a necessary control point, I would argue that had we not had a shared understanding that developed over time, had we not shared in the breakdown of our old patterns, and had we not adapted our new working practices together, then the prescriptive and deterministic process driven by governance and contracts (Winch 2010), understood theoretically as necessarily incomplete (but not often treated as such in practice) (Pryke 2017), could have severely disrupted any emerging patterns of action at the start of the delivery stage. It was the ability of the demand and supply chain actors to overcome our 'transactional' and 'relational' difficulties, with incomplete information, that created the perceptions of control and stability, not the sole fact that there was a fixed date and a fixed schedule of deliverables.

I return here to the theoretical position of the temporal paradox and the transactional and relational difficulties that have been described above and where I have argued that despite such work by Holti et al. (1999), and more recently the ICWG Project 13 report (ICE 2017), the industry continues to search for practical ways of dealing with the difficulties associated with transition. Perhaps one of the potential solutions may lie in looking beyond the ubiquitous life cycle model (Winter et al. 2006) and towards an understanding of '*how*' we manage 'transitions' between life cycle stages.

I made reference above in Section 5.2 to the IPA document 'Transforming Infrastructure Performance' and the need to develop smarter commercial relationships. They state that: 'Smarter commercial relationships between clients and their suppliers help align objectives, support the delivery of improved outcomes, boost the productivity of the industry and deliver infrastructure assets that better meet the needs of society and users. Such relationships enable a deeper engagement with the supply chain, with earlier participation in the investment life cycle where suppliers are incentivised and rewarded for finding better solutions' (IPA 2017b, p. 32).

Yet the temporal paradox of temporary organising exists, and we continue to search for ways to manage that paradox. The transactional and relational difficulties associated with short- and long-term relationships, or early/late engagement of the supply chain are what I have suggested is 'pervasive' uncertainty in the construction process. I would like to propose that in seeking to manage this pervasive uncertainty, through the lens of organisational routines and from my 'recursive process model of transitioning', *transitions* (most specifically procurement models) can be designed in a way that better connects the demand and supply chain of a project. They can be designed in such a way that looks beyond deterministic and prescriptive checklist outputs of life cycle stages and towards ways that (re)create organisational routines, reduce uncertainty, and therefore (re)create organisational capabilities. So, what might be the starting point for looking beyond the ubiquitous life cycle model of construction project management?

Morris (2013) makes it clear to us that the purpose of projects is the creation of value and if we understand value as benefit divided by cost, then it is in the business case that we should ground our thinking. If a value model is created from the business case of a 'concept' design which can be used for evaluating the tender and decision making during developing and delivering the scheme, or designing an innovative governance process, then we always return to the business case and the value we sought. Perhaps doing so might perhaps cause us to rethink the cost, time, and quality triangle so prevalent in project management and move us towards a triangle more focused on requirements, benefits, and uncertainties, with the value model at its centre.

I would also like to suggest that it is here in the value model that the primary connection between the demand and the supply chain can be found. This is set out in the benefits (social and monetary) that the demand side will receive as a result of the project and it is the supply chain that has the most opportunity to influence the extent of benefits, the costs, and the uncertainties associated with delivering the built asset. By sharing and analysing all appropriate and available information between the demand and supply side participants, trading off requirements, benefits, and uncertainties through the value model, arriving at a position of shared 'necessary incompleteness' or 'sufficient completeness' and with sufficient recognisability and repetition in our interdependent patterns of action, then perhaps construction projects can be better managed and greater value created.

Inherent in this is understanding the business case for projects and programmes as an important responsibility collectively for both the demand and the supply chain. In this way an appropriate definition of value is created and the 'ongoing' and 'dialogical' relationships between demand and supply chain actors can be harnessed to create value.

5.7 Summary

The life cycle model has become a prevalent form of organising and governing construction projects. In this chapter I have suggested that from a process ontology perspective of project organising, the efficacy of this life cycle model can be challenged as it has the potential to disrupt the ongoing and dialogical nature of the organising process.

Recognising the multiple forms that the life cycle can take, in this chapter I looked at the characterisation by Morris (2013) of the two primary stages of definition and delivery, focusing on the transition between these two stages. I presented how this transition marks a shift and a potential breakdown in the ongoing structure and governance of the project organisation and a place where the demand chain and supply chain become connected together.

To explore this *transition* phenomenon theoretically, I drew attention to the temporal paradox of temporary organising established through the lens of a theory of the temporary organisation (Lundin and Söderholm 1995) and a new theory of organisational routines (Feldman and Pentland 2003). I added to this the 'transactional' and 'relational' difficulties of organising, both from a theoretical perspective (Söderlund 2012) and from a practical perspective, which I drew from a number of reports into the construction industry over the last half century and more (Murray and Langford 2003).

I discussed these theoretical perspectives before presenting some empirical findings – with its five stages – and discussed how the predefined and fixed date for transition influenced how patterns of interdependent action were influenced in such a way that it created the five stages. I explored how these five stages were the '*ongoing*' and '*dialogical*' adaptation of our organisational routines, from design to construction that provided sufficient repetition and recognisability to be able mitigate and manage the transactional and relational difficulties as new patterns of working between the demand and supply chain at the start of the new stage of the project were sought.

Therefore, while not rejecting the life cycle as an appropriate model and recognising that it has multiple variants on many different projects, I questioned how we might better manage *transitions* across the predetermined time boundaries of life cycle stages and in doing so I returned to the work of Morris (2013) and suggested that the answer may lie in the clarity and understanding of a projects 'value model'. Elsewhere in this book we look at value and its creation (see for example chapters by Mills et al, Manu and Knight, and Broft).

It is here then that I return to the quotation at the start of this chapter and perceive how when project participants are '*transitioning*' through a project's time delimited life cycle, it is this value model that is able… 'to arouse in each other *transitory understandings* of "where" so far in their activities they have "got to", and *action guiding anticipations* of "where" or "how" next they are likely "to go on". In other words, it is only *in* the course of their actions that they can *organize* their conduct of them, not before by planning them, nor after by criticizing them' (Shotter 2008, p. 510).

References

Abbott, A. (2001). *Time Matters: On Theory and Method*. University of Chicago Press.

Ahola, T. and Davies, A. (2012). Insights for the governance of large projects: Analysis of Organization Theory and Project Management: Administering Uncertainty in Norwegian Offshore Oil by Stinchcombe and Heimer. *International Journal of Managing Projects in Business* 5 (4): 661–679.

Bakker, R.M., DeFillippi, R.J., Schwab, A., and Sydow, J. (2016). Temporary organizing: promises, processes, problems. *Organization Studies* 37 (12): 1703–1719.

Becker, M.C. (2004). Organizational routines: a review of the literature. *Industrial and Corporate Change* 13: 643–677.

Becker, M.C. (ed.) (2008). *Handbook of Organizational Routines*. Edward Elgar Publishing.

Becker, M.C. and Knudsen, T. (2005). The role of routines in reducing pervasive uncertainty. *Journal of Business Research* 58 (6): 746–757.

Brady, T. and Davies, A. (2004). Building project capabilities: from exploratory to exploitative learning. *Organization Studies* 25 (9): 1601–1621.

Bresnen, M., Goussevskaia, A., and Swan, J. (2005). Organizational routines, situated learning and processes of change in project-based organizations. *Project Management Journal* 36 (3): 27.

Cohen, D; Burkhart, R; Dosi, G; Egidi, M; Marengo, L; Warglien, M; Winter, S. (1996) Routines and Other Recurring Action Patterns of Organizations: Contemporary Research Issues. Santa Fe Institute Working Paper, page 3.

Cunliffe, A.L., Helin, J., and Luhman, J.T. (2014). Mikhail Bakhtin (1895–1975). In: *The Oxford Handbook of Process Philosophy and Organization Studies* (eds. J. Helin, T. Hernes, D. Hjorth and R. Holt). Oxford University Press.

Cyert, R.M. and March, J.G. (1963). *A Behavioral Theory of the Firm*. Englewood Cliffs, NJ: Prentice Hall as the publisher.

Davies, A. and Brady, T. (2016). Explicating the dynamics of project capabilities. *International Journal of Project Management* 34 (2): 314–327.

Davies, A. and Hobday, M. (2005). *The Business of Projects: Managing Innovation in Complex Products and Systems*. Cambridge University Press.

Davies, A., Dodgson, M., Gann, D., and MacAulay, S. (2017). Five rules for managing large, complex projects. *MIT Sloan Management Review* 59 (1): 73.

Dionysiou, D.D. and Tsoukas, H. (2013). Understanding the (re)creation of routines from within: a symbolic interactionist perspective. *Academy of Management Review* 38 (2): 181–205.

Edkins, A., Geraldi, J., Morris, P., and Smith, A. (2013). Exploring the front-end of project management. *Engineering Project Organization Journal* 3 (2): 71–85.

Egan, J. (1998). *Rethinking Construction: Report of the Construction Task Force on the Scope for Improving the Quality and Efficiency of UK Construction*. London: Department of the Environment, Transport and the Regions.

Emirbayer, M. and Mische, A. (1998). What is agency? *American Journal of Sociology* 103 (4): 962–1023.

van den Ende, L. and van Marrewijk, A. (2014). The ritualization of transitions in the project life cycle: a study of transition rituals in construction projects. *International Journal of Project Management* 32 (7): 1134–1145.

Engwall, M. (2003). No project is an island: linking projects to history and context. *Research Policy* 32 (5): 789–808.

Eriksson, T. (2015). Developing routines in large inter-organisational projects: a case study of an infrastructure megaproject. *Construction Economics and Building* 15 (3): 4–18.

Eriksson, T. and Kadefors, A. (2017). Organisational design and development in a large rail tunnel project—influence of heuristics and mantras. *International Journal of Project Management* 35 (3): 492–503.

Feldman, M.S. (2000). Organizational routines as a source of continuous change. *Organization Science* 11 (6): 611–629.

Feldman, M.S. and Orlikowski, W.J. (2011). Theorizing practice and practicing theory. *Organization Science* 22 (5): 1240–1253.

Feldman, M.S. and Pentland, B.T. (2003). Reconceptualizing organizational routines as a source of flexibility and change. *Administrative Science Quarterly* 48 (1): 94–118.

Flyvbjerg, B. (2014). What you should know about megaprojects and why: an overview. *Project Management Journal* 45 (2): 6–19.

Frederiksen, L. and Davies, A. (2008). Vanguards and ventures: projects as vehicles for corporate entrepreneurship. *International Journal of Project Management* 26 (5): 487–496.

Galbraith, J.R. (1973). *Designing Complex Organizations*. Addison-Wesley Longman Publishing Co., Inc.

Gersick, C.J. (1988). Time and transition in work teams: toward a new model of group development. *Academy of Management Journal* 31 (1): 9–41.

Goodman, R.A. and Goodman, L.P. (1976). Some management issues in temporary systems: a study of professional development and manpower—the theatre case. *Administrative Science Quarterly* 21 (3): 494–501.

Hernes, T. (2014). *A Process Theory of Organization*. Oxford University Press.

Hobday, M. (2000). The project-based organisation: an ideal form for managing complex products and systems? *Research Policy* 29 (7): 871–893.

Holti, R., Nicolini, D., and Smalley, M. (1999). *Building Down Barriers: Prime Contractor Handbook of Supply Chain Management*. Ministry of Defence.

Howard-Grenville, J.A. (2005). The persistence of flexible organizational routines: the role of agency and organizational context. *Organization Science* 16 (6): 618–636.

ICE, (2017) Institute of Civil Engineers Project 13 – From Transactions to Enterprises. www.ice.org.uk/knowledge-and-resources/best-practice/project-13-from-transactions-to-enterprises (accessed 13 August 2019).

IPA (2017a) About IPA. https://www.gov.uk/government/uploads/system/uploads/attachment_data/file/652340/IPA_Narrative_document_WEB.pdf (accessed 13 August 2019).

IPA (2017b) Transforming Infrastructure Performance. https://www.gov.uk/government/publications/transforming-infrastructure-performance (accessed 13 August 2019).

IPA (2017c) National Infrastructure and Construction Pipeline 2017. https://www.gov.uk/government/publications/national-infrastructure-and-construction-pipeline-2017 (accessed 13 August 2019).

Jarzabkowski, P.A., Lê, J.K., and Feldman, M.S. (2012). Toward a theory of coordinating: creating coordinating mechanisms in practice. *Organization Science* 23 (4): 907–927.

Jarzabkowski, P., Bednarek, R., and Lê, J.K. (2014). Producing persuasive findings: demystifying ethnographic text work in strategy and organization research. *Strategic Organization* 12 (4): 274–287.

Jones, C. and Lichtenstein, B.B. (2008). Temporary inter-organizational projects: how temporal and social embeddedness enhance coordination and manage uncertainty. Chapter 9. In: *The Oxford Handbook of Inter-Organizational Relations* (eds. S. Cropper, M. Ebers, C. Huxham and P. Smith Ring), 231–255. Oxford: Oxford University Press.

Murray, M. and Langford, D.A. (eds.), (2003). *Construction Reports 1944-98*. Chichester: Blackwell Science.

Langley, A. (1999). Strategies for theorizing from process data. *Academy of Management Review* 24 (4): 691–710.

Langley, A.N.N., Smallman, C., Tsoukas, H., and Van de Ven, A.H. (2013). Process studies of change in organization and management: unveiling temporality, activity, and flow. *Academy of Management Journal* 56 (1): 1–13.

Latham, M. (1994). *Constructing the Team: Joint Review of Procurement and Contractual Arrangements in the UK Construction Industry*. London, UK: Department of the Environment.

Lorino, P. and Tricard, B. (2012). The Bakhtinian theory of chronotope (time-space frame) applied to the organizing process. In: *Constructing Identity in and around Organizations*, vol. 2 (eds. M. Schultz, S. Maguire, A. Langley and H. Tsoukas), 201–234. Oxford University Press.

Lundin, R.A. and Söderholm, A. (1995). A theory of the temporary organization. *Scandinavian Journal of Management* 11 (4): 437–455.

Lundin, R.A. and Steinthórsson, R.S. (2003). Studying organizations as temporary. *Scandinavian Journal of Management* 19 (2): 233–250.

Manning, S. (2008). Embedding projects in multiple contexts – a structuration perspective. *International Journal of Project Management* 26 (1): 30–37.

March, J.G. (1991). Exploration and exploitation in organizational learning. *Organization Science* 2 (1): 71–87.

March, J.G. and Simon, H.A. (1958). *Organizations*. New York: Wiley.

Miller, R. and Hobbs, J.B. (2005). Governance regimes for large complex projects. *Project Management Journal* 36 (3): 42–51.

Miller, R. and Lessard, D., with Floricel, S. and the IMEC Research Group (2001). *The Strategic Management of Large Engineering Projects: Shaping Institutions, Risk, and Governance*. Cambridge, MA: MIT Press.

Mintzberg, H. (1979). *The Structuring of Organizations: A Synthesis of the Research*. University of Illinois at Urbana-Champaign's Academy for Entrepreneurial Leadership Historical Research Reference in Entrepreneurship.

Morris, P.W. (1997). *The Management of Projects*. Thomas Telford.

Morris, P.W. (2013). *Reconstructing Project Management*. Wiley.

Morris, P.W. and Hough, G.H. (1987). *The Anatomy of Major Projects: A Study of the Reality of Project Management*. Chichester: Wiley.

Nelson, R.R. and Winter, S.G. (1982). *An Evolutionary Theory of Economic Change*. Cambridge, MA: Belknap Press of Harvard University Press.

Packendorff, J. (1995). Inquiring into the temporary organization: new directions for project management research. *Scandinavian Journal of Management* 11 (4): 319–333.

Parmigiani, A. and Howard-Grenville, J. (2011). Routines revisited: exploring the capabilities and practice perspectives. *The Academy of Management Annals* 5 (1): 413–453.

Pentland, B.T. and Feldman, M.S. (2008a). Designing routines: on the folly of designing artifacts, while hoping for patterns of action. *Information and Organization* 18 (4): 235–250.

Pentland, B.T. and Feldman, M.S. (2008b). Issues in empirical field studies of organizational routines. In: *Handbook of Organizational Routines* (ed. M.C. Becker), 281–300. Edward Elgar Publishing.

Pentland, B.T. and Reuter, H.H. (1994). Organizational routines as grammars of action. *Administrative Science Quarterly*: 484–510.

Pryke, S. (2009). *Construction Supply Chain Management: Concepts and Case Studies*, vol. 3. Wiley.

Pryke, S.D. (2017). *Managing Networks in Project-Based Organisations*. Wiley Blackwell.

Rescher, N. (1996). *Process Metaphysics: An Introduction to Process Philosophy*. Suny Press.

Salvato, C. and Rerup, C. (2017). Routine regulation: balancing conflicting goals in organizational routines. *Administrative Science Quarterly* 63 (1): 170–209.

Samset, K. and Volden, G.H. (2016). Front-end definition of projects: ten paradoxes and some reflections regarding project management and project governance. *International Journal of Project Management* 34 (2): 297–313.

Sandberg, J. and Tsoukas, H. (2011). Grasping the logic of practice: theorizing through practical rationality. *Academy of Management Review* 36 (2): 338–360.

Sanderson, J. (2012). Risk, uncertainty and governance in megaprojects: a critical discussion of alternative explanations. *International Journal of Project Management* 30 (4): 432–443.

Shotter, J. (2008). Dialogism and polyphony in organizing theorizing in organization studies: action guiding anticipations and the continuous creation of novelty. *Organization Studies* 29 (4): 501–524.

Söderlund, J. (2012, 2012). Theoretical foundations of project management: suggestions for a pluralistic understanding. In: *The Oxford Handbook of Project Management* (eds. P.W. Morris, J.K. Pinto and J. Söderlund), 37–64. Oxford University Press.

Söderlund, J. and Borg, E. (2017). Liminality in management and organization studies: process, position and place. *International Journal of Management Reviews* 00: 1–23.

Stinchcombe, A.L. and Heimer, C.A. (1985). *Organization Theory and Project Management: Administering Uncertainty in Norwegian Offshore Oil*. USA: Oxford University Press.

Sydow, J., Lindkvist, L., and DeFillippi, R. (2004). Project-based organizations, embeddedness and repositories of knowledge: editorial. *Organization Studies* 25 (9): 1475–1489.

Tavistock Institute of Human Relations (1966). *Interdependence and Uncertainty: A Study of the Building Industry, Digest of a Report from the Tavistock Institute to the Building Industry Communication Research Project*. Tavistock Publications.

Tsoukas, H. and Chia, R. (2002). On organizational becoming: rethinking organizational change. *Organization Science* 13 (5): 567–582.

Van de Ven, A.H. (2007). *Engaged Scholarship: A Guide for Organizational and Social Research: A Guide for Organizational and Social Research*. Oxford University Press.

Van Maanen, J. (1979). The fact of fiction in organizational ethnography. *Administrative Science Quarterly* 24 (4): 539–550.

Winch, G.M. (2010). *Managing Construction Projects*. Wiley.

Winch, G.M. (2015). Project organizing as a problem in information. *Construction Management and Economics* 33 (2): 106–116.

Winter, M., Smith, C., Morris, P., and Cicmil, S. (2006). Directions for future research in project management: the main findings of a UK government-funded research network. *International Journal of Project Management* 24 (8): 638–649.

Wolstenholme, A., Austin, S.A., Bairstow, M. et al. (2009). *Never Waste a Good Crisis: A Review of Progress since Rethinking Construction and Thoughts for our Future*. Loughborough University Institutional Repository.

Zerjav, V., Edkins, A., and Davies, A. (2018). Project capabilities for operational outcomes in inter-organisational settings: the case of London Heathrow terminal 2. *International Journal of Project Management* 36 (3): 444–459.

6

Construction Supply Chain Management through a Lean Lens

Lauri Koskela, Ruben Vrijhoef, and Rafaella Dana Broft

6.1   Introduction

The concept of lean continues to be poorly understood. Although lean and its predecessor concepts, such as the Toyota Production System and World Class Manufacturing, have been discussed for quite some time in the West from the 1980s, the division between mainstream management thinking and lean thinking still persists. Since the 1990s, the concept of supply chain management has been introduced to the area of lean, coined as lean supply (Lamming 1996). However, at the same time, supply chain management has stemmed from other domains including logistics and information management.

From a theoretical standpoint on lean, we could classify supply chain management approaches and techniques either in a lean or non-lean perspective. Non-lean approaches to supply chain management coexist besides lean approaches. Often the problem is a one-sided and exclusive subscription of supply chain management either to the lean or non-lean domain. This problem has two angles: a lack in the understanding and explanation of lean, and a characteristic misinterpretation of supply chain management approaches. The result is that these approaches are sometimes considered lean, sometimes non-lean.

This chapter aims to answer the following questions, bundled around four topics:

1. How is lean different from mainstream thinking? And in what way is it different or similar to traditional thoughts in mainstream management? Here, the starting point is theoretical, epistemological, and ontological.
2. What is the specific characterisation of supply chain management? Which practices fall into supply chain management? Which are the theoretical approaches specific to supply chain management?
3. How are the two conceptions of lean and mainstream management reflected in supply chain management, both practically and theoretically?
4. In which way is supply chain management, especially in its lean form, contingent on the characteristics of the construction industry and firms operating within the industry?

Successful Construction Supply Chain Management: Concepts and Case Studies, Second Edition.
Edited by Stephen Pryke.

The approach in this chapter is theoretical, even historical. However, in the last part of the chapter, practical methods and current approaches are used to illustrate the conceptual and theoretical points discussed.

6.2 Theoretical and Philosophical Grounding of Lean

The Japanese scholar Fujimoto (2007) describes the development of the Toyota Production System as follows: it 'emerged as the unplanned and unexpected result of ... seemingly unrelated innovations, improvements, and initiatives'. Based on this view, it is tempting to think that there are no grand theories behind the Toyota Production System, or lean, as it derived from the Toyota Production System as its Western interpretation (Womack et al. 1990). Many see them as the same and this would be compatible with the increasingly popular view of lean as 'just' an eclectic bundle of practices or tools.

However, it is contended here that the theoretical and philosophical foundations of lean can be precisely pinpointed. Such foundations have unfortunately remained invisible as they are located in domains that are hardly discussed in the academic discipline of management, namely theory of production, epistemology, and ontology. Moreover, a treatise on the foundations of lean goes against the grain in contemporary management studies, which take it as granted that new management ideas are the primary driver for evolution of management generally and production management specifically (Koskela 2017). Instead, a discussion on the foundations of lean implies that management and organising generally are affected by the evolution of ideas and models of production (Bartezzaghi 1999).

For discussing the theoretical and philosophical grounding of lean, it is opportune to start from the corresponding grounding of the mainstream approach to production management and organising in general. The theories of production as well as epistemology and ontology of lean are then discussed.

6.2.1 Theoretical and Philosophical Grounding of the Mainstream Approach to Production Management

What could be considered the central idea of the mainstream approach to production management? It is the idea of transformation, where production is seen as a transformation of inputs into outputs (Starr 1989). However, this idea is not often recognised as foundational, perhaps for two reasons. First, it is so simple that scholars may be reluctant to pay attention to it. Second, it is so deeply ingrained that it may not be understood as a conceptualisation of production but rather as the phenomenon itself (i.e. Grubbström 1995).

How is this idea of transformation used in practice? The operational method connected to the idea of transformation is decomposition (Koskela 2000). Decomposition refers to the breaking down of the total productive tasks into smaller entities, until they can be assigned to a worker, or to a subcontractor or supplier. Here, two assumptions are underpinning the procedure of decomposition. First, that the smaller entities into which a transformation is decomposed, are by their nature similar to the original transformation. Second, that the subtransformations emerging through decomposition are mutually independent. Based especially on the latter assumption, management of production will become admirably easy: by optimising each subcomposed (and independent) transformation, the optimum for the total transformation is reached. However, the focus on subtransformations, without taking their mutual dependencies into account, has in practice the effect of suboptimising the total or aggregate transformation.

Although simple, here we have a theory of production, consisting of:

- A conceptualisation of production: transformation;
- Central principle: decomposition;
- Inherent ends and means: optimisation of decomposed subtransformations with the intention of achieving the global optimum.

Although not immediately visible, this theory of production is compatible and supported by certain assumptions on the world, which fall into the branch of philosophy called metaphysics or ontology (Rooke et al. 2007). The question addressed in these fields is: what is out there in the world? Since antiquity, one answer to this has been: the world consists of things and substances (Koskela and Kagiouglou 2005). This seems perfectly sensible and is compatible with our everyday observations. Actually, this has been the mainstream Western worldview since antiquity, and also the basis for natural science. Taking a relatively stable 'thing' that can be clearly separated from other things as a starting point, the scientific approach has been to decompose it into component parts and find the explaining causality at the decomposed level. In production, the direction of the process is reversed, the 'thing' is created from its component parts, but similarly to science, the description of production is through stable 'things' in the form of inputs and outputs.

Another relevant area of philosophy is epistemology, which is a domain to address the nature and acquisition of knowledge. An implicit assumption in the mainstream view is that knowledge is a thing too. It is an input to production, along with material inputs. Actually, such a view is aligned to the early epistemological position taken by Plato (Losee 2001). According to him, knowledge is created and held in the mind – reason – and used through deduction for any purposes. Similarly, in production, knowledge is held by the controller of production (management), and deductively pushed out into the world. If new knowledge is needed, it is acquired from elsewhere, similarly to other inputs of production.

These theoretical and philosophical starting points have influenced the way production is organised. The division of work, reflecting the procedure of decomposition, is one key idea. Another is the principle of command and control, that especially reflects the Platonic epistemology: the brain (management) gives commands to the limbs (workers).

The mainstream view on management is thus based on internally coherent, although largely implicit, foundations. Unfortunately, the shortcomings of this model started to be seen in the 1970s (Schonberger 1996). Just looking at the theoretical and philosophical explanation presented, it is easy to pinpoint some potential reasons. The assumption of independence of decomposed subtransformations is a problematic idealisation: after all, such dependencies exist and have to be taken into account in practice. Further, neither time nor customer are part of the model although both would seem relevant regarding the phenomenon of production.

6.2.2 Theoretical and Philosophical Grounding of Lean

6.2.2.1 Theory of Production

While the abovementioned transformation model of production is surely applied in specific instances in lean, there are two other models of production which are dominant, namely flow and value generation models (Koskela 2000).

The central feature of the flow model of production is time, or in other words: what is happening to objects of production in time and what do the subjects of production, workers and machines, do in time. When observing this, it becomes evident that not all time is used for transformation. Objects of production are waiting, or they are transferred, or inspected. How should such nontransformation stages be characterised? From early on, the term waste has been used, as such stages do not add value for the customer.

Again, we have here a theory of production, with its conceptualisation of production, which leads to a widely different prescription for action, namely waste reduction. Time-wise, the biggest waste is made up by waiting, in the form of storage. Thus, by compressing lead times, this waste can be reduced. These insights led first to just-in-time production and further to the Toyota Production System. Later, through mathematical analysis it has been shown that one common cause for waste is (temporal) variability, in terms of unpredictability of when a job at a workstation will start, and how long it will take (Hopp and Spearman 1996).

Although less conspicuous, there is a third model of production in use, namely that of value generation. This model, seminally presented by Shewhart (1931), adds the customer to the picture. Production is seen as happening between the customer and the supplier: the former has requirements and wishes, and the latter converts these into products that fulfil them. The central stages are design, converting requirements and wishes into a product specification, and production, creating the product to be as close to the specification as possible. This conceptualisation of production again leads to a certain prescription for action, especially how design and quality management should be carried out.

6.2.2.2   Epistemology of the Lean Concept

Plato's pupil, Aristotle, developed an epistemology that was starkly different from Plato's. Aristotle thought that theoretical knowledge can be extracted from the world, through induction (Losee 2001). That knowledge can further be proved or brought into use through deduction, similarly to Plato's scheme.

Lean subscribes to the Aristotelian epistemology in many ways (Koskela et al. 2018). The starting point of lean, waste, exists in the material world, and has to be observed empirically and analysed for its root causes, to be eliminated. In practical terms, the Plan-Do-Check-Act cycle, suggested originally by Shewhart (1931), realises this epistemology.

Epistemological choices exist also at the most practical level in production management. Pushing equals to Platonism: deduction from ideas (plan) to action, implying production is controlled 'top-down' and pushed forth by management and planning. Whereas pulling, emphasised in lean, equals to Aristotelianism, basing action on the state of the world, implying production is controlled 'bottom-up' and planned by autonomous mechanisms within production itself based on the status of it, and pulling other resources and management when needed.

6.2.2.3   Ontology of the Lean Concept

Regarding ontology, there has already been an alternative to thing metaphysics from antiquity onwards, namely process metaphysics (Rescher 1996). In this approach, the world is conceived to consist of temporal processes. The natural mode of inquiry, when examining a phenomenon, is then to identify and address the relations that this phenomenon has to other things and processes. Another corollary is to pay attention to small changes over time, rather than one-off changes.

Process metaphysics is visible in lean in two key ways. Continuous improvement represents a focus on small changes. In turn, the emphasis on collaboration is compatible with process metaphysics: it is through collaboration that the dependencies between tasks in design and construction, often invisible in advance, are handled.

6.2.3   Implications for Management and Organising

These theoretical and philosophical foundations influence in many ways how business activities, especially production, are organised and managed. To illustrate this, in Table 6.1, we summarise, based on the discussion above, what basically distinguishes mainstream and lean, or in other words, the fundamental aspects and differences between mainstream and lean in basic theory and philosophy. One important conclusion that can be drawn is that mainstream and lean ideas and concepts are not mutually exclusive or diametrically opposite. Rather, the lean approach is wider, and contains the concepts and ideas belonging to the mainstream as its parts, both theoretically, epistemologically, and ontologically.

It is noteworthy that the different elements depicted in Table 6.1 can be directly connected to specific management approaches, as illustrated in Table 6.2 (with examples mainly from the supply chain area).

However, the analysis needs to be deepened in two senses. First, the supply chain management theory has to be outlined for comprehensively discussing the instantiation of lean and mainstream thinking in this general domain. Secondly, the peculiarities of construction have to be examined, to assess specific features of construction supply chain management from lean and non-lean viewpoints.

Table 6.1 Basic theoretical and philosophical framework: mainstream versus lean elements of basic theory from a theoretical/philosophical perspective.

Theoretical or philosophical foundation	Mainstream elements	Lean elements
Production theory	Transformation: production is conceptualised as transformation, and associated principles are used for managing production	Production is conceptualised both as transformation, flow and value generation, and associated principles are used in a balanced manner for managing it
Epistemology	Platonic: knowledge is held by managers and engineers, and it is pushed towards the world deductively through plans and designs and their implementation	Aristotelian: the deductive push of knowledge is acknowledged; however, knowledge is also inductively created, seeing production as a scientific laboratory
Ontology	Thing: management addresses clear-cut, stable things. Changes are abrupt, realised by introducing new things into the setting	Process: management concerns interrelated, fluid processes (embracing also things). Changes may be abrupt or continuous

Table 6.2 Examples of management approaches supported by different theoretical and philosophical bases.

Aspect of philosophical foundation	Specific theoretical approach	Example of management approach
Production theory	Transformation	Supplier selection based on bidding
	Flow	Joint development of the logistical chain
	Value generation	Supplier prequalification and early supplier involvement
Epistemology	Platonic	Using an optimisation model for designing a supply chain
	Aristotelian	Joint continuous improvement along the supply chain
Ontology	Thing	Transactional contracts
	Process	Relational contracts

6.3 Theoretical Background and Characterisation of Supply Chain Management

Although a monotheoretical perspective would be sufficient in order to understand sole aspects of supply chain management, a full understanding of the various aspects in the exchange between organisations and humans within the supply chain requires multiple perspectives on the supply chain (Cousins et al. 2006). Several theories and concepts apply to the phenomena existing in the supply chain. These theories and concepts can be arranged in larger theoretical domains. Since supply chain management is aimed at the integrated management and governance of actors, processes, and activities through the supply chain, corresponding topics from multiple theories need to be applied in a joint framework.

Previously, four sets of theoretical topics have been put into perspective from four respective theoretical domains: organisational topics; human resource management and social topics; technological and production topics; and procurement and economical topics (Vrijhoef 2011). In each of the four theoretical perspectives existing concepts and constructs have been identified that jointly need to be applied for management in the supply chain, that is, supply chain management (Figure 6.1).

6.3.1 Production Perspective

From a production perspective, particularly lean production, any form of waste, i.e. non-value-adding activities existing in the production system of the organisations within the supply chain will need to be reduced or removed. This view is compatible with the flow model of production (Koskela 2000), as introduced above, where production is conceived as a flow of materials and semiproducts leading to an integrated final product. Instead of minimising a specific type of waste associated with one activity type in production, the objective is rather to design, manage, operate, and improve the entire production system through the supply chain as a whole in such a way that total flow is optimised and the aggregate amount of waste is minimised.

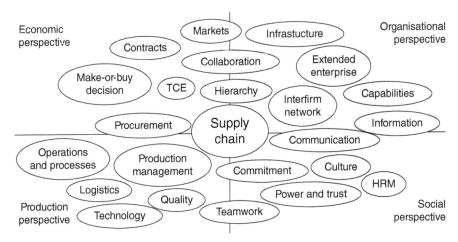

Figure 6.1 Four perspectives on supply chain management. HRM, human resource management. TCE, transaction, cost, economics. Source: Vrijhoef et al. (2003), p. 290.

As a result, in essence, production management and supply chain management become integrated. Christopher (1992) observes that 'supply chain management covers the flow of goods from supplier through manufacturing and distribution chains to the end user'. This means that the independent organisations within the supply chain agree upon the way in which production and information flows are organised. A typical consequence of this kind of agreement is an integrated planning and organisation of activities and logistics among the group of organisations involved (Bowersox and Closs 1996).

However, besides this flow aspect, supply chain management also covers the other conceptualisations and aspects of production and production management (Vrijhoef 2011). Especially, the selection of an organisational form to organise the production is focused on the general objective of value creation for end customers, rather than just waste minimisation, which may be regarded as a subgoal of production management (Galbraith 1995). Seeking a general explanation for the selection of an organisational form of production on the basis of minimising waste or production costs only is not justified. Instead, the supply chain, defined as 'the network of organisations that are involved, through upstream and downstream linkages, in the different processes and activities', must produce 'value in the form of products and services in the hands of the ultimate customer' (Christopher 1992, p. 13) (Figure 6.2). The integrated control of the supply chain must therefore have a positive effect on the delivery of end products and services in the hands of end customers (Hobday et al. 2005).

6.3.2 Economic Perspective

From an economic perspective, the aim of supply chain management is reducing production costs and particularly transaction costs, by searching for cost level reductions and aligning coordination and communication in the supply chain (Hobbs 1996). Rather than treating each transaction separately, benefits are to be gained from organising clusters of related transactions within the supply chain (Williamson 2008). Following the logic of transaction cost economics, a change in transaction costs that arises from the exchange of a product may lead to a change within the governance of the supply chain.

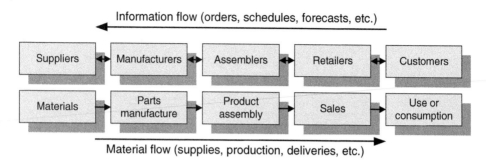

Figure 6.2 Generic configuration of a supply chain in manufacturing. Source: Vrijhoef and Koskela (2000), p. 170.

This depends on the degree of uncertainty, asset specificity, and frequency. Generally high levels of uncertainty, asset specificity, and infrequency would lead to more formal types of supply chain management, and possibly vertical integration, that is, internalisation. Often this implies manufacturers acquiring parts of the supply chain, e.g. taking over suppliers and placing them within their organisational boundaries (Blois 1972; Díez-Vial 2007).

6.3.3 Organisational Perspective

From an organisational perspective on the level of the supply chain and the constituting individual organisations, distributed business activities raise the governance issue of how to bundle and align competences of individual organisations within the supply chain (Prahalad and Hamel 1990). The coordination of networks requires organisational alignment of the multiple strategies, as well as cultural alignment to manage people working along extended business processes (Bititci et al. 2003). Supply chains being networks imply a form of governance that is distinctly different from market and hierarchy (Sydow and Windeler 1998). The balance between control and emergence is important, in the sense that control tends to detract from innovation and flexibility, and emergence decreases predictability and manageability of operations (Choi et al. 2001; Dooley 1997). Therefore, a major challenge is to devise a governance model for the supply chain that leaves room for the 'adaptive, collective behaviour of the supply chain' (Surana et al. 2005). A long-term outlook encourages organisations to search for ways of jointly accomplishing tasks, learning and promoting trust and stability. The 'shadow of the future' in long-term relations decreases opportunistic behaviour (Axelrod 1984). Trust is thereby generated, which is, as Arrow (1974) noted, an 'efficient lubricant to economic exchange' between organisations in supply chains.

6.3.4 Social Perspective

From a social perspective, supply chain management can be described as 'the coordination of efforts of people working on a collaborative task broken down into a set of specialised activities. Coordination is then achieved through communication' (Taylor 1993). Next, this communication needs to lead to commitment to get the jointly coordinated activities accomplished. One idea representing this is the language/action perspective (Van Reijswoud 1996). Regarding the practical significance of this perspective, two directions were pinpointed by Winograd and Flores (1986). First, the process of requesting, creating, and monitoring commitments can be facilitated by systems for constructing and coordinating conversation networks. Second, people can learn to communicate for action by improving their skills in understanding requests and making promises and commitments. This depends on various social phenomena and undercurrents such as patterns of exchange power that underpin and influence networks of relationships (Cousins and Menguc 2006; Cox 2001; Ireland and Webb 2007). It is the nature of exchange power as exercised in organisations that influences the possibilities and tendencies of individuals to act or not, and as an effect determines the culture whether individuals can and will engage and thrive in work relationships with others beyond their own organisation and in the supply chain too. In this respect, trust and transparency are key issues that need to be fostered, for aims of socialisation, interpersonal exchange, joint learning, and ultimately the performance and improvement of supply chains (Kwon and Suh 2005; Sense and Clements 2006). In addition, Pryke (2017) identified empathy as an important network relationship enabling factor.

6.4 Analysis of Supply Chain Approaches and Conceptualisations through a Lean Versus Mainstream Lens

The next step is to view the concept of supply chain management through the lean lens, based on the above presentations of both lean and supply chain management. We will use the notions of what is lean and what is not lean (mainstream) from Section 6.2. Based on this we will look into the four perspectives of supply chain management from a lean and mainstream viewpoint, and underline those elements that make a difference. An outcome of this analysis is presented in Table 6.3.

In summary, both the comparison of mainstream versus lean elements in basic theory and philosophy as well as in supply chain management theory show a characteristic difference between the two. The mainstream view mostly applies to the 'thing', the 'artefact' being produced based on mere transaction, and separation of demand and supply, respectively customer and supplier. The mainstream interpretation of supply chain management on interactions between organisations and individuals is that these are predominantly based on bargaining, financial transactions, transaction cost reduction, control by exercising power, and judgement by customers over suppliers as a mechanism in relations.

In supply chain management through the lean lens view, however, the emphasis is on the 'value' produced, mapping the 'value stream' to produce the value, and channelling the corresponding 'flow' of activities through the supply chain. To this end the lean view propagates alignment of organisations, individuals, processes, activities, and respective organisation cultures, approaching the supply chain as a single 'virtual organisation', as one 'extended enterprise' (Boardman and Clegg 2001). It is considered as an integrated whole, where interfaces between different organisations are seen as artificial (Lamming 1996). As a result (Broft and Koskela, 2018), one may conclude that lean elements in the supply chain often have a relational focus, including long-term relations and commitments, alignment of engagements and commitments, continuous improvement to increase value and eliminate waste, and suppliers and clients interacting based on mutual trust, respect, and learning.

Table 6.3 Understanding supply chain management from two viewpoints: mainstream versus lean interpretations of the four views on supply chain management.

Four theoretical views on supply chain management	Mainstream interpretation	Lean interpretation
Production view	Getting the product produced; with all organisations in the supply chain. The production systems of individual organisations do not necessarily have to be aligned in flow	Controlling and improving the complete flow of production through the supply chain of organisations, aimed at reducing variability of production processes, and as a result reducing waste and increasing value
Economic system	One-sided focus on transaction costs, often assuming production cost as given. Governance structure determined by bargaining lowest buying costs through the supply chain. This may be a result of bargaining power or deal making. Mostly relying on short-term transactional contracts	Continuous improvement aimed at cost level reductions, also the costs of production itself. Efficiencies through modelling and innovative design of the supply chain. Long-term arrangements with suppliers as a basis for joint investments and achieving continuous improvement, such as framework agreements and other types of relational contracts
Organisational system	Centralised models for controlling supply chains, formal information exchange. Decreasing opportunistic behaviour by control and exercising power by the focal organisation in the supply chain. Supplier rating	Common understanding of the joint organisation of the supply chain. Joint innovation and collaborative improvement based on the plan-do-check-act cycle. Commitment to elimination of waste. Collaboration for customer value. Supplier development
Social system	Organising for communication; sending and receiving orders between individuals and the organisation; commitment based on exercising power; judging on performance	Avoiding breakdowns in conversation and commitment through communication; creating trust and transparency; fostering learning and socialisation among organisations and individuals

6.5 Contingency of Supply Chain Management in Construction through a Lean Lens

In the last decades, the construction industry has shown a growing interest in both the concepts of supply chain management and lean construction, including corresponding approaches and techniques. The implementation of supply chain management, however, has often been limited to project-specific approaches as opposed to other industries, where supply chain management has become a central strategy (Broft et al. 2016). The concept of supply chain management originated in manufacturing, whereas construction is considered a project-based industry with many characteristics different to manufacturing. The most differentiating ones are often summarised as: 'one-of-a-kind nature of projects, site production, and temporary multiorganisation' (Koskela 1992).

The organisation of production and the supply chains is strongly adapted to these basic characteristics (Koskela 2000; Broft and Koskela 2018). Next, following the same perspectives as used in the previous section, we try to describe the impact and implications of these peculiarities on the general organisation and management of the construction supply chain. Thereafter, based on the lean elements of supply chain management as presented in Table 6.3, we make a crossover of this theory to the practical context of construction.

6.5.1 Construction from a Production Perspective

Construction could be related to prototype production due to its temporary and non-repetitive nature – it includes one-off construction projects (Vrijhoef and Koskela 2005). Its low levels of repetition make continuous improvement difficult and subsequently limits innovation to projects. This is often presented as one reason for a lack of performance in the industry. The project itself resembles an assembly-like process, conducted on site, where the labour (subject) and its services flow through the construction project (object) (Ballard and Howell 1998; Friedrich et al. 2013).

6.5.2 Construction from an Economic Perspective

Subcontracting has been adopted as the dominant procurement strategy in construction, resulting in a high supplier involvement. Nowadays, the main contractor, the principal construction organisation that manages a construction project, executes only a small part of the product using its own personnel and its own production facilities (Dubois and Gadde 2000). From an economic system perspective, almost all construction projects are divided into parts that are subcontracted to individual enterprises – again, this is the decomposition principle, where a production process constitutes subprocesses (Broft and Koskela 2018). Its organisations cooperate in ever changing patterns, decided mainly by the lowest bids for the project in question. Procurement strategies, often pursued by clients, promote such competitive pricing, and feature purchasing transactions (Gadde and Dubois 2010), in order to get every new project executed at the lowest possible cost. As a result of the industry's fragmentation and prevalent competitive tendering, supply chains become disjointed (Eriksson 2015). The industry has been criticised for missing a relational component in its largely transactional relationships. There had been a move towards the formation of strategic partnerships and collaborative agreements between supply chain actors (Briscoe and Dainty 2005).

6.5.3 Construction from an Organisational Perspective

From an organisational system perspective, construction involves temporary interorganisational collaborations of coalitions as a result of constant reconfiguration. The project is the primary business mechanism in construction with 'a number of independent organisations coming together for the purpose of undertaking a single building project' (Winch 1989). The organisations are legally autonomous and need to evolve systems for joint project delivery (Sydow and Staber 2002). The supply chain is also interwoven, as practically every organisation at the same time participates in more than one project, utilising the same production capacity (Bertelsen 2003).

Table 6.4 Construction supply chain management: crossover from supply chain management theory through a lean lens to the practical context of construction.

Four views on the supply chain	Typical characterisation of the construction supply chain	Application of supply chain management to construction in a lean view	Practical considerations of the application
Production system	Prototype production within one-off construction projects involving large quantities and varieties of components and materials assembled on site by many different specialists and trades	All tasks, either in-house or subcontracted, are aligned for optimised flow in the subsequent processes of individual suppliers of materials to site and subcontractors involved in the final assembly on site	(Partial) off-site prefabrication; creating stable conditions on site through balancing and shielding the assembly. Just in time delivery of materials and services. Supplier and subcontractor involvement
Economic system	Parts of the construction are subcontracted to individual enterprises, based on decomposition of the work and the expertise and added value and/or lowest price subcontractors are able to provide per decomposed part	Economic agreements are considered to be transactional, but based on value/cost considerations, with a relational component and long-term focus in principle. Joint mechanisms and incentives are in place for minimisation of total costs within the entire supply chain	Long-term agreements with suppliers and subcontractors. Fixed arrangements for stable cost levels. Reward schemes for achieving joint cost reductions
Organisational system	Temporary interorganisational collaboration with a constant reconfiguration of firms organised around a project. The organisations involved are legally and operationally autonomous	The organisations are organisationally connected through both collaborative and integrated processes, and joint product development. Improvement and control of supply chain performance is a joint obligation and effort. Long-term perspective mitigates a constant reconfiguration of team members	Creating a project entity that operates as if it were a single organisation (Integrated Project Delivery). Early supplier involvement. Long-term relationships enabling strategic collaboration with subcontractors and suppliers
Social system	A highly transient social system with a varying lifecycle of relationships leading to divergent loyalty and differing responsibilities, interests, and priorities	Long-term relationships increase common understanding and alignment in the supply chain. Organisations and individuals become more involved and intrinsically motivated to look through projects and find joint opportunities for improvement. Teamwork is fostered to increase mutual commitment	Fixed teams through projects. Alignment of the goals on an organisational and project level. Using joint planning solutions (such as Last Planner) for improving the social cohesion and joint commitment. Early involvement of supply chain actors

6.5.4 Construction from a Social Perspective

Socially, the project and the construction site is a working place for humans and a place for cooperation and social interaction, which – because of the temporary character – forms a highly transient social system (Bertelsen 2003). In fact, for every project a new coalition is established with a varying lifecycle of relationships between individuals and groups (Tavistock Institute 1966; Tuomela-Pyykkönen et al. 2015). Following from the high status of projects, two cultural identities exist: the corporate formal culture of organisations, and a distinctive, informal culture within each separate project (Vrijhoef and Koskela 2005) resulting in divergent loyalty, and differing responsibilities, interests, and priorities. While the main contractor coordinates several subcontractors on one site, a subcontractor often coordinates its workforce on several sites (Pennanen and Koskela 2005).

6.5.5 A Crossover of Supply Chain Management and Lean in the Context of Construction

Based on the previous section we will now make a crossover between supply chain management and lean in a construction context taking into account the above-mentioned characterisation of construction. Table 6.4 presents this crossover and the considerations to be taken into account when supply chain management is applied to construction in a lean manner. What are the key issues for the application of lean supply chain management to construction? How does the application work out in practice?

6.6 Discussion

Lean and mainstream management theories are basically different. However, they are not diametrically opposite. Rather, they seem complementary in many aspects. The lean approach is wider and contains more relational and social concepts and ideas than the mainstream approach. Mainstream approaches seem to be more dominated by economics and control. Epistemologically and ontologically, the approaches differ regarding their preference of either pull and emergence (Aristotelianism) or push and control (Platonism). In a construction context, which by nature is relatively unpredictable, arguably both push and pull are needed; in particular, there need to be pull mechanisms to cope with emergent unpredictability at the level of design and production. Supply chain management as a concept can be viewed and analysed based on both theories. Each of these theories emphasise different elements of supply chain management.

Supply chain management has originated partly from the lean domain. However, both in theory and in practice supply chain management seems to reflect and combine both lean and mainstream elements. However, one could argue that a lean view on supply chain management is more beneficial for the supply chain because of the more comprehensive and integrative stance it takes toward the supply chain as a collective organisation.

In summary, we postulate that most or all supply chain management practices, and in construction in particular, could be analysed and explained through one of these theories. The lean elements in supply chain management theory have been presented, through the framework in this chapter, as a 'lean lens'. However, the definitions and explanations of supply chain management presented in the chapter are still limited and subject to further extension, from within the lean domain, the mainstream domain, or beyond. So, the potential of the framework is greater than presented in this chapter.

6.7 Conclusion

The chapter contributes to the conceptual and theoretical understanding of supply chain management both generally and philosophically, as well as specifically and practically applied to construction. It puts in perspective the approaches and methods used in supply chain management that actually are lean, and sometimes mainstream, which often are misunderstood or looked at in isolation. By putting the lean lens on supply chain management, the chapter defies mainstream explanations of construction supply chain management and triggers further explanation and exploration of lean applications to construction supply chain management that has been lacking in theory and not sufficiently applied in construction practice so far.

References

Arrow, K.J. (1974). *The Limits of Organisation*. New York: Norton.

Axelrod, R. (1984). *The Evolution of Cooperation*. New York: Basic Books.

Ballard, G. and Howell, G.A. (1998), What kind of production is construction? In: *Proceedings 6th Annual Lean Construction Conference* (IGLC-6), 13–15 August 1998, Guarujá, Brazil.

Bartezzaghi, E. (1999). The evolution of production models: is a new paradigm emerging? *International Journal of Operations & Production Management* 19 (2): 229–250.

Bertelsen, S. (2003) Complexity – Construction in a new perspective. In: *Proceedings 11th Annual Conference IGLC*, Blacksburg.

Bititci, U.S., Martinez, V., Albores, P., and Mendibil, K. (2003). Creating and sustaining competitive advantage in collaborative systems: the what and the how. *Production Planning & Control* 14 (5): 410–424.

Blois, K.J. (1972). Vertical quasi-integration. *Journal of Industrial Economics* 20: 253–272.

Boardman, J.T. and Clegg, B.T. (2001). Structured engagement in the extended enterprise. *International Journal of Operations & Production Management* 21 (5/6): 795–811.

Bowersox, D.J. and Closs, D.J. (1996). *Logistical Management: The Integrated Supply Chain Process*, 3e, 730. New York: McGraw-Hill.

Briscoe, G. and Dainty, A. (2005). Construction supply chain integration: an elusive goal? *Supply Chain Management: An International Journal* 10 (4): 319–326.

Broft, R.D. and Koskela, L. (2018). Supply chain management in construction from a production theory perspective. In: *26th Annual Conference of the International Group for Lean Construction (IGLC)*, Chennai, India, 18–20 July 2018, 271–281.

Broft, R.D., Badi, S., and Pryke, S. (2016). Towards SC maturity in construction. In: *Built Environment Project and Asset Management*. Emerald Special Issue.

Choi, T.Y., Dooley, K.J., and Rungtusanatham, M. (2001). Supply networks and complex adaptive systems: control versus emergence. *Journal of Operations Management* 19: 351–366.

Christopher, M. (1992). *Logistics and Supply Chain Management: Strategies for Reducing Cost and Improving Service*. London: Pitman/Prentice Hall.

Cousins, P.D. and Menguc, B. (2006). The implications of socialization and integration in supply chain management. *Journal of Operations Management* 24 (5): 604–620.

Cousins, P.D., Lawson, B., and Squire, B. (2006). Supply chain management: theory and practice - the emergence of an academic discipline? *International Journal of Operations & Production Management* 26 (7): 697–702.

Cox, A. (2001). Understanding buyer and supplier power: a framework for procurement and supply competence. *Journal of Supply Chain Management* 37: 8–15.

Díez-Vial, I. (2007). Explaining vertical integration strategies: market power, transactional attributes and capabilities. *Journal of Management Studies* 44 (6): 1017–1040.

Dooley, K.J. (1997). A complex adaptive systems model of organization change. *Nonlinear Dynamics, Psychology and Life Sciences* 1 (1): 69–97.

Dubois, A. and Gadde, L.E. (2000). Supply strategy and network effects – purchasing behaviour in the construction industry. *European Journal of Purchasing & Supply Management* 6 (3–4): 207–215.

Eriksson, P.E. (2015). Partnering in engineering projects: four dimensions of supply chain integration. *Journal of Purchasing and Supply Management* 21: 38–50.

Friedrich, T., Meijnen, P., and Schriewersmann, F. (2013). *Praxis des Bauprozessmanagements – 2 Lean Construction*, 47. Ernst & Sohn GmbH & Co. KG.

Fujimoto, T. (2007). *Competing to Be Really, Really Good: The behind-the-Scenes Drama of Capability-Building Competition in the Automobile Industry*, vol. 22. International House of Japan.

Gadde, L.E. and Dubois, A. (2010). Partnering in the construction industry – problems and opportunities. *Journal of Purchasing and Supply Management* 16 (4): 254–263.

Galbraith, J.R. (1995). *Designing Organizations*. San Francisco: Jossey-Bass.

Grubbström, R.W. (1995). Modelling production opportunities - an historical overview. *International Journal of Production Economics* 41: 1–14.

Hobbs, J.E. (1996). A transaction cost approach to supply chain management. *Supply Chain Management: An International Journal* 1 (2): 15–27.

Hobday, M., Davies, A., and Prencipe, A. (2005). Systems integration: a core capability of the modern corporation. *Industrial and Corporate Change* 14 (6): 1109–1143.

Hopp, W. and Spearman, M. (1996). *Factory Physics: Foundations of Manufacturing Management*, 668. Boston: Irwin/McGraw-Hill.

Ireland, R.D. and Webb, J.W. (2007). A multi-theoretic perspective on trust and power in strategic supply chains. *Journal of Operations Management* 25 (2): 482–497.

Koskela, L. (1992). *Application of the new production philosophy to construction (CIFE Technical Report 72)*. Stanford, CA: Stanford University.

Koskela, L. (2000) An exploration towards a production theory and its application to construction. PhD Thesis. VTT Publications 408. VTT Technical Research Centre of Finland, Espoo.

Koskela, L. (2017). Why is management research irrelevant? *Construction Management and Economics* 35 (1–2): 4–23.

Koskela, L.J. and Kagioglou, M. (2005). On the metaphysics of production. In: *Proceedings of 13th International Group for Lean Construction Conference*, 37–45.

Koskela, L., Ferrantelli, A., Niiranen, J. et al. (2018). Epistemological explanation of lean construction. *Journal of Construction Engineering and Management* 145 (2): 04018131.

Kwon, I.G. and Suh, T. (2005). Trust, commitment and relationships in supply chain management: a path analysis. *Supply Chain Management: An international Journal* 10 (1): 26–33.

Lamming, R. (1996). Squaring lean supply with supply chain management. *International Journal of Operations & Production Management* 16 (2): 183–196.

Losee, J. (2001). *A Historical Introduction to the Philosophy of Science*, 4e. Oxford University Press.

Pennanen, A. and Koskela, L. (2005) Necessary and unnecessary complexity in construction. First International Conference on Complexity, Science and the Built Environment, 11–14 September 2005 in Conjunction with Centre for Complexity and Research, University of Liverpool, UK.

Prahalad, C.K. and Hamel, G. (1990). The core competence of the corporation. *Harvard Business Review* 68 (3): 79–91.

Pryke, S.D. (2017). *Managing Networks in Project-Based Organisations*. Oxford: Wiley.

Rescher, N. (1996). *Process Metaphysics: An Introduction to Process Philosophy*. Suny Press.

Rooke, J.A., Koskela, L., and Seymour, D. (2007). Producing things or production flows? Ontological assumptions in the thinking of managers and professionals in construction. *Construction Management and Economics* 25 (10): 1077–1085.

Schonberger, R.J. (1996). *World Class Manufacturing: The Next Decade*, 275. New York: The Free Press.

Sense, A.J. and Clements, M.D.J. (2006). Ever consider a supply chain as a community of practice?: embracing a learning perspective to build supply chain integration. *Development and Learning in Organizations* 20 (5): 6–8.

Shewhart, W.A. (1931). *Economic Control of Quality of Manufactured Product*, 501. New York: Van Nostrand.

Starr, M.K. (1989). *Managing Production and Operations*, 647. Prentice-Hall International Editions.

Surana, A., Kumara, S., Greaves, M., and Raghavan, U.N. (2005). Supply-chain networks: a complex adaptive systems perspective. *International Journal of Production Research* 43 (20): 4235–4265.

Sydow, J. and Staber, U. (2002). The institutional embeddedness of project networks: the case of content production in German television. *Regional Studies* 36 (3): 215–227.

Sydow, J. and Windeler, A. (1998). Organizing and evaluating interfirm networks: a structurationist perspective on network processes and effectiveness. *Organization Science* 9 (3): 265–284.

Tavistock Institute (1966). *Interdependence and Uncertainty: A Study of the Building Industry*. London: Tavistock Publications.

Taylor, J.R. (1993). *Rethinking the Theory of Organizational Communication*. Norwood: Ablex.

Tuomela-Pyykkönen, M., Aaltonen, K., and Haapasalo, H. (2015). Procurement in the Real Estate and Construction Sector (RECS) – preliminary context-specific attributes, 8th Nordic Conference on Construction Economics and Organization. *Procedia Economics and Finance* 21: 264–270.

Van Reijswoud, V.E. (1996) The structure of business communication. Dissertation. Delft University of Technology, Delft.

Vrijhoef, R. (2011). *Supply Chain Integration in the Building Industry: The Emergence of Integrated and Repetitive Strategies in a Fragmented and Project-Driven Industry*. Technische Universiteit Delft. Amsterdam: IOS/Delft University Press.

Vrijhoef, R. and Koskela, L. (2000). The four roles of supply chain management in construction. *European Journal of Purchasing and Supply Management* 6 (3–4): 169–178.

Vrijhoef, R. and Koskela, L. (2005). Revisiting the three peculiarities of production in construction. In Proceedings 13th International Group for Lean Construction Conference, 19–27. Unitec, New Zealand.

Vrijhoef, R., Koskela, L., and Voordijk, H. (2003). Understanding construction supply chains: a multiple theoretical approach to inter-organizational relationship in construction. In: *Proceedings 11th Annual Conference on Lean Construction* (eds. J.C. Martinez and C.T. Formoso), 280–292. Blacksburg: Virginia Polytechnic institute and State University.

Williamson, O.E. (2008). Outsourcing: transaction cost economics and supply chain management. *Journal of Supply Chain Management* 44 (2): 5–16.

Winch, G. (1989). The construction firm and the construction project: a transaction cost approach. *Construction Management and Economics* 7 (4): 331–345.

Winograd, T. and Flores, F. (1986). *Understanding Computers and Cognition: A New Foundation for Design*. Norwood: Ablex.

Womack, J.P., Jones, D.T., and Roos, D. (1990). *The Machine that Changed the World*, 323. New York: Rawson Associates.

7

Supply Chain Management and Risk Set in Changing Times: Old Wine in New Bottles?

Andrew Edkins

7.1   Introduction and Overview

Over the last decade, the general level of interest in the topic of supply chain man-agement has continued to mature and it has now become part of the established way of thinking and practising around many parts of the world. This is true of the general production of goods and services, as well as in the construction sector, which is the focus of this chapter. After considering the current state of supply chain management and the more general developments in this area, this chapter will dwell on some of the emergent 'big ticket' risks that are requiring management thought and attention and the issue of the different forms of power structure, incentives, and behaviours that are required or result.

While reference to supply chains and indeed supply chain management itself is fre-quently used throughout industry, academe, and governments, this chapter reiterates the more strict definition of supply chain management as proposed by Cox (1999, 2003) and the work of Sanderson and Cox (2008). According to these authors, for full sup-ply chain management to occur, there is the need to accomplish a lot more than simply *sourcing* and *selecting* the supplier or even to then go on to *develop* the supplier rela-tionships. These three activities are unquestionably important, but they pose different challenges. While it will be only some readers who remember the volumes of 'Yellow Pages' where businesses and service providers listed themselves in a paper directory, supply chain management involves more than picking details from these pages and then seeking prices and terms from these suppliers; even having to develop a supplier's capa-bility and or capacity may only be a direct and immediate relationship. As argued by Sanderson and Cox (2008), true supply chain management is a rich and deep set of rela-tionships that travel up and down the supply chain, creating opportunities for mutual advantage and enhancement.

An indication of the maturing of the topic of supply chain management is the devel-opment of supply chain risk management. The principal author on this is Finch (2004) with other useful contributions coming from Ritchie and Brindley (2004, 2007), MIT and PWC (PWC and MIT and Forum for Supply Chain Innovation 2013) and a useful review of the supply chain risk management literature as produced by Ho et al. (2015).

Successful Construction Supply Chain Management: Concepts and Case Studies, Second Edition.
Edited by Stephen Pryke.
© 2020 John Wiley & Sons Ltd. Published 2020 by John Wiley & Sons Ltd.

While there is now far more widespread and indeed worldwide familiarity with the concepts of supply chains and supply chain management, the world has moved on in many ways. In the area of supply chain management the ubiquity of the internet and the embracing of globalisation now make it not only possible, but actually relatively cheap and easy for procurers to work with suppliers around the world to develop ever more customised and bespoke solutions, whether they be in the form of goods, services, or hybrids of both. Feedback from these suppliers is easier to transmit and, where the management culture is appropriate, the resulting two-way flow of information and understanding can lead to rapid and significant improvement in the design, production, and other associated tasks involved. But in establishing and operating these relationships and contracts there will always be risks. The word 'risk', which is a focus of this chapter, is seen as both negative and positive. In the literature of supply chain risk management, the tendency is for risk to be seen negatively, with the focus on the concerns on the probability and consequence associated with supply chain disruptions or shocks. This makes a lot of sense as our reliance on supply chains makes us conditioned to successful operation and when an event such as factory fire, volcanic eruption, or financial problem causes a key supplier or supply route to become problematic, the shockwave can travel along the supply chain and be amplified. Indeed, this is known in supply chain management as one of the causes of the so called 'bullwhip' effect (see for example, Lee et al. 1997).

However, it is always useful to point out that risk is balanced – with upside opportunities as well as the downside problems. This is something that is implicit at many levels: from the personal and organisational right through to the societal and national level. For example, when we as individuals decide to focus on an area to gain a job or develop a career we are risking that we might make the wrong choice – what economists refer to as the opportunity cost. Indeed, for many of us, this can stem from choices we made at school: how hard we studied, what we studied, who we associated with as role models, etc. This extends well beyond our personal lives; the choices made for commercial enterprises have to involve some form of risk if the opportunity for profit is the goal. Politicians and political parties take risks when they propose manifesto or similar election ideas, or develop policies that they believe to be the most appropriate. It is only with the passage of time that the decisions made, and actions taken, will be revealed to be good or successful. In such cases the risk was handled well, but this is not assured.

In the context of the project, all projects are risky because there is a lag between the decision to proceed with a project and the delivery of the project and all that it promised. Of the many authors who write on project risk management, the more recent work of Chapman and Ward, with their recognition of risk as well as opportunity management, is pertinent as it emphasises the need to balance the 'fear' of risk with the 'excitement' of risk (Chapman and Ward 2011). They, among many others, also recognise that there is a difference between risks and uncertainties, where the distinction is effectively the presence or lack of reasonable data and appropriate data processing techniques. The more uncertainty is present, the more reliance there is on the humanistic ability to judge and assess. With the rise of machine learning and artificial intelligence it may well be that the innate skill of human assessment and judgement that is recognised and rewarded today, may in the future be something that will be performed better by computers. However, this is conjecture and speculation and thus presents itself as an uncertainty now and a risk for a future version of this chapter to review!

While, therefore, we may have a constantly evolving context, the existing challenges of supply chain management have not gone away – the issues being discussed during the formative years of supply chain management in construction (from the 1980s through to the early 2000s in the case of the UK) remain relevant today. As with other sectors, IT-enabled commerce in construction has no doubt made a large difference to the task of tendering and the procurement of construction projects. The internet's global reach has expanded market limits, and the routine digitisation of design and manufacture information and processing has lowered cost and improved quality. This has been helpful, but has meant that now there is more widespread use of complex, digitally-reliant supply chains. While the choice-set of potential suppliers has increased and the ways of interacting with these suppliers has become more digital, cheaper, and easier, there is a set of new concerns that are emerging.

Cataloguing the many and varied issues that need to be considered in the context of the modern and IT-enabled supply chain management world could become both a significant deviation for this chapter and, because this is fast-moving area, one that may well cause the reader unease. Therefore, to avoid this risk and add some structure, the chapter will instead consider the challenges that we all need to be cognisant of and, ideally, have a response to, that arise from the work of the United Nations in its development of the 17 'Sustainable Development Goals'.[1] Within these Sustainable Development Goals are principal concerns about planetary impact, fairness, equity, and equality. The significance of these issues is not to be underestimated, nor is the trickle-down impact on supply chains and their management. If we take the 'big ticket' items captured within the Sustainable Development Goals and interweave them with the relentless march of technological progress, we have more than enough fodder to feed this chapter on supply chain management and risk.

The strategic intention of this chapter is therefore revised to reflect the varying forces of change. It will look at how the percolation of a richer 'sustainability' agenda has affected supply chain management and, as a separate but clearly interrelated topic, consider the potential impact of what can be considered as a possible 'technology enabled cultural shift' in construction project supply chain management through the rise of interest and deployment in building information modelling (BIM). However, before this is discussed, there was a development in the UK at the beginning of 2018 that had the potential to strategically reawaken the interest and discussion on supply chains and their management in the more commercial sense: in January 2018 the UK newspapers and news reports across all forms of media were dominated for some days by the collapse of the construction-centred firm Carillion plc.

7.2 The Collapse of Carillion: Consequences for Consideration – Implications for Construction Supply Chains

At the time there was much discussion about the collapse of this colossus of UK construction and the collateral damage of its collapse on other companies with similar types

---

1 The 17 UN Sustainable Development Goals can be found here: http://www.undp.org/content/undp/en/home/sustainable-development-goals.html (accessed July 2019).

of operation. The failure of Carillion triggered a reflective discussion about the approach taken in countries such as the UK where the public sector has transferred a great deal of reliance on the private sector in the form of delivery of public sector assets and services.

Carillion, like many other major UK construction contractors, recognised back in the 1990s that there were opportunities for both improved profit and diversification of risk if these construction contracting companies moved into the delivery of services, often related to both hard and soft facilities management. The origins of this expansion in remit can be argued as being the result of the creation of the Private Finance Initiative (PFI) and its subsequent evolution into first, Public Private Partnerships (PPP), and most recently to PF2 (National Audit Office 2018). Carillion's collapse impacted on major construction and infrastructure projects, the operational performance of important public sector facilities such as schools and hospitals, and it deeply impacted those whose jobs were dependent on Carillion, either directly as employees or indirectly as they worked for other companies that were tied into Carillion projects. It is this 'ripples on the pond' issue that is of particular interest as the consequence of Carillion's collapse had more far reaching effects than those simply linked to the company itself. Indeed, the many companies that were contracted with Carillion, either directly or indirectly, may ultimately mean that there was far greater collateral damage outside Carillion than was felt within the company directly. The major press coverage may move on from Carillion, but the consequences – the ripples on the pond – may travel and be felt for a longer time than many of us might realise. While it is in the nature of the corporate world for individual companies to fail, the Carillion collapse was exceptional. Indeed, such was the significance of the Carillion collapse that it required government and bank action to ensure continuity of both public sector services and the financial survival of companies embedded within its supply chain. The damage done by this one company's failure will be felt well beyond those who had invested or lent money to it. Employees will have lost jobs, those expecting to deliver services from the projects that Carillion were constructing will experience delays – as in the cases of the hospitals being built – and the government will have had to recognise that transferring risk to the private sector is a risk that is real and causes harm. It is therefore to be noted that in the budget held towards the end of 2018, the Chancellor of the Exchequer focused on the delivery of infrastructure projects and during his speech noted:

> Mr Deputy Speaker, Half of the UK's £600 billion infrastructure pipeline will be built and financed by the private sector. And in financing public infrastructure…
> I remain committed to the use of public–private partnership where it delivers value for the taxpayer… and genuinely transfers risk to the private sector.
> But there is compelling evidence that the Private Finance Initiative does neither. We will honour existing contracts. But the days of the public sector being a pushover, must end.
> We will establish a centre of excellence to actively manage these contracts in the taxpayers' interest starting in the health sector. And we will go further.
> I have never signed off a PFI contract as Chancellor…and I can confirm today that I never will. I can announce that the Government will abolish the use of PFI and PF2 for future projects.
> Source: https://www.gov.uk/government/speeches/budget-2018-philip-hammonds-speech (accessed July 2019)

Carillion's failure and the cascade effects that radiated out from it have inevitably led to a fundamental questioning of the practice of fragmenting construction projects for delivery by many organisations linked together through contracts and relationships. The alternative to these supply chains and networks of independent organisations linked together through the use of explicit commercial contracts is to have more vertically-integrated organisations that offer more 'cradle to grave' services in-house. This is not new and has been tried in the past, both in the public and private sectors. In the UK public sector, for example, there have, in the past, been a variety of ministries of central government and a plethora of local councils employing their own designers, constructors, and specialist tradespeople. In the private sector, major construction companies would have offered similar full in-house services.

The substantial move away from such vertically integrated organisations to a situation of where companies are more 'atomised' and the routine use of 'contracting out', was promoted on the basis of the economic and managerial efficiencies gained from both specialisation and ability to use contracts to align incentives. This was particularly the case where the public sector wished to transfer risk to the private sector that was itself prepared to accept these risks – in return for higher margins or increased fees. The recognition that the public sector could move risk to the private sector was a key development for construction supply chain management, as a recognised problem for the public sector became an opportunity for the private sector. This occurred where the private sector could see a way of accepting that transferred risk, treating the risk in various ways (such as using innovative practices and technologies to handle the risk directly, onward transferring of the risk through a supply chain, or arranging insurance policies to protect in the case of risk manifestation). Thus, we can see the temporally viewed situation as one where a pendulum swings slowly through time, moving from fully in-house provision to complete out-sourcing – and possibly back again.

In the UK construction sector, in this century and in the last half of the twentieth century, the pendulum has swung towards considerable and established use of such risk transfer, supply chains, and subcontracting. The case of Carillion is powerful for those considering and studying supply chain management and supply chain risk management as it reveals the many and varied problems that arise when a single organisation takes on the portfolio of risks that Carillion did. The nature of Carillion's operations meant that there were risks driven by construction projects it had undertaken to deliver that had a significant propensity to go wrong and lose money or were related to the more operational management challenge of providing various facility management services at very low margins to parts of the public sector. Post Carillion, the extent to which there has been a shift to more interest in 'in-housing' rather than 'out-sourcing' is as yet unclear. However, if such a shift were to occur it could be seen as a 'sea change' in thinking and practice. This is true in the UK as well as in many other countries that continue to use many different types of organisations from the private and third sectors linked together via webs of contracts and relationships to deliver construction projects. The size and diversity of global construction markets, coupled with their fickleness as they are moved by forces arising from economics, politics, public pressure, and technology development has encouraged the use of many firms working on construction projects in chains that can be many links long and webs that can involve multiple connection points.

Given this context, the collapse of Carillion during the first half of 2018 caused alarm and concern as this UK construction firm's prominence in many construction projects and service delivery operations affected other organisations that were linked to them via supply chains. What this case has revealed is that even though supply chains are extensively used, there remains an open question as to how many organisations are actively 'managing' their supply chains and by this, the specific issue is the ability for organisations using or involved in these supply chains to know they are functioning well.

Some may argue, with some justification, that a single company failure is no reason to conduct some form of introspectively philosophical critique of supply chain management and its association with risks. However, the counter to this is that if the approach to commercial and financial risks arising in and from the use of supply chains is not being fully and actively managed – as the Carillion case would suggest – then it raises questions around the rigour used to manage other important issues, such as approaches to sustainability and ethical practice. Indeed, it questions the extent of what can be considered as appropriate or proper management control in supply chains. And if such controls are not present, then a further interesting question is what lies beneath this lack of management control? Part of the answer is concerned with different power structures (Emerson 1962) between organisations in supply chains, the incentives that they create, and the behaviours they facilitate or constrain.

7.3 Risk, Power Structures, and Supply Chains

This chapter will now consider and then use these power structures within supply chains as the conceptual hook to explore some of the emerging 'hot' areas in the construction industry. The first of these is the UN's Sustainable Development Goals' influence. Implicit within these goals is a focus on circular supply chains. Second is the rise of new technologies such as block-chain crypto-currencies and ledgers, and BIM. Our proposition is that construction industry supply chains and the risks inherent in them should be seen as systems in need of strategic coordination. Following this line of argument, we then need to discuss what creates incentives for firms to work harmoniously together. It is this reasoning that led to the subtitle for the chapter – old wine (power and incentives in supply chains) in new bottles (industry hot topics).

We begin with some fundamentals. The construction project, and its inherent processes, is traditionally and normally imagined as linear – indeed we often refer to the 'life cycle' of the project. Pedants will already have spotted the inappropriateness of this term, because projects are nonsentient and do not live the way you or I do. What is true is that projects are started and completed. For traditional buildings to be created, materials are extracted from the earth or provided by nature, processed and manipulated, and then assembled to form the final construction artefacts (buildings and similar), which then go on to provide varying forms of usefulness or utility. At end of 'life', these buildings are demolished. This process is far from simple and contains within it myriad opportunities for harm – to both the planet and specifically to humans. Where unfettered commercial incentives dominate, we can see case after case of destructive and harmful actions, from illegal forest felling for timber, through dangerous and harmful mining for materials, the exploitation of vulnerable individuals – including children – to produce construction products, to the illegal dumping of construction waste (see Figure 7.1 for a personal encounter with such nefarious activity).

Figure 7.1 Encountering illegal 'fly-tipping' that is clearly the result of construction activities. Source: Copyright Andrew Edkins.

Much of this type of activity is clearly abhorrent and morally wrong, yet the presence of global supply chains in combination with power structures and incentives makes this risk real and present. The lack of effective enforcement that leads to the prospect of not getting caught explains cases such as shown in Figure 7.1, but poor legislation, lax regulation, and the presence of corruption are all reasons why incidents that should not happen continue to occur. However, there are countervailing forces at work to give us hope, and these come in the form of heightened awareness – arguably one of the main reasons for producing the UN's 17 Sustainable Development Goals – and the ability to gather and process substantial amounts of data and information.

7.3.1 Commercial Power and the Role of Law and Regulation

As more information flows over the internet, it is far easier and cheaper to trace materials and labour supplies, but it is also easy to fool and forge. With greater awareness and information we are all able to put the welfare of both people and the planet more centre stage in the construction sector through our choices on types of material, sources of supply, or recruitment of labour. Since the advent of the 1974 Health and Safety at Work Act in the UK, we have seen great strides being taken to improve the welfare of those employed – and not just in construction. The introduction in the UK of the Construction (Design and Management) Regulations in 2007 reinforced the need to consider the safety of those involved in the construction process and, critically, provided what is in effect a *duty of care* on those designing the building to consider how it will be built safely and then to ensure that safe methods of construction are used. This may appear obvious, but formal regulations were needed, that are still in force today (although updated), to

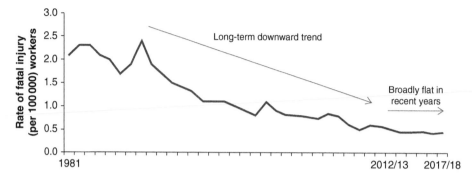

Figure 7.2 UK Construction industry fatal injury rates. Source: Health and Safety Executive: Fatal injuries in Great Britain – http://www.hse.gov.uk/statistics/fatals.htm (accessed July 2019).

ensure that the power structure between the designer, acting as the articulator of the client's demands or requirements for the building, and constructor is fair and balanced. The 'constructor' in the previous sentence is actually the entire supply chain that will deliver the project. Thus, we have a situation where the law and regulation are used to counterbalance the potential for the commercial and informational power held by the client and their designers. This is not to say that all clients and their designers are rapacious and immoral – far from it. Many clients and designers are at the forefront of concerns for the planet and the construction workforce, but there was still a problem, and the law and regulation (in the UK at least) have made a positive difference. The evidence for this is manifest; the construction industry still has the potential to hurt, maim, and kill its workers, but the year on year results show rates falling, as illustrated in Figure 7.2.

Our second area of interest is linked to the point made in the last paragraph and concerns the actions of those who are in the vanguard of construction-related best practice in terms of environmental considerations. This builds on the ideas of those who are advocating the principles and practices of the 'circular economy' (see for example Ghisellini et al. 2016) and involves the recognition of the need to be less linear in our conceptualising of success; more is not necessarily better. This has to be put in historical context because for the majority, improvement in our lives is about success and progress that can be measured in terms of economic and financial success. As a species we have the intelligence and tenacity to solve our problems and create new materials, things, and opportunities for new experiences – and we have. From rare jewels worn as human adornment, to mega-tall buildings in which we desire to live and work, there is a yearning for more – bigger and better. From the industrial revolution to the present day, we have witnessed remarkable improvements in our health, wealth, and lifestyles. This has direct consequences for construction as the focus for the need and/or desire for more built environment is coupled with the traditional construction production processes of extract from nature, craft by human hand using technology where available, use until demise, and then discard and destroy. This has led to many problems and issues, including: the impact on nature as we take more from it; the harm done to the construction workers who may have been treated poorly; and the damage caused when we throw away or destroy all the waste materials that construction produces. All this results in a growing view of the dangers that this linear growth paradigm presents, and not only in construction. This matters in construction because we have developed supply chains and we manage them to accord with this linear growth paradigm. The reason why we have to challenge this persuasive argument of linear and material growth is stark and powerful in the extreme: we are operating within a planetary and climatic system that has limits – hard limits.

This is the world of both intended and 'unintended' consequences. Whether comfortable or not, we are increasingly having to have discussions as we deal with the challenges to our planet of climate change, global warming, widespread pollution, and resource degradation in for the form of deforestation, soil erosion, and damage to the ozone layer. If this were not enough to contend with, we also have the issues of human trafficking, modern day slavery, skills shortages, and ageing populations. These are all the negative consequences or the prices we have unwittingly paid for the improvements that many of us have benefited from – including, to a great extent, those of the construction sector. We enjoy the many benefits that result from the construction production processes we have in place, but we are rapidly recognising that, in an individual and collective sense, we may have underestimated the accumulating full consequences of these improvements.

While it is beyond the remit of this chapter to explore fully the ramifications of considering a limit to the material growth we seek, what is clear is that there is merit in looking to move from 'linearity in production' to more of a 'circularity' in approach – which is the view that we must seek to throttle back on all that we may wish for materially, reuse as much as we can, before recycling or even upcycling the remainder. This may seem too bohemian for some, but the evidence is mounting daily that we simply cannot afford to continue to extract from nature knowing that the ultimate activity will be to throw away, bury, and burn 'stuff' (including construction-related materials) as we have done and continue to do. This has resulted in some interesting developments within supply chains as a result of technology and it introduces the next of our organisational power structures – where commercial power vies with technological power.

7.3.2   Technology-Based Power Structures: Cases of Construction Waste and BIM

Construction waste is a major issue. In the UK the latest official estimate is that around 55 million tonnes of construction and demolition waste were produced in 2014. To give this some context, the official estimate for commercial and industrial waste in the same year was around 41 million tonnes and the waste from households was around 27 million tonnes. However, whereas the recovery rate for household waste was around 45% in 2014, remarkably, for UK construction waste, almost 90% was recovered – see Figure 7.3.

While the process of recovery will include sifting and sorting to extract material streams such as paper, wood, certain plastics, glass, metal, and cementitious and similar materials, there will still be materials that cannot be easily or economically extracted for recycling. With combustible material forming some of this nonrecyclable waste from construction, it is relevant to look at how the disposal of construction waste becomes one of the important inputs into the emerging 'waste to energy' market (Science for Environment Policy 2010). This relatively recent innovation in energy generation has resulted from the concern about the historically crude disposal solution of organised dumping, sometimes out at sea but most frequently in landfill sites (Tam and Tam 2006). In parts of the UK, for example, there is a current and growing

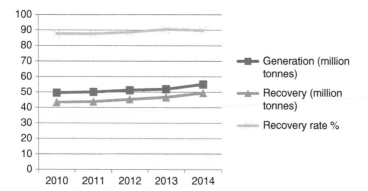

Figure 7.3 UK Construction Waste Output 2010–2014. Source: Derived from Government Statistical Service, 2018, UK Statistics on Waste, p. 7, URL: https://assets.publishing.service.gov.uk/government/uploads/system/uploads/attachment_data/file/746642/UK_Statistics_on_Waste_statistical_notice_October_2018_FINAL.pdf (accessed July 2019).

problem with exposure of old landfill sites through erosion caused by locating such sites next to watercourses and near the coast.[2] As a result of the various concerns, there was political recognition that there had to be some form of safeguarding. Politicians found themselves under pressure from the environmental science base as surveys and monitoring exercises revealed problems, as well as from relevant and often expert nongovernmental organisations, and from the media and electorate. The result was a compelling case for legislative and regulatory action – as we have noted previously.

The introduction of fines and financial charges for unfettered disposal not only immediately changed behaviours, it also created one of the key requirements for innovation, which is that there was a problem to be solved and an opportunity to be exploited. This has travelled from the 'tail' end of the end of the chain where construction waste is handled, to the 'top' end where design and specification for waste elimination are fundamentally factored in. Coupled with the absolute requirement to reduce the problem of waste generation there was the direct and indirect opportunity for scientific and technological invention and innovation that resulted in various offerings of potential alternatives to treating this waste. Lastly, there are the commercial incentives resulting from this combination of regulatory and legal restrictions together with scientific and technological opportunity. This involves two sources of potential revenue: the first is being paid to take waste away and then, after converting this waste into either re-sellable materials (e.g. ferrous and nonferrous metals) or turning the waste into energy and possibly other commercially valuable by-products such as incinerator fly-ash, there is the prospect of the sale of such energy and by-products into new markets. The result is a context in which laws and regulations demand better solutions than simply burying or dumping waste, the scientists and technologists develop ways of taking waste as an input and converting this waste into useful energy and possibly other sellable materials, and the potential for cash profit or surplus to be generated arises. The result of these interventions and developments has been profound on construction supply chains and their

---

2 See this section of Friends of the Earth website: https://friendsoftheearth.uk/climate-change/leaking-landfills-ticking-timebomb-climate-change-hits?origin=d7 (accessed July 2019).

management. What we can say has been learned is that the introduction of the problem of better waste handling and management needs to be seen as a negative risk to be handled as well as a positive risk to be exploited.

In considering this waste-based construction challenge, the public sector could have taken this on itself – collecting waste, creating the technology to treat this waste to obtain energy from it, and selling the resulting energy – probably back to itself as it is a large user of such energy. However, in countries where the public sector has shrunk, and where the role of the private sector has been expanded, this is not how this happens. The significance of laws and regulations may compel the public sector to find an acceptable waste handling solution that may include a waste-to-energy process, but because of its diminution, it will simply have to work with organisations in other sectors, as these organisations have some attributes and resources that the public sector may struggle to possess. Within this list is scientific and technological prowess. As already noted, the most obvious waste-to-energy solution is to simply burn rubbish in incinerators to create steam to drive turbines that are used to generate electricity, with the by-products of steam and heated water. Eliminating or substantially reducing the resulting toxic and noxious gases is technologically challenging and expensive, but it is possible. Economies of scale may mean that such a plant would only be needed for quite large populations and their operation would lend itself to a specialist organisation. Such a facility would also be expensive to build and maintain. These two factors alone may make it too difficult for the public sector to finance and make it appealing for the private sector to organise and arrange. It is here that we have an irony, as some of the players entering this market are themselves major construction companies, who have built supply chains to exploit this new form of construction. Amongst these is a UK company called Interserve, and at around the same time as the collapse of Carillion, Interserve's share price fell as it revealed its move into the incinerator construction business was not going to plan. This demonstrates how the risks present in managing supply chains in sectors such as construction are constantly varying and subject to external forces and factors that are themselves difficult to predict and forecast.

The second issue that illustrates technology-based power structures is very much of construction: BIM. Just like the case of the public sector dealing with waste and rubbish, large construction companies can take on the challenge of BIM via direct in-house investment, but there is the opportunity for existing smaller players to embrace and adopt BIM and effectively become new entrants and could be considered as 'BIM natives'. This is something noted by the architect David Miller in a UK government report of 2012:

> We believe BIM represents a fantastic opportunity for smaller organisations to change the odds; there is even a real opportunity for SMEs [small and medium enterprises] to lead rather than follow ....
>> Source: David Miller quoted in (HM Government 2012, p. 8)

The potential for BIM to be a 'game-changer' is real because, for it to deliver its full promise, it has to be a common technological thread that not just runs through supply chains, but unites them. The exciting possibilities for BIM are also the reasons why it may take a long time, if ever, to fundamentally change construction systems, as it presents myriad risks that will need to be managed both within individual enterprises and throughout project supply chains. Professional and dedicated supply chain management-BIM managers may be needed in the future to ensure both the promise of harmonious construction delivery and the step changes in improved performance that BIM offers.

The great challenge, and therefore the great opportunity, lies in the future 'roll-out' of BIM to the very many smaller construction companies that inhabit many national construction sector universes. The specifics of both challenge and opportunity are to make BIM ubiquitous, cheap, and easy to use. Mandating its use, as is the case with UK government construction procurement (HM Government: Cabinet Office 2011), will stimulate some action, but the challenge (i.e. risk) is that BIM fails to permeate throughout the full construction market. It is far more likely to be adopted on the bigger projects and by the bigger players, but the extent to which BIM will percolate down the many layers of supply to the smaller organisations is not clear. However, one does not have to look very far to see how some technologies can be both adopted and lead to improved performance. The most obvious case is the use of mobile phone technology and integration within the handset of access to the Internet. BIM is recognised as a more niche technology, but if in the future it can be made to be as user-friendly as our smartphones are now, then BIM could deliver on the promises made of it.

While BIM is a recognised 'hot' topic for construction, a more embryonic topic that is also technology centred is the potential of crypto-currencies based on the use of blockchain. This is effectively described by Koutsogiannis and Berntsen (2017):

> The Blockchain is a new way to store and record transactions. To put it simply, we could define the Blockchain as a peer-to-peer controlled distributed transactional database.
> [It is] A digital ledger where different types of agreements (e.g. contracts, financial transactions) are recorded and confirmed as completed.
> Its main difference to traditional databases is that it lacks the need for a central authority. There is no middleman, such as a bank transferring money or a lawyer to confirm the conditions of a contract. In that sense, there is no single database or company on which it hinges.

The linking of design and production through BIM and the contracts and payments through crypto-currencies could revolutionise the process of construction and the way the supply chain operates. The consequences in terms of power distribution amongst the supply chain members would alter, with clarity, fairness, and integrity being implicit to the function of the process and thus help to eradicate perennial issues such as late payment to the smaller and lower tier construction supply chain companies, which has been a problem recognised for many years (Akinsiku and Oluwaseyi 2016; Dainty et al. 2001) and which was still being discussed in the UK construction industry at the time of writing (Bounds 2018; Garner-Purkis 2018). The noteworthy conclusion is that if the risk of the mastery of technology can be made successful, then this provides a form of clear power structure that can bring totally new players and the rise of existing players into both existing and new markets.

7.4 Conclusions

The purpose of this chapter was not to provide a compendium of all construction-related risks that beset the members of construction supply chains. Nor was the chapter seeking to provide any form of toolkit or magic bullet solution for managing the risks that are inherent in such supply chain situations. The hoped for contribution is to ask the reader to consider some of the more strategic or 'higher level' risks that are entering into the discussion, and the chapter has identified some examples, but there are more. Before commencing on this, however, a significant and deliberate diversion was undertaken to consider the case of Carillion. This was a necessary deviation from the main thrust of the chapter as its relevance could not be overstated to the chapter's focus. The collapse of Carillion has sent various forms of shockwave through both the UK construction industry as well as those nations that have adopted UK-style structures and practices for their own construction markets. The consequences for companies connected to Carillion, either directly or indirectly, will continue to be felt for some time to come. The UK banking sector offered help, but this was limited. The concern arising is the latent financial vulnerability of lower supply chain order companies who are surviving thanks in part to the UK having experienced an extended period of unprecedentedly low interest rates. The ending of period of extraordinarily low interest rates, or some other shock, could lead to the mass demise of many construction-based SMEs. Construction companies, and especially contracting companies, operate on very narrow margins and with often precarious cash flows, so they are acutely sensitive to issues like interest rate rises. The full extent of the damage done by Carillion's failure is unlikely to ever be recorded entirely as the shockwave will take time to travel and take time to have its effect. But this episode reminds us both of the fundamental use of supply chains in delivering our built assets and the risks that are being run in using them to the extent we do.

As well as these risks, this chapter has introduced additional areas that present risk – both upside and downside. The United Nations' 17 Sustainable Development Goals may seem unrelated to the cut-and-thrust world of everyday supply chain operation and management, but these goals identify and signify an expanding set of criteria on which supply chain performance is measured and judged. This means that supply chain performance is no longer only assessed on the obvious parameters of technical performance, commercial performance, and timeliness. We now look at the way staff and other people associated with the firms are treated and looked after, how these companies make sure that their choices are factoring in ethical and moral positions, and the sustainability implications measured in terms of both our environment and our societies. And then we have to consider the risk (again upside and downside) of emergent and emerging technologies such as BIM and crypto-currencies. The chapter could also have brought into play other strategic-level challenges and possibilities, such as the potential for the Internet-of-Things (IoT) as a further development, but the point being made is not the choice of example, but the way these changes are affecting supply chains and their management.

For those in academe studying this area, or those practitioners in leadership and management positions, it is hoped that the chapter will reinforce the need to stay vigilant to the challenges and opportunities facing organisations forming construction supply chains. This will almost certainly include being affected by new technologies that may come from within the construction sector – as BIM has – or maybe by a more universally adopted technology as was the case with mobile phones and looks set to be the case for both crypto-currencies and the IoT.

So what does this all add up to? To see the principal points one has to fly above the day-to-day issues that beset and test us. This is easier said than done as we are weighed down with the day-to-day challenges. Managing construction supply chains properly and completely is a challenge as they are constantly forming, adapting, and ending. This fluidity and level of fluctuation sets out a challenging landscape that was already testing those in construction-related supply chain management. However, there are more challenges emerging – and some of these are big challenges. Construction companies that have the willingness and ability to look ahead will spot these challenges and this is an essential first part of effective risk management. For some, these challenges will represent tremendously exciting opportunities (i.e. upside risks) and if they can invest the time, effort, and no doubt money, then they will not just be better prepared for the future, they will be more likely to be one of the leaders that creates a better future.

The scholars and scholarly minded practitioners and policymakers reading this chapter are asked to consider the proposition that, having had some decades now to solve the many operational and managerial challenges involved in creating and using supply chains, we are now increasingly identifying systemic risks requiring our careful thinking and management action. This is where conceptual models and theoretical thinking, such as the power structures used as the intellectual underpinning of this chapter, can be effective. Using such an open mind-set allows us to make sense of all the data and other informational 'noise' that surrounds us on a daily basis. Such practice-to-inform-theory-to-inform-practice has proven useful in the past and looks to continue to be valuable in the future. The ability to see things such as the opportunity for new theory, to develop, test and then deploy this new theory or conceptual thinking to practice is also an illustration of systems thinking (Senge 2006). Such a systems view is obviously useful when it comes to considering construction supply chains, as they are indeed solution systems, comprising different organisational entities, driven by different forces and operating in different ways. Successful construction supply chains counter the difficulties and obstacles by combining careful selection of supply chain members, appropriate contractual arrangements, and the development and management of relationships. These established mediating elements (the old wine) are themselves affected by issues such as global concerns and technological development (our new bottles), which can have the most substantial impact on approaches, attitudes, behaviours, and actions.

References

Akinsiku, O. and Oluwaseyi, A. (2016) Effects of Delayed Payment of Contractors on Construction Project Delivery. The Construction, Building and Real Estate Research Conference of the RICS, September 2016, Toronto, Canada. Tag as a conference proceeding.

Bounds, A. (2018). *Carillion Failure Adds to Subcontractors' Case Against Late Payment*. London: Financial Times.

Chapman, C. and Ward, S. (2011). *How to Manage Project Opportunity and Risk [Electronic Resource]: Why Uncertainty Management Can Be a Much Better Approach than Risk Management*. Translated from English by, 3rd edition. Chichester: Wiley.

Cox, A. (1999). Power, value and supply chain management. *Supply Chain Management: An International Journal* 4 (4): 167–175.

Cox, A.W. (2003). *Supply Chain Management : A Guide to Best Practice*. London: FT Prentice Hall.

Dainty, A.R.J., Briscoe, G.H., and Millett, S.J. (2001). Subcontractor perspectives on supply chain alliances. *Construction Management and Economics* 19 (8): 841–848.

Emerson, R.M. (1962). Power-dependence relations. *American Sociological Review*: 31–41.

Finch, P. (2004). Supply chain risk management. *Supply Chain Management: An International Journal* 9 (2): 183–196.

Garner-Purkis, Z. (2018) Carillion failure adds to subcontractors' case against late payment. *Construction News*.

Ghisellini, P., Cialani, C., and Ulgiati, S. (2016). A review on circular economy: the expected transition to a balanced interplay of environmental and economic systems. *Journal of Cleaner Production* 114: 11–32.

HM Government (2012). *Industrial Strategy: Government and Industry in Partnership: Building Information Modelling*. London: HM Publishing Service https://assets .publishing.service.gov.uk/government/uploads/system/uploads/attachment_data/file/ 34710/12-1327-building-information-modelling.pdf (accessed August 2019).

HM Government: Cabinet Office (2011). *Government Construction Strategy*. London: https://assets.publishing.service.gov.uk/government/uploads/system/uploads/ attachment_data/file/61152/Government-Construction-Strategy_0.pdf [accessed August 2019].

Ho, W., Zheng, T., Yildiz, H., and Talluri, S. (2015). Supply chain risk management: a literature review. *International Journal of Production Research* 53 (16): 5031–5069.

Koutsogiannis, A. and Berntsen, N. (2017) *Blockchain and construction: the how, why and when*. Available online: www.bimplus.co.uk/people/blockchain-and-construction-how-why-and-when [accessed August 2019]

Lee, H.L., Padmanabhan, V., and Whang, S. (1997). The bullwhip effect in supply chains. *Sloan Management Review* 38 (3): 93.

National Audit Office (2018). *PFI and PF2*. London.

PWC and MIT and Forum for Supply Chain Innovation (2013). *Making the Right Risk Decisions to Strengthen Operations Performance*. Boston, US: PWC.

Ritchie, B. and Brindley, C. (2004). Risk characteristics of the supply chain – a contingency framework. In: *Supply Chain Risk* (ed. C.S. Brindley), 28–42. UK: Ashgate Publishing.

Ritchie, B. and Brindley, C. (2007). Supply chain risk management and performance: a guiding framework for future development. *International Journal of Operations and Production Management* 27 (3): 303–322.

Sanderson, J. and Cox, A. (2008). The challenges of supply strategy selection in a project environment: evidence from UK naval shipbuilding. *Supply Chain Management: An International Journal* 13 (1): 16–25.

Science for Environment Policy (2010). *Best Options for Disposing Construction Waste*. Brussels: EU: European Commission DG Environment News Alert Service.

Senge, P.M. (2006). *The Fifth Discipline: The Art and Practice of the Learning Organization*. Broadway Business.

Tam, V.W. and Tam, C.M. (2006). A review on the viable technology for construction waste recycling. *Resources, Conservation and Recycling* 47 (3): 209–221.

8

Linkages, Networks, and Interactions: Exploring the Context for Risk Decision Making in Construction Supply Chains

Alex Arthur

8.1   Introduction

This chapter explores the risk management systems within construction supply chain management. The emphasis is on the processes applied in identifying and managing risk events generated through the linkages, networks, and interactions within construction supply chain relationships. The chapter has been organised into five sections covering the following themes: the evolution of the UK construction industry and supply chain relationships; the concept of risk; the construction risk management system; risk generation in construction supply chain relationships; and risk management decision-making systems in construction supply chain relationships.

The discussion begins by tracing the evolution of the UK construction industry from the era of sole master mason profession (Walker 2007) to the contemporary state of multiple differentiated organisations, and professional disciplines specialising in aspects of the construction process (Vrijhoef and Koskela 2000; Edkins 2009). There is further review of how the resultant organisational metamorphosis has stimulated the adoption of relational project governance approaches including supply chain management. The section concludes with a theoretical review of how the multiplicity of organisations and specialist roles involved in construction supply chain relationships creates variability in project risk communication.

The second section reviews the conceptual meaning of risk including the origin of the terminology and the factors accounting for the recent heightened public risk discussions. This is followed by a theoretical review of the different conceptual meanings of risk including: real and objective; socially mediated; real and socially constructed; socially transformed; socially constructed; and subjectively biased. The section concludes with an analytical review of: the conceptual relationship between risk and uncertainty, and risk and probability; risk as a potential future event; and the impact of a risk event on an objective or interest.

The third section explores the different ideological perspectives of risk management. This is followed by systems analysis of the risk management system within construction supply chain relationships. The analysis employs schematic models in describing the risk management interactions within the construction transformational processes.

Successful Construction Supply Chain Management: Concepts and Case Studies, Second Edition.
Edited by Stephen Pryke.
© 2020 John Wiley & Sons Ltd. Published 2020 by John Wiley & Sons Ltd.

There is further theoretical review of the internal subsystems within the construction risk management system, including the risk identification subsystem, the risk analysis subsystem, and the risk response subsystem.

The systems analysis is extended in the fourth section by reviewing risk identification and treatment in construction supply chain relationships. The review focuses on the theoretical relationship between a 'risk event' and its 'impact on an objective or interest'.

Finally, the fifth section discusses the different approaches to risk management decision making within construction supply chain relationships including subjective and objective techniques. Following on, there is a review of the prevailing empirical evidence which suggests a high incidence of intuitive construction risk management practices. This is followed by a review of the behavioural patterns of the intuitive risk management decision-making system.

The chapter ends with a concluding discussion on the key theoretical and conceptual issues raised. This is followed by a theoretical review of recommendations for enhancing the risk management systems in construction supply chain relationships, including integrated risk management practices, and constant harmonisation of the variations in technical and behavioural stimulus emanating from the coalition of multiple professionals and organisations.

8.2   The Evolution of the UK Construction Industry and Supply Chain Relationships

Walker (2007) traces the evolution of the UK construction organisational structure to the medieval period when master masons were the most prominent actors in the construction industry. The master masons were responsible for organising material and labour for construction projects on behalf of their rich clients. The exceptionally skilled master masons were appointed by Kings to build their castles, and were given the title, 'King's masons'. There were also instances of client representatives being appointed to manage projects, but this was purely an administrative role without technical or professional duties. The relatively stable construction environment in the medieval period did not necessitate the subdivision of the master mason's role.

During the sixteenth century, the construction industry experienced some changes; military installations were constructed, the architect's role became more prominent, and projects started to be procured through contracts (Walker 2007). These changes introduced additional specialist roles into the construction industry.

Between the sixteenth century and the industrial revolution era, there was increased mobility, with transnational travels to established cities including Rome and Athens, which boosted the demand for architectural services in the UK (Bowley 1966). The great London fire incident in September 1666 (Robinson 2011) also accounted for significant changes in the structure of the construction industry. The need to quantify and pay indemnities for the fire destructed construction projects, triggered the emergence of professional measurers who have subsequently evolved into the present quantity surveying professional discipline. The rebuilding of London also increased the demand for architectural and engineering services, resulting in the subdivision of the engineering role into civil, mechanical, and electrical. Within this era, the specialist subgroups began consolidating themselves through the formation of clubs and associations. In 1771, the society of engineers was set up (Derry and Williams 1960), followed in 1791 by the architects' club (Allinson 2008), and later the surveyors club in 1792 (RICS 2018a).

In the nineteenth century, the professional clubs and associations developed into institutions with charters restricting their members to the specific functions of their sub-groupings (Bowley 1966). This had a major influence on the construction industry with the members within the various professional clubs developing higher allegiance to their respective subgroups instead of the construction industry. Differentiation along professional subgroups was further consolidated and entrenched when between the First and Second World Wars, the professional institutions introduced professional qualification tests, and codes of ethics and conduct for their members (Walker 2007). According to Bowley (1966), the differentiation then was formed along class structures, with the architects at the top, followed by engineers, then quantity surveyors, and the builders who were seen as trades.

Between the Second World War and the present era, environmental changes have continued to account for the emergence of additional specialist roles, and the subdivision of the established professional groups into multiple differentiated specialist disciplines. There are 22 specialist surveying groups listed on the RICS website (RICS 2018b); there are also specialist architects, builders, and engineers for the different construction sectors, types, and project delivery stages. Recent emphasis on sustainable construction, and integration of building information modelling (BIM), have also seen the emergence of specialist sustainability consultants (GreenBook live 2018; BREEAM 2018) and BIM managers (RICS 2018c). The cumulative effect of this increasing specialisation has been fragmentation in the structure of the construction industry with multiple niche organisations. According to the records from the Office for National Statistics, there were 296 093 construction companies registered in Britain in 2016 (Office for National Statistics 2017).

The evolution of the multiple differentiated construction disciplines has resulted in industrial transformation, from an era of self-sufficient firms, to a state where most firms now specialise in aspects of the construction process and rely on contribution from other firms for project execution (Edkins 2009). Vrijhoef and Koskela (2000) have confirmed an increasing trend in labour and service outsourcing within the construction industry, with most main contractors now relying on subcontractors and material suppliers for their project execution. This has resulted in the adoption of relational project management approaches including supply chain. Kahraman and Oztaysi (2014: vii) describe supply chain as the systems of individuals, human and material resources, firms, processes, and technologies involved in the production and sale of a project. Khalfan et al. (2010) have also suggested two ideological strands for supply chain comprising logistics management of labour and materials for site production, and lean thinking of the construction production processes.

According to Vrijhoef and Koskela (2000) the application of construction supply chain management originated from the manufacturing industry, where it was first applied as a management tool for promoting efficiency, through reduction in inventory, and improvement in the relational management of the production line and their suppliers. Morledge et al. (2009) have also suggested that supply chain management was introduced into the UK construction industry as a strategic project governance initiative for managing the nonformal relationships promoted by the noncontractual procurement options such as partnering (Egan 1998), and also the operational and managerial inefficiencies (CBI 2010) caused by the fragmented industrial structure (Aloini et al. 2012). Pryke (2009) has further suggested the prime attraction for construction supply chain relationships to be the desire for companies to align themselves with other companies that possess complimentary skills and functionalities. Construction supply chain relationships tend to be product centred, temporal, and bespoke to the specific project context (Vrijhoef and Koskela 2000). The management approach focuses on the administration of construction processes from a central point originating from the client organisation or an external consultant organisation acting on behalf of the client. Depending on the procurement strategy, the supply chain relationship could comprise the client and their key consultants entering into a first tier relationship with the main contractor. The contractor subsequently forms second-tier relationships with their sub-contractors, who also rely on third-tier relationships with their suppliers (Pryke 2009).

Khalfan et al. (2010) believe the construction industry, unlike the other industrial sectors, has three unique interconnected supply chain systems comprising professional services, site construction production, and supply of materials. Vrijhoef and Koskela (2000) have also described four roles of construction supply chain management including:

- Improvement in site efficiency through improved relationship with suppliers to facilitate constant material flow.
- Promotion of efficiency in the material production and delivery through improved supply logistics management.
- Feeding back expectations of the site construction base to their suppliers to facilitate improved coordination in site production and material specification and delivery.
- Integrated management of the site production with the supply of materials to facilitate total efficiency across the supply chain processes.

According to Aloini et al. (2012), there have been multiple theoretical initiatives on the implementation of construction supply chain management. The practical evidence nevertheless suggests that the perceived benefits are yet to be fully realised. This has been attributed to the lack of holistic application of the concept across the construction production processes (Aloini et al. 2012). The other reasons include the empirical differences between conceptual rhetoric and reality, and the lack of deeper understanding of the risk management systems within construction supply chain relationships and the associated behavioural patterns. This often results in failure to identify and treat project risk events robustly, and consequently impedes the realisation of the expected process efficiency.

The prevailing theoretical evidence suggests that the multiplicity of organisations and professional disciplines involved in construction supply chain relationships creates variability in project risk communications (Japp and Kusche 2008; Arthur and Pryke 2014; Arthur 2018), in relation to how risk is perceived and responded to. The variations result from the difference in behavioural stimulus (Kahneman 2011; Slovic 2010) emanating from the diverse organisational, professional, and personal backgrounds (Arthur and Pryke 2013; Arthur 2017). The failure to recognise and manage these differences results in the introduction of additional risk to process integration, which consequently affects the efficiency of the project transformational system. The link between differentiation in supply chain relationships and risk becomes significant when the variability in behavioural stimulus is so intense that it influences the members within the network of organisations and professional disciplines to exhibit higher allegiance to their micro business and personal objectives, to the detriment of the achievement of the macro objectives of the supply chain relationship (Walker 2007).

8.3   The Concept of Risk

Risk has always been a part of the human condition both within the events that happen spontaneously and the regular activities which we are familiar with (Lock 2003). According to Smith et al. (2006) the term 'risk' originated from the French word 'risqué' which was introduced in England around 1830 when it was used in the insurance industry. Flyvbjerg et al. (2003) have suggested that contemporary sociologists such as Ulrich Beck and Anthony Giddens believe that modern society functions in risk and therefore excluding risk from social, economic, political, and environmental deliberations is likely to result in failures. While we may not be able to completely eliminate risk (Beck 2002, 2007; Loosemore 2006), an informed risk management approach will nevertheless go a long way in helping to diagnose the sources of risk and formulating strategies to respond to their threats and opportunities (Arthur 2017).

According to Loosemore et al. (2006), increased public awareness of the risk factors associated with both personal and corporate activities, coupled with increased media reporting on the impact of risk events, have accounted for heightened discussions on risk, and influenced public attitude to risk management. Media reporting on the European *E.coli* outbreak in 2011 led to the National Health Service in the UK assessing the risk impact on vegetable importation, processing, and consumption, and issuing public health advice in June 2011 (NHS 2011). The use of the Blackberry Instant Messenger system by the perpetrators of the August 2011 British riots to elude police intelligence, triggered discussions on the security risk associated with smart phones (The Economic Times 2011). The previous political uprising in the Arab nations also influenced many western governments to assess the travel risk to their nationals and warn against nonessential travels to countries which they considered as high risk, including Tunisia, Egypt, and Libya (Guardian 2011).

The tragic deaths of a six-year-old girl at Carnival Place, Moss Side, Manchester, UK in June 2010, and a five-year-old girl at Brook Court, Bridgend, South Wales, UK in July 2010, after they became trapped within automatic sliding doors at blocks of flats, also ignited discussions on the potential risk associated with powered controlled door entry systems, and prompted investigations and risk assessment on electrically controlled door entry systems (The Independent 2011). The impact from the recent global economic downturn on the construction industry which resulted in companies including Connaught, Rok, John Laing Partnership, Greenacre Homes, and Carillion Group going into administration in the middle of project execution has also increased client's awareness of potential risks associated with project procurement, and the need to carry out robust financial audits before contracts are awarded.

The recent fire incidence and fatalities in high rise buildings (28-storey tower block in Shanghai, China in 2011; 25-storey tower block in Krasnoyarsk, Russia in 2014; 79-storey skyscraper in Dubai in 2015; 17-storey commercial building in Tehran, Iran in 2017; and the 24-storey Grenfell Tower in Kensington, UK in 2017) have also heightened public awareness on safety and fire risk in high rise buildings, and cladding materials (Arthur 2017; Ponniah 2017).

Table 8.1 Different risk conceptual interpretations.

Risk as …	Viewpoint	Related disciplines
Real and objective	Appreciates risk as a real event that can be analysed	Toxicology, epidemiology, actuarial science, engineering,
Socially mediated	Appreciates the influence of social institutions in determining risk perceptions and responses	Sociology
Real and socially constructed	Appreciates both the reality of risk and the influence of social institutions in defining risk perceptions and responses	Sociology
Socially transformed	Appreciates how social institutions can transform threats into risk events	Sociology, anthropology, criminology, cultural and media studies
Socially constructed	Appreciates social construction of risk through the actions of multiple differentiated subsystems	Systems, governmentality
Subjectively biased	Appreciates human subjectivity in defining risk perceptions and responses	Psychometric research: objective/subjective utility analysis

Source: Adapted from Zinn (2008, p. 8).

Zinn (2008) describes six different conceptual interpretations of risk, as illustrated in Table 8.1. Taking the 'real and objective' ideology, risk is understood as a real event which can be analysed (Renn 1992). The realist perspective is usually applied in the toxicology, epidemiology, actuarial science, engineering, and probability disciplines (Zinn 2008).

The socially mediated risk conception appreciates the influence of social institutions in determining risk perceptions and risk responses. This perspective is typically applied in the sociology discipline (Zinn 2008).

The real and socially constructed risk conception combines the real and socially constructed perspectives. A risk conception based on this model appreciates both the reality of risk, and the influence of social institutions and culture in defining risk perceptions and judgements (Beck 2002, 2007).

The concept of risk as socially transformed as explained by Zinn (2008, p. 8) is where *real threats are transformed into risk for sociocultural boundaries*. This conceptual interpretation draws ideas from a wide range of social science disciplines including sociology, anthropology, criminology, culture, and media studies (Tulloch 2008).

A conceptual model of socially constructed risk is applied in this chapter, employing social calculative techniques to conceptualise social systems to discover their embedded risk variables (Zinn 2008). The underlining analytical thinking is the social construction of risk, through social communications and decision making, by the multiple horizontal differentiated functional subsystems which have displaced the traditional stratified vertical social system (Japp and Kusche 2008). A major criticism of this approach is the lack of correlation which sometimes exists between the virtual risk conceptualisation and the real threats posed in societies, as observed in the erroneous risk prediction of the level of Bulgarian and Romanian migration into the UK, following the lifting of their European Union migration restrictions on 1 January 2014 (Kirkup and Barret 2014). The socially constructed perspective is applied in systems and governmentality disciplines (Japp and Kusche 2008; O'Malley 2008).

A conceptual examination of risk as subjectively biased, also as applied in this chapter, recognises the influence of perceptions inspired by emotions and previous experiences of risk judgement and decision making (Renn 1992; Taylor-Gooby and Zinn 2006; Slovic et al. 2010). The subjective perspective of risk is usually applied in psychometric research and subjective analysis.

Risk, as defined by Loosemore et al. (2006), is an unpredictable event that might occur in the future whose exact likelihood and outcome is uncertain but could potentially affect interest or objectives in some way (normally adversely). Analytical review of the above definition suggests a theoretical relationship between risk, and uncertainty, and probability judgement. It also emphases risk as a future potential occurrence with the propensity to impact on an objective or interest.

8.3.1   Uncertainty

The absence of information on future events creates uncertainties (Winch 2010), which makes it difficult to confidently predict the future. Until an event has taken place it is usually difficult to predict the outcome accurately, especially where there are external factors which are beyond the control of the actors involved in executing the event. It has been said that nothing in life is certain except taxes and death; however, employment and financial status do excuse certain people from paying taxes. Also, not all people who are required to pay taxes honour that obligation. The lack of information on a future event owing to the fact that the event has not occurred, suggests that potential risk events can only be predicted with certain caveats.

Edkins (2009) makes a distinction between risk and uncertainty by arguing that risks are uncertainties that have been identified and recorded to be given further treatment, while uncertainties exist in obscurity, relegated to the project background and may not attract formal attention because they may not be tangible enough to warrant recording and treatment. This view is also supported by Flyvbjerg et al. (2003), who in analysing risk management practices in mega projects, identified that some of the negative uncertainties which normally affect project success are known at the project organisational level, but not properly managed because they may not be tangible enough at the time of their discovery to warrant analysis.

An important point to note is that the impact of uncertainty is not always negative. There are instances where uncertainties actually do present opportunities to individuals and organisations (Dallas 2006). This should, however, be analysed in the context of the individual's or organisation's objectives (Loosemore et al. 2006), since difference in objectives could mean that an uncertain event will present an opportunity to a particular organisation, while at the same time pose a threat to another organisation. For instance, the money.co.uk website published an article on 9th October 2008 reporting that, while the banks and financial institutions were suffering the effects of the recent credit crunch, fast food companies (including McDonalds and Burger King), and other businesses producing home entertainment gadgets and maternity products, as well as bargain retail companies, experienced a boom in their business activities (Money.co.uk 2011).

8.3.2 Probability

The fact that an event has occurred in the past does not in any way confirm that it will reoccur in the future; however, analysis based on past observations of similar occurrences can help us in forecasting future occurrences under similar circumstances with a degree of certainty termed probability (Aven 2004). The value of a probability ranges between zero and one. A value of one represents that an event is certain to occur, while a value of zero means an event is certain not to occur.

Adams (2008) has suggested that the classification of construction project risk as either *objective* or *subjective* determines the method of risk analysis, including probability evaluation. Objective risk which involves the actual observation and evaluation of risk occurrence, and impact, usually adopts quantitative techniques in the probability evaluation. Subjective risk, on the other hand, employs qualitative tools and techniques such as perceptions and personal judgements. According to Kahneman (2011), personal judgement of the probability of an event occurring is influenced by how easily a similar past occurrence can be recalled, and the relationship between the event and its parent population. Bateman et al. (2010) also explained that the value of probability which we assign to a potential uncertain event is influenced by our pictorial image of the event based on our past experience of the level of risk of a similar past event.

8.3.3 Risk as a Potential Future Event

Loosemore et al. (2006) argue that risk should be analysed in terms of the events that give rise to the risk outcome rather than the impact of the risk. Risk analysis based on the risk impact becomes reactive and ineffective, and more of a crisis management than risk management, which seeks to mitigate the potential events that give rise to the risk impacts.

Aven (2010) has challenged the conceptualisation of risk as an 'event' on the basis that such a definition limits the ability to qualify risk as being high or low, severe or mild. He has therefore suggested that risk analysis and quantification should focus on the antecedent uncertainties which give rise to the actualisation of potential risk events.

8.3.4 The Impact of a Risk Event on an Objective or Interest

One important principle emphasised in the 'risk' definition offered by Loosemore et al. (2006) is that a potential event becomes risk to a subject where there is a likelihood for its occurrence to impact on the objectives or interests of the subject. This implies that not all potential events can be classified as risk to all subjects. As an illustration, the potential occurrence of a specific bridge collapsing will be classified as a risk event to just the individuals and organisations whose personal and/or business objectives are likely to be affected.

8.4 The Construction Risk Management System

Risk management involves the identification and treatment of events that have the potential to impact on future activities. Loosemore (2006) has also described risk management as a field of competing ideologies between the *homeostatic* and *callibrationist* perspectives (Hood and Jones 1996). The *homeostatic* perspective proposes a scientific approach to risk management through structured risk identification and rational decision making, while the *callibrationist* viewpoint on the other hand advances human subjectivity in risk interpretation, identification, and treatment (Loosemore 2006). Another significant ideology of the *callibrationist* view is the subjectivity and biases of the personal perceptions responsible for guiding the risk identification and treatment processes (Tversky and Kahneman 1982; Slovic 2010; Kahneman 2011).

The subsequent discussion employs principles of system thinking in analysing the construction risk management processes. Systems thinking and analysis is an approach to studying a subject by employing concepts and principles similar to those of classical natural sciences (Checkland 1999). Checkland (1999) explains that systems thinking and analysis should be seen as an approach to studying a subject rather than being a distinct discipline. Walker (2007) also describes systems thinking and analysis as 'thinking about complex processes so that the interrelationships of the parts and their influence upon the effectiveness of the total process can be better understood, analysed and improved' (Walker 2007, p. 36). The theory underlying systems thinking and analysis originated from the biological sciences through the discovery of general systems theory in the 1960s by Ludwig Von Bertalanffy (Walker 2007). This was as a result of his studies of the interdependencies between the different science disciplines, which were then studied as distinct subjects. Bertalanffy (1968) has suggested that application of the concepts within systems thinking existed prior to the discovery of general systems theory and might have been applied in preliminary studies including Paracelsus' studies in medicine and Leibniz's studies in philosophy, although the term 'systems' was not specifically used. In addition, Apostle Paul's comparison of the mission philosophy of the Christian church to the collective functioning of the different body parts can also be argued as a form of systems analysis (Romans 12:4–8; 1 Corinthians 12:12–31 – Holy Bible, King James Version). The concepts and principles underlining general systems theory are therefore not restricted to the natural sciences only, but can also be applied in other fields (Bertalanffy 1968).

According to Smith et al. (1999), the standard risk management model comprises three stages namely, risk identification, risk analysis, and risk response. Dallas (2006) also believes the risk management stages exist as a cycle consisting of interconnected activities, as illustrated in Figure 8.1.

Drawing from the concepts and principles in systems thinking and analysis (Checkland 1999; Bertalanffy 1968; Cole 2004; Carmichael 2006; Walker 2007), each of the risk management stages can be viewed as a system having input, transformation, and output stages. Each stage also exhibits both inter- and intradependencies (Carmichael 2006;

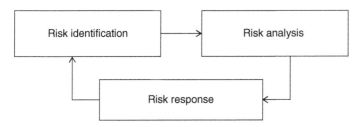

Figure 8.1 Risk management cycle. Source: Adapted from Smith et al. (1999) and Dallas (2006: p. 41).

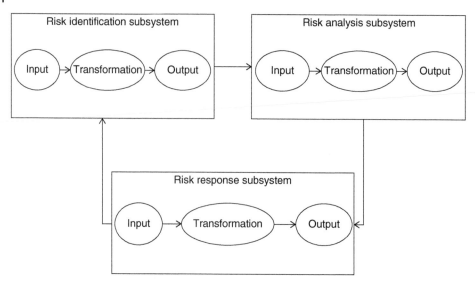

Figure 8.2 Risk management subsystems. Source: Adapted from Walker (2007: p. 113), Smith et al. (1999), and Dallas (2006: p. 41).

Walker 2007), as illustrated in Figure 8.2. The output of a stage becomes the input for the succeeding stage.

The construction industry being an open system further suggests that each of the subsystems will interact with both their immediate and broader environments (Cole 2004) to form a complex model as illustrated in Figure 8.3.

The subsystems receive inputs from both their immediate subenvironment and the broader construction system's environment (Cole 2004; Walker 2007), and these are processed into outputs which, together with other inputs from the environment, forms the input for the succeeding stage of the risk management process. Each subsystem is composed of different parts which interact amongst themselves and their environmental forces during the transformational processes. The reciprocal nature of interactions enables forward and backward interactions between the different stages of the risk management system (Walker 2007).

The complex interactions coupled with the changing nature of environmental forces suggest that the construction risk management system rather than being static, is very dynamic, requiring constant revision in the light of changing internal and environmental interests and activities.

8.4.1   Risk Identification Subsystem

The first stage of the construction risk management system is risk identification. Loosemore et al. (2006) have suggested that contracts are the principal tools for identifying and allocating project risks. However, before risk events either in expressed or implied forms are identified and documented in contracts, there will have to be a decision to undertake a project, which will lead to the design of a brief (Latham 1994) upon which the potential risk events are generated.

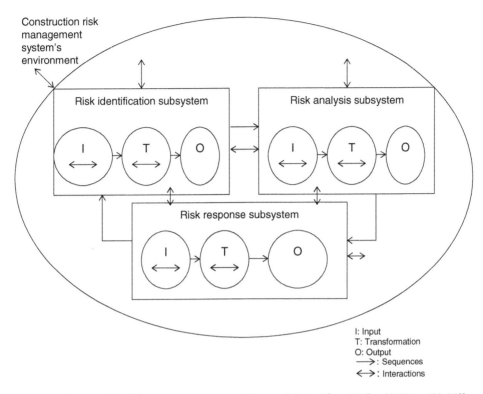

Figure 8.3 Construction risk management system. Source: Adapted from Walker (2007: pp. 76, 113), Smith et al. (1999), and Dallas (2006: p. 41).

Schieg (2006) has suggested a classification of six construction project risk events: quality, personnel, cost, programme, strategic decision, and external. Mustafa and Al-Bahar (1991) ,while acknowledging the nonexistence of an exhaustive list, have also suggested a categorisation of six events including: acts of God, physical, financial and economic, political and environmental, design, and job site-related.

8.4.2   Risk Analysis Subsystem

The risk analysis stage is where we evaluate the reliability of the identified uncertainties developing into future risk events. It also involves the classification of risk events based on the probability of occurrence and the likely impact of occurrence, using rankings such as low, medium, and high (Lock 2003).

8.4.3   Risk Response Subsystem

The risk response stage is the last phase of the risk management system. The finding from the risk analysis stage serves as the basis for deciding whether to accept or reject the potential risk event, revise the project objectives to eliminate the likelihood of the potential risk event occurring, or delay decision making until additional information becomes available (Winch 2010). Where a project risk is to be accepted, a decision will have to be made to either manage it internally or transfer it to another company using procurement and contractual tools, and insurance policies (Akintoye and Macleod 1997; Latham 1994).

8.5 Risk Generation in Construction Supply Chain Relationships

This section expands the systems analysis in evaluating project risk identification and treatment processes in construction supply chain relationships. The discussion centres on the theoretical relationship between a 'risk event' and its 'impact on an objective or interest' (Loosemore et al. 2006).

The interactions between the components and transformational stages of a system and the wider environment suggest that the success of a construction supply chain relationship will depend on how well the activities of the various processes, interactions, and networks are coordinated. Any event that breaks the chain of coordination causes the system to malfunction and poses risk to the achievement of the supply chain system's objectives (Walker 2007). Potential risk events could be generated through the interactions between the project delivery processes and also through the network of relationships of the different professionals and organisations involved in supply chain relationships.

8.5.1 Project Risk Events Generated through the Project Delivery Processes

Systems' objectives established at the project conception stage define the expected project outputs at the completion stage. Between these two cardinal stages are alternative transformational processes which could be selected to link 'objective' to 'expected output', as illustrated in Figure 8.4. Loosemore et al. (2006) have suggested that in practice, objectives and outputs are usually defined independently at the project onset, with the transformational processes grafted into the project delivery programme, without consideration to any comprehensive review of the project environmental forces. The result normally is vast differences between the projections at the project conception stage and the actual at the project completion stage. Flyvbjerg et al. (2003) report that, in major transport infrastructure projects, the differences between estimated and actual investment costs ranges from 50 to 100%.

The direct reciprocal relationship existing between an objective and its expected output implies that a change in one variable will result in corresponding change in the other variable, as depicted in Figure 8.5. Failure to recognise this relationship poses a risk to project success.

Figure 8.4 Project delivery system.

Figure 8.5 The relationship between project objectives and output.

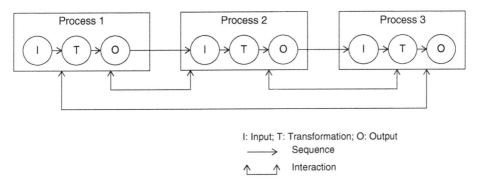

I: Input; T: Transformation; O: Output

———→   Sequence

↑   ↑   Interaction

Figure 8.6 Interaction between construction systems' transformational processes.

The changing nature of supply chain stakeholders' interests and preferences during the project delivery stages suggests that any revision to the expected project output without a corresponding revision in the project objectives, will result in variances between the initial projections and the actuals, thereby generating project risk.

Inherent in the output of each of the construction system's transformational processes are potential opportunities and risks, which are fed as inputs into the succeeding processes (see Figure 8.6). Where these processes are not coordinated, there is the risk of supply chain organisations and professionals involved in the succeeding processes failing to identify the potential risks and opportunities and designing inappropriate response strategies.

Loosemore et al. (2006) have argued that the segregation of the project delivery activities into phases, and the resultant exclusion of property management professionals such as facility managers from initial design development processes, usually results in the production of buildings that fail to meet the aspirations of the end users. According to Bea (1994), the bulk of the initial project design errors later mature into project risk events, with approximately 42% and 50% actualising during the construction and operational phases respectively.

8.5.2 Project Risk Events Generated through the Network and Interactions within Construction Supply Chain Relationships

Systems thinking confirms that every system has an objective that navigates its transformational processes (Walker 2007). The success of a system therefore depends on the integrative working of the parts (Arthur and Pryke 2013). Any event that interferes with the collaborative working of the components has the potential to affect the transformational processes and ultimately hinder achievement of the system's objectives. Similarly, the success of a construction supply chain relationship will greatly depend on the integrative working of the networks of professionals and contracting organisations involved in the project production processes.

Carmichael (2006) has also argued that systems exist within their broader environment as subsystems interacting with other subsystems to form a network that defines the main system's boundaries. Applying the principles of systems thinking enables further analysis of the components within subsystems; subsystems reflect different micro objectives. The failure to identify and integrate these micro objectives has the potential to affect the main system's transformational processes.

The analysis described previously would suggest that the differentiation in micro objectives between the networks of professionals and contracting organisations involved in construction supply chain relationships will result in differences in impact from environmental forces, leading to difference in risk interpretation and response (Arthur and Pryke 2013; Arthur 2018). Therefore, in a situation where the different professionals and organisations adopt different risk management systems, the variations in strategies may lead to cross-purpose working practices which could affect the project transformational processes (Pryke and Smyth 2006; Cagno et al. 2007). Even in a state of an established joint risk management system, the failure to identify and integrate the different micro objectives could introduce additional risk events to affect the integration of the risk management decision processing subsystem (Mullins 2005).

8.6   Risk Management Decision-Making Systems in Construction Supply Chain Relationships

Following on from the discussions on the structure and processes of the risk management system within construction supply chain relationships, we now evaluate the behavioural patterns underlying the associated decision-making processes. Behavioural scientists believe that there are two systems of thinking and decision making; the intuitive system guided by personal perceptions and qualitative decision-processing techniques, and the rational system which operates based on logic and cognitive analysis (Kahneman 2011). According to Kahneman (2011), the intuitive system is always active, and effective in the processing of risk events associated with reflex actions and emotions, but ineffectual with logical risk management decision settings. The rational system, on the other hand, requires conscious effort to activate, through recall of relevant cognitive analytical programmes, and decomposition of decision task into structured sequential order. This invariably makes the rational system suitable for logical risk management decision-making settings, but ineffective with the processing of sentimental risk events. Decision processing usually commences in the intuitive system, and then rolls onto the rational system when there is the need for further cognitive analysis.

Inbar et al. (2010) have suggested that the key features of a decision task and its environmental setting are what determines the appropriate decision-processing system. Decision tasks capable of being simplified into organised sequential patterns are most suited for rational processing, whereas fragmented decision tasks tend to thrive under the intuitive decision processing system. In light of construction supply chain risk management decision tasks exhibiting evidence of both structured order, and fragmentation, the logical expectation would have been the application of an equal degree of rationality and intuition. The empirical evidence, however, confirms minimal application of rational risk management decision-making systems. Analysis of the findings from a questionnaire survey of 100 general contractors and project management organisations in the UK identified low competency in the application of rational risk management decision-making techniques (Akintoye and Macleod 1997). A review of the findings of another study involving senior managers drawn from the Queensland engineering industry in Australia likewise identified low levels of training and application of rational risk management tools (Lyons and Skitmore 2004). The evidence from a similar study conducted in Malawi also identified limited application of structured risk management techniques amongst the large-scale construction organisations, with the risk management decision processing of the medium and small-scale construction organisations dominated by intuitive practices (Kululanga and Kuotcha 2010).

The prevailing theoretical evidence suggests that intuitive decision making relies on personal perceptions acquired through physiological make-up, social conditioning, and environmental exposures (Benthin et al. 1993; Maytorena et al. 2007; Slovic et al. 2010). As we experience live activities, our environmental exposures are stored in our brains in the form of perceptual images (Damasio 2006) relating to the human senses. Attached to the sensory files are behavioural stimulus corresponding to the feeling state experienced at the time of the recorded exposure, termed 'affective heuristics' (Slovic et al. 2010). Intuitive decision-making processing of a risk event commences with the consultation of the human image library for evidence of available sensory files with exhibits similar to the risk decision setting. Where a related sensory file is identified, the associated affective heuristics induces a corresponding physiological state which inspires the modelling of the risk decision event outcome to reflect the pattern of the related previous exposure (Slovic et al. 2010). The absence of a related sensory file impedes intuitive risk management decision processing, leading to recall of the rational decision-making system for structured analysis (Arthur and Pryke 2013; Arthur 2018).

This can be illustrated hypothetically by evaluating the possible outcomes of an intuitive risk analysis of a thermal insulation material by a project team drawn from diverse professional and organisational backgrounds (see Figure 8.7).

The specialists with previous positive experience of the thermal insulation material are likely to model the project outcome in patterns reflecting the key attributes of their previous related project experiences, and consequently classifying the application of the material as a positive risk event. The specialists with previous negative experience of the thermal insulation material, such as the 2017 Grenfell fire disaster, are likely to model the project outcome to reflect patterns similar to their previous related project experiences, leading to the classification of the application of the material as a fire risk event. The other specialists without previous working experience or technical knowledge of the fire risk associated with thermal insulation material are likely to exhibit psychological difficulties in their intuitive risk evaluation of the thermal material. Consequently, they are likely to resort to deliberate activation of their cognitive reasoning system for analytical risk evaluation of the thermal insulation material.

The differences in personal exposures confirms difference in personal affective heuristics, and hence differences in perception generation in intuitive risk management decision making, as illustrated. The variability in personal perceptions inspired by the differences in personal exposures confirms the need for care in the formation of construction project teams, by ensuring compatibility between project specific contexts

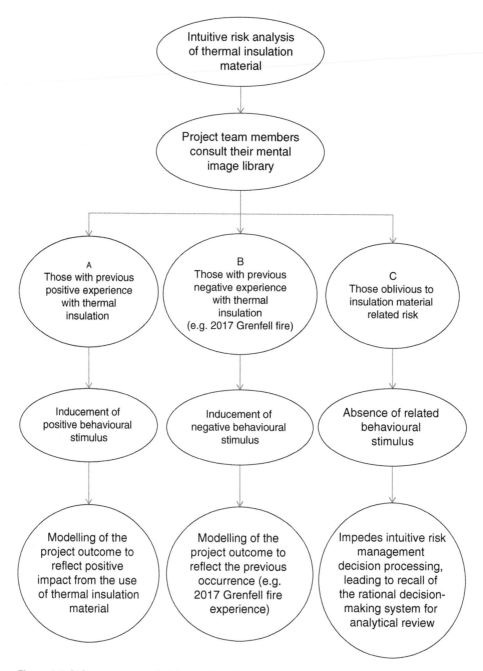

Figure 8.7 Risk management decision making. Source: Original.

and the technical and behavioural backgrounds of assembled project coalitions. There is also the need for a wide range involvement of specialists from diverse technical backgrounds to facilitate effective intuitive risk management systems (Arthur and Pryke 2013; Arthur 2018). The theoretical evidence described in summary gives credence to recognising behavioural differences whenever an intuitive risk management system is applied within a construction supply chain relationship.

8.7   Conclusion

This chapter has explored the risk management systems within construction supply chain relationships. The discussions have centred on the processes applied in identifying and managing risk events generated through the linkages, networks, and interactions among the professionals and organisations within construction supply chain relationships. The deliberations were structured under four themes covering the evolution of the UK construction industry and supply chain relationships: the concept of risk; the construction risk management system; risk generation in construction supply chain relationships; and risk management decision-making systems in construction supply chain relationships.

The transformation of the UK construction industry from the era of sole master mason profession to the contemporary state of multiple differentiated organisations and professional disciplines (Walker 2007) confirms the influence of environmental forces in creating emerging professional disciplines, and also subdividing the established disciplines into multiple differentiated specialist roles. The differentiation in organisational structure has led to a situation where most firms now specialise in aspects of the construction processes and rely on contributions from other firms for their project execution (Vrijhoef and Koskela 2000; Edkins 2009). This has further inspired the adoption of relational project management approaches including supply chain management.

Construction supply chain management originated from the manufacturing industry where it was first applied as a management tool for promoting efficiency in production and supply logistics (Vrijhoef and Koskela 2000). The introduction of the concept into the construction industry was expected to generate similar levels of efficiency, but the practical evidence suggests that the perceived benefits are yet to be fully realised (Aloini et al. 2012). This has been attributed to the lack of holistic concept application (Aloini et al. 2012), the variances between conceptual rhetoric and reality, and the lack of deeper understanding of the multiplicity of risk communications within construction supply chain relationships.

The recent surge in risk discussions has been credited to the rise in media reportage on disasters, which has increased the public's awareness of the impact of risky activities (Loosemore et al. 2006). Zinn (2008) has described six different conceptual interpretations of risk including: real and objective; socially mediated; real and socially constructed; socially transformed; socially constructed; and subjectively biased. Further analysis of the concept of risk using a working definition by Loosemore et al. (2006) suggests the following attributes: risk is an uncertainty which has been identified to be managed (Edkins 2009); risk can be expressed in probability format to predict the degree of likelihood of a potential future event becoming actualised (Aven 2004; Adams 2008; Bateman et al. 2010); risk relates to an occurrence in the future (Loosemore et al. 2006);

the establishment of a risk event requires prior establishment of potential impact on an objective (Loosemore et al. 2006).

The approaches to risk management can be categorised into two main ideological perspectives: the *homeostatic* viewpoint which proposes rationality, and the *callibrationist* standpoint which advances human subjectivity in risk identification and treatment (Hood and Jones 1996; Loosemore 2006). Theoretical analysis of the construction risk management processes using the concept of systems decomposition (Carmichael 2006) has revealed three interconnected cyclical parts, beginning with the risk identification subsystem where potential risk events are identified and allocated, mainly through contracts. The next stage is the risk analysis subsystem which evaluates the reliability of the identified risk events becoming real. The last stage is the risk response subsystem where the decision is made whether to accept, reject, delay, or mitigate the impact of the established risk events.

Systems analysis of the internal structure of the construction risk management system confirms the relevance of integrative working practices (Walker 2007). Any event that breaks the chain of coordination amongst a system's components has the potential to generate a risk event that could consequently impede the efficiency of the product transformational processes. Similarly, the failure to synchronise the dynamism in construction supply chain stakeholders' interests with the related project's specifications at the various execution stages may result in the production of a building that is not fit for purpose. Also, the failure to coordinate the activities of the different supply chain management stages may lead to the transfer of a risk event from an initial product transformational stage to the subsequent stages, without corresponding appropriate risk response strategies. The micro differences in objectives of the different professional disciplines and organisations involved in construction supply chain relationships, and the associated multiplicity in risk communication, also suggest that the application of separate risk management systems may result in divergent strategies leading to cross purpose working practices (Pryke and Smyth 2006; Cagno et al. 2007). Even in a state of an established joint risk management system, the failure fully to identify and integrate the diverse micro objectives could generate additional risk events to affect the integration of the risk management decision processing subsystem (Mullins 2005).

The selection of an appropriate risk management decision-making system is defined by the applied management ideological perspective (Inbar et al. 2010). Decision settings capable of being decomposed into structured patterns are most suited for the rational system, while disjointed and subjective decision settings thrive under intuition (Kahneman 2011). Construction supply chain decision tasks involving both structured and subjective decision settings would have suggested an equal measure of rationality and intuition in the risk management decision processing systems. The empirical evidence, however, reveals a high incidence of intuition (Akintoye and Macleod 1997; Lyons and Skitmore 2004; Kululanga and Kuotcha 2010).

The science of intuitive decision processing confirms that the thought generation is inspired by personal perceptions acquired through physiological make-up, social conditioning, and environmental exposures (Benthin et al. 1993; Maytorena et al. 2007; Slovic et al. 2010), which are stored in the human image library in the form of perceptual images (Damasio 2006). Attached to these sensory files are affective heuristics corresponding to the emotional state experienced at the time of the related exposure (Slovic 2010). Intuitive decision processing of a risk decision task begins with the scanning of the human image library for evidence of available similar sensory files. The presence of a related record influences the modelling of the risk decision task to reflect the patterns of the previous exposure, and subsequently classification as either positive or negative risk event. The absence of a related record, however, impedes progression of the intuitive decision processing, leading to recall of the rational system for structured analysis.

The differences in physiological make-up, social conditioning, and environmental exposures of the different professionals and organisations involved in construction supply chain relationships confirms differences in storage and recall of affective heuristics and, therefore, differences in personal perceptions, during intuitive risk management decision processing. The variability in personal perceptions in effect suggests the need for prudence in the formation of construction supply chain relationships, by ensuring compatibility between the project specific context, and the technical and behavioural backgrounds of the assembled project coalition. The theoretical evidence on the difficulty in risk perception generation without having related previous exposure also suggests the necessity for wide range involvement of specialists from diverse technical backgrounds to facilitate robust intuitive risk identification and treatment (Arthur and Pryke 2013; Arthur 2018). Bringing the above together confirms the need for constant harmonisation of both the behavioural and technical differences of the professionals and organisations involved in construction supply chain relationships to ensure integrative risk management processes.

References

Adams, F.K. (2008). Risk perception and Bayesian analysis of international construction contract risks: the case of payment delays in a developing economy. *International Journal of Project Management* 26: 138–148.

Akintoye, A.S. and Macleod, M.J. (1997). Risk analysis and management in construction. *International Journal of Project Management* 15 (1): 31–38.

Allinson, K. (2008). *Architects and Architectures of London*. Oxford: Elsevier.

Aloini, D., Dulmin, R., Mininno, V., and Ponticelli, S. (2012). Supply chain management: a review of implementation risks in the construction industry. *Business Process Management Journal* 18 (5): 735–761.

Arthur, A. C., (2017) Is It Time to Learn the Lessons? Fire Incidence at Atomic Junction. *Ghanaian Daily Graphic*, October 16, 2017. No. 20510, 7.

Arthur, A.C. (2018). Do construction project managers need training in Behavioural sciences to effectively manage project risk? In: *Proceedings of the Construction, Building and Real Estate Research Conference of the Royal Institution of Chartered Surveyors, London, April 2018*. London: RICS.

Arthur, A.C. and Pryke, S.D. (2013). Instinctive thinking; analysis of construction risk management systems. In: *Proceedings of the Construction, Building and Real Estate Research Conference of the Royal Institution of Chartered Surveyors, New Delhi, September 2013*. London: RICS.

Arthur, A.C. and Pryke, S.D. (2014). Are we adding risk to our projects by mixing objective assessments of compound conjunctive and disjunctive project risks with intuitive approaches? In: *Proceedings of the 30th Annual ARCOM Conference, 1–3 September*

2014 (eds. A.B. Raiden and E. Aboagye-Nimo), 1399–1408. Portsmouth, UK: Association of Researchers in Construction Management.

Aven, T. (2004). On how to approach risk and uncertainty to support decision making. *Risk Management* 6 (4): 22–39.

Aven, T. (2010). *Misconceptions of Risk*. Chichester: Wiley.

Bateman, I., Dent, S., Peters, E. et al. (2010). The affect heuristic and the attractiveness of simple gambles. In: *The Feeling of Risk, New Perspective on Risk* (ed. P. Slovic), 3–19. London: Earthscan Ltd.

Bea, R.G. (1994). *The Role of Human Error in Design, Construction, and Reliability of Marine Structures*, Ship Structure Committee SSC-378,. Washington, DC: U.S. Coastguard.

Beck, U. (2002). The terrorist threat, world risk society revisited. *Theory, Culture and Society. Sage* 19 (4): 39–55.

Beck, U. (2007). *World at Risk*. Cambridge: Polity Press.

Benthin, A., Slovic, P., and Severson, H. (1993). A psychometric study of adolescent risk perception. *Journal of Adolescence* 16: 153–168.

Bertalanffy, L.V. (1968). *General System Theory, Foundations Development Applications*. Norwich: Fletcher and Son Ltd.

Bowley, M. (1966). *The British Building Industry*. Cambridge: Cambridge University Press.

BREEAM (2018). *What is BREEAM?* (online). Available from http://www.breeam.com (accessed August 2019).

Cagno, E., Caron, F., and Mancini, M. (2007). A multi-dimensional analysis of major risks in complex projects. *Risk Management* 9 (1): 1–18.

Carmichael, D.G. (2006). *Project Planning, and Control*. Oxon: Taylor & Francis.

CBI (2010). *Procuring in a Downturn*. CBI.

Checkland, P. (1999). *Systems Thinking, Systems Practice*. Chichester: Wiley.

Cole, G.A. (2004). *Management Theory and Practice*. Singapore: Seng Lee Press. 78910-11 10 09, 27.

Dallas, M. (2006). *Value and Risk Management, a Guide to Best Practice*. Oxford: Blackwell Publishing Ltd.

Damasio, A. (2006). *Descartes' Error, Emotion, Reason and the Human Brain*. London: Vintage Books.

Derry, T.K. and Williams, T.I. (1960). *A Short History of Technology, From The Earliest Times to A.D. 1900*. Oxford University Press.

Edkins, A. (2009). Risk management and the supply chain. In: *Construction Supply Chain Management, Concepts and Case Studies* (ed. S. Pryke), 115–136. Oxford: Blackwell Publishing.

Egan, S.J. (1998). *Rethinking Construction–The Report of the Construction Task Force*. London: Department of Trade and Industry.

Flyvbjerg, B., Bruzelius, N., and Rothengatter, W. (2003). *Megaprojects and Risk, an Anatomy of Ambition*. Cambridge: Cambridge University Press.

GreenBook Live (2018) *A-Z of Green Book Live Schemes* (online). Available from http://www.greenbooklive.com/atoz.jsp (accessed August 2019).

Guardian News and Media Limited, (2011) *FCO Travel Advice Mapped: The World According to Britain's Diplomats* (online). Available from www.guardian.co.uk/news/datablog/2011/mar/23/fco-travel-advice-map (accessed August 2019).

Hood, C. and Jones, D.K.C. (1996). *Accident and Design: Contemporary Debates in Risk Management*. Abingdon: UCL Press.

Inbar, Y., Cone, J., and Gilovich, T. (2010). People's intuitions about intuitive insight and intuitive choice. *Journal of Personality and Social Psychology* 99 (2): 232–247.

Japp, K. and Kusche, I. (2008). Systems theory and risk. In: *Social Theories of Risk and Uncertainty: An Introduction* (ed. J.O. Zinn), 52–75. Oxford: Blackwell Publishing.

Kahneman, D. (2011). *Thinking Fast and Slow*. London: Penguin Books Ltd.

Kahraman, C. and Oztaysi, B. (2014). Supply chain management under fuzziness. *Recent Developments and Techniques* 313 Springer. ISSN 1860-0808.

Khalfan, M.M.A., Kashyap, M., Li, X., and Abbott, C. (2010). *Knowledge Management in Construction Supply Chain Integration. International Journal of Networking and Virtual Organisations* 7 (2/3): 207–221.

Kirkup, J., and Barrett, D. (2014) *Bulgarian and Romanian arrivals down after open door as Tories admit missing target for net migration (online)*. Available from www.telegraph.co.uk/news/uknews/immigration/10832220/Bulgarian-and-Romanian-migrants-down-after-open-door-as-Tories-admit-missing-target-for-net-migration.html (accessed on August 2019).

Kululanga, G. and Kuotcha, W. (2010). Measuring project risk management process for construction contractors with statement indicators linked to numerical scores. *Engineering, Construction and Architectural Management* 17 (4): 336–351.

Latham, S.M. (1994) *Constructing the Team*: Joint Review of Procurement and Contractual Arrangements in the United Kingdom Construction Industry, Final Report., London: HMSO. ISBN 0–11-752994-X.

Lock, D. (2003). *Project Management*, 8e. Hampshire: Gower Publishing Limited.

Loosemore, M. (2006). Managing project risks. In: *The Management of Complex Projects, a Relationship Approach* (eds. Pryke and Smyth), 131–146. Oxford: Blackwell Publishing.

Loosemore, M., Raftery, J., Reilly, C., and Higgon, D. (2006). *Risk Management in Projects*. Oxon: Taylor & Francis.

Lyons, T. and Skitmore, M. (2004). *Project risk management in the Queensland engineering construction industry: a survey. International Journal of Project Management* 22 (2004): 51–61.

Maytorena, E., Winch, G.M., Freeman, J., and Kiely, T. (2007). *The influence of experience and information search styles on project risk identification performance. Transactions on Engineering Management, IEEE Transactions on* 54 (2): 315–326.

Money.co.uk, (2011) *Which Companies are Benefiting from the Credit Crunch* (online). Available from www.money.co.uk/article/1001644-which-companies-are-benefiting-from-the-credit-crunch.htm (accessed August 2011).

Morledge, A., Knight, A., and Grada, M. (2009). The concept and development of supply chain management in the UK construction industry. In: *Construction Supply Chain Management, Concepts and Case Studies* (ed. S. Pryke), 115–136. Oxford: Blackwell Publishing.

Mullins, L.J. (2005). *Management and Organisational Behaviour*, 7e. Harlow: Prentice Hall.

Mustafa, M.A. and Al-Bahar, J.F. (1991). Project risk assessment using the analytic hierarchy process. *IEEE Transactions on Engineering Management* 38 (1): 46–52.

NHS, (2011) *Deadly E. coli Outbreak hits Germany* (online). Available from http://www.nhs.uk/news/2011/05May/Pages/cucumbers-german-e-coli-infections.aspx (accessed August 2019).

Office for National Statistics (2017) *Construction statistics:* Number 18, 2017 edition (online). Available from www.ons.gov.uk/businessindustryandtrade/constructionindustry/articles/constructionstatistics/number182017edition#toc (accessed August 2019).

O'Malley, P. (2008). Governmentality and risk. In: *Social Theories of Risk and Uncertainty: An Introduction* (ed. J.O. Zinn), 52–75. Oxford: Blackwell Publishing.

Ponniah, K., (2017) *How are fires fought in high rise blocks?* BBC News online, 4 August 2017. Available from www.bbc.co.uk/news/uk-england-london-40273714 (accessed August 2019).

Pryke, S. (2009). *Construction Supply Chain Management Concepts and Case Studies – Introduction*. Wiley-Blackwell.

Pryke, S. and Smyth, H. (2006). *The Management of Complex Projects, a Relationship Approach*. Oxford: Blackwell Publishing Ltd.

Renn, O. (1992). Concepts of risk: a classification. In: *Social Theories of Risk* (eds. S. Krimsky and D. Golding), 53–79. Westport: Praeger Publishers.

RICS (2018a). *Who We Are, And What We Do, History* (online). Available from https://www.rics.org/uk/about-rics/ (accessed August 2019).

RICS (2018b). *Your Dashboard* (online). Available from https://www.rics.org/uk/knowledge/latest-on (accessed August 2019).

RICS (2018c). *BIM Manager Certification* (online). Available from http://www.rics.org/uk/join/member-accreditations-list/bim-manager-certification (accessed August 2019).

Robinson, B. (2011) *London's Burning: The Great Fire* (online). BBC online. Available from www.bbc.co.uk/history/british/civil_war_revolution/great_fire_01.shtml (accessed August 2019).

Schieg, M. (2006). Risk Management in Construction Project Management. *Journal of Business Economies and Management* Vii (2): 77–83.

Slovic, P. (2010). *The Feeling of Risk, New Perspective on Risk Perception*. London: Earthscan Ltd.

Slovic, P., Finucane, M.L., Peters, E., and MacGregor, D.G. (2010). Risk as analysis and risk as feelings: some thoughts about affect, reason, risk and rationality. In: *The Feeling of Risk, New Perspective on Risk* (ed. P. Slovic), 21–36. London: Earthscan Ltd.

Smith, N.J., Merna, T., and Jobling, P. (1999). *Managing Risk in Construction Projects*. Oxford: Blackwell Science.

Smith, N.J., Merna, T., and Jobling, P. (2006). *Managing Risk in Construction Projects*, 2e. Oxford: Blackwell Publishing.

Taylor-Gooby, P. and Zinn, J.O. (2006). Current directions in risk research: new developments in psychology and sociology. *Risk Analysis* 26: 2.

The Economic Times, (2011) *"Blackberry riots" in Britain, Addition to Smart phones* (online). Available from http://economictimes.indiatimes.com/news/international-business/blackberry-riots-in-britain-addiction-to-smartphones/articleshow/9601104.cms (accessed August 2019).

The Independent (2011). *Five Year Old Girl Killed by Automatic Gate* (online). Available from www.independent.co.uk/news/uk/home-news/fiveyearold-girl-killed-by-automatic-gate-2018572.html (accessed August 2019).

Tulloch, J. (2008). Culture and risk. In: *Social Theories of Risk and Uncertainty: An Introduction* (ed. J.O. Zinn), 52–75. Oxford: Blackwell Publishing.

Tversky, A. and Kahneman, D. (1982). Judgement under uncertainty: heuristics and biases. In: *Judgement under Uncertainty: Heuristics and Biases* (eds. D. Kahneman, P. Slovic and A. Tversky), 3–20. Cambridge: Cambridge University Press.

Vrijhoef, R. and Koskela, L. (2000). The four roles of supply chain management in construction. *European Journal of Purchasing & Supply Management* 6: 169–179.

Walker, A. (2007). *Project Management in Construction*. Oxford: Blackwell Publishing Ltd.

Winch, G.M. (2010). *Managing Construction Projects*, 2e. Chichester: Blackwell Publishing Ltd.

Zinn, J.O. (2008). *Social Theories of Risk and Uncertainty: An Introduction*. Oxford: Blackwell Publishing.

9

Culture in Supply Chains

Richard Fellows and Anita Liu

9.1 Introduction – Context

'Essentially, business is about appropriating value for oneself … only by having the ability to appropriate value from relationships with others … can business be sustained … [there] … must … be conflicts of interest between vertical participants in supply chains, just as there are between those competing horizontally … In Western … culture most suppliers are basically opportunistic' (Cox 1999, p. 171). Most clients, consultants, and constructors are businesses and, consequently, operate under business performance imperatives. The businesses are subject to regulation and, particularly for (design) consultants, need to behave professionally and on a moral/ethical basis for perceived social good.

While the concepts of competition are applied to horizontal situations (competitors in similar positions within other supply chains) most commonly, vertical competition between members of a supply chain (those at different positions, or tiers, within the supply chain) may have much greater consequences for the effectiveness of the output and the efficiency of the supply processes. Indeed, an important trend is that '…competition is becoming less "firm vs. firm" and more "supply chain vs. supply chain"' (Hult et al. 2007, p. 1047).

A systems model of the typical realisation process of a construction project is shown in Figure 9.1. The larger and the more complex the project, the greater is the number and diversity of specialist participants in each of the functional categories. Internationalisation and globalisation compound the ensuing difficulties. Usually, there is a great diversity of participants and specialist functions. Therefore, it is difficult, if not impossible, to identify all the participants at the early stages of a project, prompting assumptions of identities and interests of parties not involved at this point; it can be very tempting, for simplicity and convenience, to ignore such parties!

Strict application of the concept of supply chains may give a somewhat limited perspective of the realities of project realisation, occupation and use, and disposal processes. This is particularly important as perspectives move from linear (a single project) to circular (cycling through many projects – for sustainability). Focusing on project realisation, it seems more appropriate to examine networks of potential participants, leading to integrations of chains of suppliers for actual realisation processes, preferably

Successful Construction Supply Chain Management: Concepts and Case Studies, Second Edition.
Edited by Stephen Pryke.
© 2020 John Wiley & Sons Ltd. Published 2020 by John Wiley & Sons Ltd.

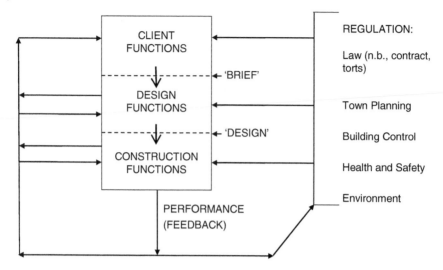

Figure 9.1 The project realisation process. *Note*: (i) Performance leads to satisfaction of participants via perspectives of project success (success is achievement of specified performance targets, as a minimum). (ii) Performance-Satisfaction-Success also produces feedforward in the 'cycling' of project data and information to aid realisations of future projects through participants' perception-memory-recall filtering ('experiences').

also incorporating 'client' purpose-orientated perspectives. This encourages use of the concept of 'value-added' by the supply chain (network) members. Some chains operate in series, others in parallel.

Supply networks operate in two primary contexts: firstly, as horizontal arrays of organisations from which particular organisations may be selected to participate in the realisation of a project, depending on factors such as expertise, resources, workload, and price; secondly, the selected participants then form a vertically orientated network of organisations with their integration aimed at realising the project effectively and efficiently. In these contexts, much of the research and literature has focused on the realisation processes with the objective of enhancing *project management performance* (time, cost, and quality measures) (see, e.g. Jin et al. 2007); the design, briefing, and buildability/constructability literature adopts a focus more towards *project performance* (functional effectiveness of the project in occupation and use) (see, e.g. Green and Simister 1999; Griffith and Sidwell 1995).

Thus, the construction project business world emphasises operating/performance imperatives relating to individual organisations – the links in the construction supply chain/network. However, the holistic performance perspectives (which are becoming ever more widespread), in order to yield successful projects with satisfied participants, need to emphasise integration of the supply network members. A key constituent is recognition of interdependencies – pooled, sequential, reciprocal (see Thompson 1967; Lazzarini et al. 2001).

Thorelli (1986) suggests that networks lie between markets and hierarchies, as quasi-firms or multiorganisations (see, Eccles 1981). This is part of a contingent perspective on firms' macrobehaviour, in that they operate in complex environments in which firms' behaviours are interdependent (Aldrich 1979); the degree of interdependence is contingent on the nature of the processes and the environment. Thus, competing becomes an issue of a firm locating itself within the network, as in a Cournot equilibrium under oligopoly – i.e. that a firm endeavours to maximise its profit, given the outputs of the other, competing firms.

Cox (1999) distinguishes supply chains, which deliver physical goods and services to customers, from value chains, which relate to the revenue streams from customers and so operate in the reverse direction. Such a perspective limits conceptualisation and analysis due to its market price orientation. While appreciation of the flows of goods/services and of the revenues generated from them is useful, understanding the fundamental causations of the flows, and amounts of those flows, is vital for systemic improvements.

The usual objective of business transactions is increasing the wealth of the participants. The chaining and networking concepts provide the linkages between the individual participants – why and how they relate to each other for realisation of their objectives. Thus, participants need to '… understand the physical resources that are required within a supply chain to create and deliver a finished product or service to a customer… understand the exchange relationship between particular supply chain resources and the flow of revenue in the value chain… understand what it is about the ownership and control of particular supply chain resources to command more of the flow of value than others' (Cox 1999, p. 174). Such appropriation of value depends, to a major extent, on the power and relational structures of the supply chain.

Usually, wealth is measured in terms of financially realisable assets. This objective is the underpinning of market capitalism, the generic economic system of most societies. However, the world retains a rich mix of 'not-for-profit organisations' which, although requiring financial break-even performance, measure wealth in various ways and, often, for several stakeholder groups. Such a situation is also occurring amongst for-profit organisations, which embrace an expanding array of nonfinancial performance measures to supplement the financial ones.

What such contextual analyses indicate is that the organisations are operating in a world of increasing pluralism and, hence, complexity which is driven by, and in turn drives, culture and cultural change. Wealth is determined by what people value and thus the pluralistic approach to wealth reflects the diversity of people's values. These values are articulated through the manifestations of culture by governing how people behave and conduct relationships and this has consequences for performance of individual and collective activities (such as project realisation).

Thus, this chapter addresses the nature of culture, both national and organisational, together with derivatives of organisational climate and the concerns of behaviour modification, organisational citizenship, and corporate social responsibility (CSR). Consideration of ethics addresses the moral bases which impact on human behaviour. Relational issues of team formation and functioning, and alliances between businesses are examined as human and organisational contexts through which supply chain participants operate and projects are realised.

Typically, culturally orientated analyses have been used to determine the appropriate dimensions for study and to detect typical behaviours and their underpinning causes (values and beliefs – see later). These studies have facilitated further work to examine interrelationships between human groups (nations, organisations, etc.), commonly featuring the examination of cultural differences and/or measures of cultural distance. Such examinations have been applied to strategic alliances between organisations and consequently are of direct importance to the inevitable joint-venture nature of construction supply chains (whether for domestic or international projects). Inevitably, analyses have concentrated on formal organisations and alliances when by far the more common encounters within construction are informal.

Thus, in construction supply chains/networks, two organisations impact on each participant directly – the 'permanent' organisation which employs that person and the project temporary multiorganisation to which that person is attached to fulfil a supply chain role. Of course, the diversity of construction project members and stakeholders (both individuals and organisations) with which any particular member of the project supply chain must interact lends further complexity to the appreciation of behavioural imperatives of the supply processes to secure successful performance; this requires an understanding of and sensitivity to diverse, often competing, interests so that they can be accommodated.

The argument advanced here is that only through appreciation of the requirements of others, as well as of self, can fragmented supply chains (individual objectives) be integrated appropriately to secure both individual and holistic successful performance of both the project management process (realisation) and the project (product in use). This appreciation relates to own and others' cultures in terms both of the manifestations of the cultures (language, behaviour, etc.) and the underpinning/determining variables and constructs (values, beliefs). Hence, this appreciation should include organisational climates and project 'atmospheres', how meaning is determined and how understanding arises, what drives behaviour, and how behaviour may be changed together with the likely outcomes of any change initiatives.

Such appreciation is essential for management of construction supply chains/networks due to the wide array of diverse stakeholders and the labour-intensive nature of the processes. Intraorganisational management, interorganisational management, and boundary management are intensified in the fluid, power-based temporary multiorganisations which are the norm for any construction project.

9.2 Culture

Culture pervades all our lives: we experience and 'do' culture every moment; culture conditions our behaviour and, in turn, our behaviour modifies culture. Culture determines how we communicate, how we relate to other people, how we regard property, our interaction with the environment, and our perspectives of time. The buildings we construct are potent symbols of culture. However, the study of culture, especially relating to the construction industry, is nascent, and so cultural processes and the consequences of or on culture are fundamental but, as yet, not fully understood.

Initially, culture may be described as 'how we do things around here' (Schneider 2000, p. 26). Of course, much more is involved: culture is not merely *how* things are done – the scope is much more extensive and includes *what* is done, *why* things are done, *when*, and *by whom* . . . . However, the description does have a behavioural focus and so draws attention to that primary manifestation of culture.

Kroeber and Kluckhohn (1952: p. 181) define culture as '... patterns, explicit and implicit of and for human behaviour acquired and transmitted by symbols, constituting the distinctive achievements of human groups, including their embodiment in artefacts;

the essential core of culture consists of traditional (i.e. historically derived and selected) ideas and, especially, their attached values; culture systems may, on the one hand, be considered as products of action, on the other as conditioning elements of future action'. Culture is a collective construct which concerns groups of people. Furthermore, culture is iteratively dynamic – culture shapes behaviour and, in turn, behaviour shapes culture; development spirals through time.

Hatch (1993) advances a model of cultural dynamics which encapsulates the cyclical processes of manifestation, realisation, symbolisation, and interpretation. The dynamism arises from the continual construction and reconstruction of culture as the context for setting goals, taking action, making meaning, constructing images, and forming identities. In construction, a vital consideration is the impact of culture on what performance is achieved and measured against predetermined, culturally bound, goals and targets.

Hofstede (1994a: p. 1) defines culture as '... the collective programming of the mind which distinguishes one category of people from another'. This definition suggests that culture is learned, rather than being innate in the person or genetic. It is inherited behaviourally through replicating and responding to the behaviour of others, most importantly close family and contacts during early life. Furthermore, as culture is a collective construct, categorisation of people may be by ethnic origin, political nationality, organisation, etc. (the important aspect of such categorisation is that 'within category' variability is significantly less than 'between category' variability; although, for certain categorisations in practice – notably, nation-states – within category variation may be large). This is very important in examining similarities and differences between cultures – where the boundaries are drawn, what dimensions are considered, and how measurements are made and used (absolute/relative).

Schein (1990) views culture as grounded in basic assumptions which constitute communal values and are 'taken for granted'. Cultures arise through the formation of norms of behaviour relating to critical incidents which are communicated through stories passed on between members of the community, as well as through identification with leaders and what they scrutinise, measure, and control.

Hence, culture seems to operate, and be modelled, as a system of meaning that is shared by members of a group. This incorporates how meaning is established (notably, via sensemaking – e.g. Weick 2009; Fellows and Liu 2016), how meanings are communicated (languages, assumptions, norms), and what actions ensue (behaviour, motivations, reasoning, and decisions).

Many models of culture employ vertical analyses: physiological instincts and beliefs at the core (survival imperatives; religion, morality, etc.); values as the intermediate layer (the hierarchical ordering of beliefs, perhaps with possible trade-offs); and behaviour at the outer layer (as in language, symbols, heroes, practices, artefacts), as in Figure 9.2.

Culture may also be analysed horizontally, which yields categories of national culture, organisation culture, organisation climate, project/workplace atmosphere, and behaviour of people (see Figure 9.3). Due to interrelationships (caused by people fulfilling different roles), the categorical boundaries are quite fuzzy. Individuals, at any one time, may be of a certain ethnicity (e.g. Chinese), live in a nation (e.g. Australia), work in a particular organisation (e.g. Bechtel), belong to a 'social interest group' (e.g. Greenpeace), and show certain behaviour (e.g. organisational citizenship behaviour, OCB). Clearly, situational variables impact on behaviour, promoting a contingency

Behaviour, Heroes, Symbols, Artifacts, Language, etc.

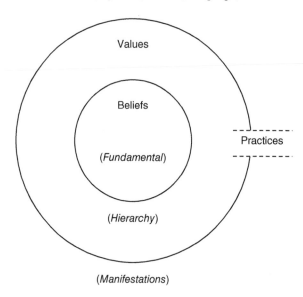

Figure 9.2 Layers of culture. Source: Derived from Hofstede (1980, 2001). Note: Schein (2004) considers levels of culture to be artefacts, espoused beliefs and values, and underlying assumptions (as the deepest level). '…beliefs are statements about reality that individuals accept as true, values are generalized principles of behavior to which people feel strong positive or negative emotional commitment, and norms are shared rules or standards regarding the extent to which specific behaviors are to be considered socially acceptable…' (Wagner and Moch 1986, p. 286).

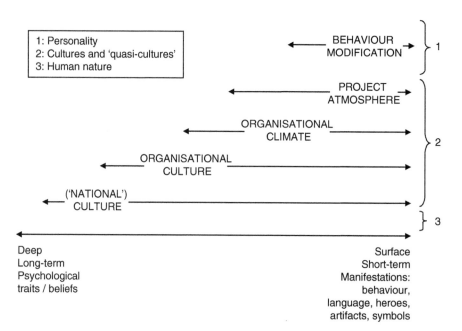

Figure 9.3 Culture spectrum. Source: Developed from Fellows 2006. Note: Boundaries between cultures, climate, atmosphere, and behavioural modification are fuzzy.

perspective. The complexity has prompted a variety of approaches for determining and analysing the categories. One consequence is a common confusion of the categories when 'managers' wish to effect change – in particular what has changed, how, why, and the permanence of any change – culture cannot be changed by use of a '40-hour workshop' (although behaviour modifications may result).

Thus, the concept of a 'project culture' is a misconception because cultures can develop and change only in the long-term (far greater than the life of all but the longest duration megaprojects). However, the concept of 'project atmosphere' (derived from 'workplace atmosphere' – Hovmark and Nordqvist 1996) may be useful – an amalgam of the organisational cultures of the (main) parent organisations, weighted by the power/influence of each (and their agents on the project) (see also Fellows and Liu 2013).

There is a fundamental philosophical difference in perspectives of culture that leads to diametrically opposed views for researching culture and for management of/in cultures. A positivist/functionalist perspective regards culture as '…something that an organization has…', whereas a social constructivist/interpretivist perspective regards culture as '…something the organization is…' (Smirchich 1983, p. 347). For positivists/functionalists, culture can be managed – especially made consistent and congruent with management's desires and, thereby, enhance performance (see Peters and Waterman 1982; Egan 1998). For constructivists/interpretivists, a holistic, circumspect view applies, so that 'The evolution of culture is shaped by agency and power, but cannot be created by fiat…' (Weeks and Gulunic 2003, p. 1315) and so, '…culture may be malleable albeit fraught with difficulties and ethical dilemmas…' (Harris and Ogbonna 2002, p. 34). Thus, programmes to change/manage (organisational) culture are likely to be resisted and yield somewhat different outcomes from those desired and intended. Furthermore, change initiatives must be long term and continuously reinforced to effect a change in culture (UK car seat belt legislation is a good example).

Culture is an all-pervading construct of human existence but its conceptualisation is contested. Hence, it is questionable to define or measure culture as different paradigms adopt radically different approaches. Emic approaches are, essentially, inward-looking and via a constructivist paradigm, assert that a culture can be investigated validly only from that culture's own perspective (idiographic). This leads to highly individual perspectives that foster deep understanding of a particular culture but do not facilitate comparisons across cultures. Etic approaches are concerned with an outside view, especially for crosscultural investigations, and so tend to adopt a positivist perspective using surveys, models, and dimensions (nomothetic) – such as in Hofstede's research. These studies tend to yield a small number of dimensions that are employed to depict individual cultures and facilitate (relative) comparisons.

9.3   Dimensions of Culture

In order to gain appreciation of culture and to understand similarities and differences between cultures, it is helpful to determine dimensions on which cultures may be 'measured' – most advisedly, employing relative positionings (Hofstede 2001). The dimensions of culture and related constructs constitute common bases for determining 'profiles' of cultures, to facilitate both appreciation of individual cultures, as well as comparisons of cultures (which, commonly, focus on differences). However, as Hofstede stresses, it is important that measurements on the dimensions are NOT treated as absolute in either individual culture or crosscultural studies; what such measurements do provide is useful, relative indicators.

In order to assemble a comprehensive perspective of culture, the levels of national culture, organisational culture, organisational climate, project/workplace atmosphere and behaviour modification are examined (i.e. progression from the more general to the particular, as in Figure 9.3). The principles and knowledge of cultures are examined to enable a supply chain manager to gain understanding of the manifestations likely to be encountered in project practice and to determine actions from an informed basis. Clearly, the number of possible combinations existing in the world is enormous!

9.3.1 National Culture

The notion of national culture is at once convenient and problematic; convenient as the boundary is clear (usually) and national statistics are available in most countries; problematic as borders change and nations may comprise many different ethnic and social groups of people. Thus, it is important to ensure that the use of national metrics is appropriate. At best, they provide useful indicators, while, at worst, such indicators may be meaningless! Thus, Au (2000) echoes Hofstede (1989) that within national boundaries, culture may have greater variability than across national boundaries. (Here, statistics of national means and standard deviations, measured using Hofstede's dimensions of national culture, are helpful – see, Hofstede 2001.)

Hofstede (1980) determines four dimensions for measuring national culture:

- *Power distance.* 'the extent to which the less powerful members of institutions and organizations within a country expect and accept that power is distributed unequally' (Hofstede 1994b, p. 28).
- *Individualism/collectivism.* 'individualism pertains to societies in which the ties between individuals are loose: everyone is expected to look after himself or herself and his or her immediate family. Collectivism as its opposite pertains to societies in which people from birth onwards are integrated into strong, cohesive groups, which throughout people's lifetimes continue to protect them in exchange for unquestioning loyalty' (ibid, p. 51).
- *Masculinity/femininity.* 'masculinity pertains to societies in which gender roles are clearly distinct (i.e. men are supposed to be assertive, tough, and focused on material success whereas women are supposed to be more modest, tender, and concerned with the quality of life); femininity pertains to those societies in which social gender roles overlap (i.e. both men and women are supposed to be modest, tender, and concerned with the quality of life)' (ibid, pp. 82–83).
- *Uncertainty avoidance.* 'the extent to which the members of a culture feel threatened by uncertain or unknown situations' (ibid, p. 113).

A fifth dimension of long-termism/short-termism was added later (Hofstede 1994b), following studies in Asia which detected important impacts of 'Confucian dynamism' (The Chinese Culture Connection 1987). Long-termism is 'the fostering of virtues orientated towards future rewards, in particular perseverance and thrift' (Hofstede 1994b, p. 261), while short-termism is 'the fostering of virtues related to the past and present, in particular respect for tradition, preservation of "face", and fulfilling social obligations' (ibid, pp. 262–263).

Following analyses of the World Values Survey (WVS) across more than ninety countries, Hofstede et al. (2010) determine that a sixth dimension should be included, named Indulgence – 'a tendency to allow relatively free gratification of basic and natural human desires related to enjoying life and having fun', versus Restraint – 'a conviction that such gratification needs to be curbed and regulated by strict social norms' (ibid, p. 281).

Furthermore, in the Values Survey Module (2008) (VSM 08) manual one other dimension is noted: Monumentalism – where a society rewards persons who are proud and unchangeable; the opposite is Self-Effacement – where a society rewards persons for humility and flexibility. This dimension remains an 'experimental' addition, although it is acknowledged to have potential to enrich the profiling of culture in conjunction with the other six national dimensions.

Triandis (1989) discusses tightness (strong social norms and sanctions, i.e. low tolerance of deviation from the norms) and looseness as a dimension of culture. Tight societies have high levels of accountability, as perceived by its members and so impact their behaviour.

Benedict (1946) suggests a typology of shame-cultures and guilt-cultures. In a shame-culture, emotion is a reaction to criticism by others and is deeply rooted in the external environment which emphasises standards imposed on the individual by society. In contrast, guilt-cultures operate through individuals developing conscience – inner standards of behaviour according to absolute moral standards of society.

Markus and Kitayama (1991) adopt a typology of interdependent-self and independent-self (where an interdependent-self is closely connected to the environment, and an independent-self is quite separate from context and values expressions of uniqueness). Witkin (1976, 1979) employs a typology of field-dependent and field-independent (where field-dependent people are compliant with the properties of the context and are less autonomous; field-independent people operate more autonomously).

Hall (1976) classifies cultures as high-context or low-context; 'in high-context culture, surrounding situations, external physical environments, and non-verbal behaviors are important for its members to determine the meanings of messages conveyed in communications' (Yamazaki 2005, pp. 524–525). Hall and Hall (1990) examine manifestations of the high/low context/content typology and add two further dimensions of cultures concerning approaches to time (monochronic – polychronic), and possessiveness of territory and space, i.e. monochronism and high possessiveness cultures are low context (high content), polychronism and low possessiveness cultures are high context (low content).

A particular concern is whether the classifications are continuous or discrete. Most are continuous and furthermore, as cultures are dynamic, the 'positioning' of a culture is likely to evolve over time due to changing social and economic conditions.

Trompenaars and Hampden-Turner (1997, pp. 123–125) recognise the culturally bound approaches to management of time, characterised as sequential or synchronic behaviour. Sequential behaviour leads people to deal with matters one at a time, often involving prioritising them by perceived importance; punctuality is vital (common in 'Western' cultures). Synchronic behaviour occurs where people deal with matters as they arise, often involving 'multitasking' (for instance, immediate greeting of a visitor is essential, even if another vital task is being carried out when the visitor arrives); punctuality is not too important (common in 'Eastern' cultures).

Gomez et al. (2000) explain that people in collectivist cultures favour in-group members but discriminate against out-group members. Chen et al. (1997) examine the cultural dimension of collectivism and determine that it is a construct which comprises vertical and horizontal components. They juxtapose those components to Hofstede's (1980) concept of individualism as '… individualism (low concern for collectivity and low concern for in-group others) at one end of the spectrum with vertical collectivism (high concern for the collectivity) and horizontal collectivism (high concern for in-group others) at the other end' (p. 64). They find that, 'Because the vertical scale items refer to work situations and the horizontal scale items primarily refer to non-work situations, one may speculate that the Chinese are becoming "organizational individualists" even though they are still cultural collectivists in other domains …' (ibid). This finding may be extended to the 'Asian Tiger economies', consequential upon their rapidly rising levels of industrialisation and wealth (e.g. Triandis 1990; Hofstede 1983, 1994b, p. 75). Hofstede (1983, p. 81) notes the correlation between wealth and Individualism in various countries but continues that '… Collectivist countries always show large Power Distances but Individualist countries do not always show small Power Distance'. These relationships are discussed extensively in Hofstede (2001).

Some studies seem to run into danger of conflating constructs. Denison (1996, 2009) employs a 'competing values' model in which flexibility and stability are juxtaposed on one dimension with organisational focus – internal juxtaposed to external – on the other dimension. The resultant quadrants comprise mission, consistency, involvement, and adaptability, each comprising three constituents. However, Denison employs a model comprising the same dimensions and quadrants, but with different constituents, to analyse leadership.

Project GLOBE (House et al. 2001; Javidan and House 2001) develops a questionnaire which was, initially, employed in the 1990s across 58 countries, and establishes '…nine dimensions of societal culture and nine isomorphic dimensions of organisational culture' (p. 493) – Uncertainty Avoidance, Power Distance, Collectivism I: Societal Collectivism, Collectivism II: In-Group Collectivism, Gender Egalitarianism, Assertiveness, Future Orientation, Performance Orientation, Humane Orientation. The aim of GLOBE is to understand '…cultural influences on leadership and organisational practices' (ibid: 489), thereby suggesting a strong organisational orientation (akin to Denison 2009), as reflected in the study by Javidan et al. (2006a) of national clustering on the GLOBE dimensions and the implications for global leaders. However, as with Hofstede's research, there are significant criticisms of the GLOBE methodology, methods, and findings (see, Graen 2006; Javidan et al. 2006b; Hofstede 2006; Minkov and Blagoev 2012). Despite their identical focuses, it is important not to conflate the dimensions of Hofstede and project GLOBE intellectually. Additionally, the weight of argument indicates that it is best to address the constructs separately – 'notional' culture, organisational culture, and leadership.

9.3.2 Organisational Culture

Organisational culture may be regarded as the way of doing business that the organisation adopts. More specifically, it is the '…shared understandings that, through subtle and complex expression, regulate social life in organizations' (Ouchi and Wilkins 1985, p. 458).

Usually, organisational cultures derive and evolve from the founders and others who have had a major impact on the organisation's development. Such people, through setting organisational objectives and influence over employment of staff, have shaped the values and behaviour of members of the organisation to develop the organisation's identity – both internally and externally. Organisational cultures are embedded in their national culture (potentially, more than one national culture for international organisations – nation of origin, headquarters location, local branches). Furthermore, they contain a nesting of organisational climate and other, subordinate levels (see Figure 9.3); for larger organisations, divisions of the organisation (functional/technical, geographical) may each have their own climates, if not their own cultures.

Organisational cultures (and climates) are, largely, self-perpetuating – persons who 'fit' are hired and they 'fit' because they are hired; errors of 'fit' are subject to resignations or dismissals. Thus, organisational cultures develop through the necessity of maintaining effective and efficient working relationships amongst stakeholders (both permanent and temporary). Pressure for cultural change commonly arises from external parties, particularly in situations of environmental turbulence, innovations, and attempts to enter new markets.

Organisational culture types and the dimensions determined for analyses (see later) show marked categorical similarity to the organisational behaviour typology of the human relations – task schools (such as Herzberg et al. 1967 – Theory X and Theory Y). This dichotomous perspective emphasises the spectrum of, often alternative, orientation of an organisation towards people ('employees', as the 'active factor' in supply chain processes which, as such, are the primary determinants of performance and, hence, success) or outcome (the product/service output as the 'supply' of the organisation). Today, the concept of 'employees', especially in a supply chain perspective, should be interpreted liberally to encompass all types and levels of persons employed in an organisation, whether a permanent organisation (e.g. a company) or a temporary (multi-) organisation (as for a construction project supply chain) and with a view to including perspectives on and behaviour towards other stakeholders – as conceived in CSR and OCB; the outward and inward manifestations of organisational ethics.

Schein (1984) determines two primary types of organisational culture: 'free flowing', an unbounded, egalitarian organisation without much formal structure, thereby encouraging debate and internal competition; or 'structured', a bounded, rigid organisation with clear rules and requirements (analogous to the organic–mechanistic typology of Burns and Stalker 1961).

Handy (1985) suggests that the main factors which influence organisational culture are: history and ownership, size, technology, goals and objectives, environment, and people. He advances four primary forms of organisational culture (echoed by Williams et al. 1989):

- *Power* is a web with the primary power at the centre; emphasis is on control over both subordinates and external factors (suppliers etc. and nature).
- *Role* involves functions/professions which provide support of the overarching top management; emphasis is on rules, hierarchy, and status through legality, legitimacy, and responsibility (as in contractual rights, duties, and recourse).

- *Task* in which jobs or projects are a primary focus, yields an organisational net (as in a matrix organisation); structures, functions, and activities are evaluated in terms of contribution to the organisation's objectives.
- *Person* in which people interact and cluster relatively freely; emphasis is on serving the needs of members of the organisation through consensus.

Hofstede (2002, p. 6) asserts that 'What holds a successful multinational together are *shared practices*, not, as the "corporate culture" hype of the early 1980s wanted it, shared values' [emphasis added]; see Figure 9.2. Hofstede (1994b, 2001) identifies six dimensions of organisational cultures:

- *Process - Results Orientation* (technical and bureaucratic routines [can be diverse]; outcomes [tend to be homogeneous]). This reflects means or ends orientation. Process cultures tend to be routine based and (developed to be) risk-avoiding. Strong (homogeneous) cultures tend to be more results orientated.
- *Job - Employee Orientation* (derives from societal culture as well as influences of founders, managers). Job cultures emphasise getting a job done (only outputs of employees matters), while employee cultures focus on concern for the welfare of the people involved (including personal matters).
- *Professional - Parochial* (one category of people identify with a profession[s], their type of job/occupation; other people identify with their employing organisation). People in parochial cultures consider that the norms of the organisation apply outside the workplace, hiring is on a holistic perspective, and the firm takes the long-term view, allowing individuals to be more short-term in focus. Professional people separate private life and work life aspects, they are hired for occupational competence, and adopt long-term perspectives.
- *Open - Closed System* (ease of admitting new people, styles and ease of internal and external communications). In open cultures, new people are fully and rapidly incorporated, while in closed systems inclusion is likely to take a very long time and such systems remain highly secretive towards people inside the organisation as well as towards those outside it.
- *Tight - Loose Control* (degrees of formality, punctuality, etc., may depend on technology and rate of change). Tight control demands extensive and rigid structuring with high levels of cost consciousness and time-keeping; stringent, if unwritten, codes of behaviour and dress follow.
- *Pragmatic - Normative* (how to relate to the environment, e.g. customers; pragmatism encourages flexibility). Pragmatic organisations are driven by markets, usually emphasising customer orientation. Normative organisations emphasise following rules and procedures and are perceived as having high standards of honesty and ethics. Depending on the nature of the business and its context, either approach may lead to good performance (although persons in pragmatic organisations see themselves as more results orientated).

Cameron and Quinn (1999) employ a 'competing values' model which juxtaposes 'flexibility and discretion' and 'stability and control' on one dimension, with 'internal focus and integration' and 'external focus and differentiation' on the other. The resultant model (see Figure 9.4) yields four quadrants, each denoting a type of organisational culture – Clan, Adhocracy, Market, and Hierarchy.

```
                    ┌─────────────────┐
                    │   FLEXIBILITY   │
                    │      AND        │
                    │   DISCRETION    │
                    └─────────────────┘
```

CLAN CULTURE	ADHOCRACY CULTURE
Leader: Facilitator; Mentor; Parent	**Leader:** Innovator; Entrepreneur; Visionary
Effectiveness Criteria: Cohesion; Morale; Development of Human Resources	**Effectiveness Criteria:** Cutting-Edge Output; Creativity; Growth
Organisation Theory Basis: Participation fosters commitment	**Organisation Theory Basis:** Innovation fosters new resources

```
┌──────────────────┐            ┌──────────────────┐
│  INTERNAL FOCUS  │            │  EXTERNAL FOCUS  │
│       AND        │            │       AND        │
│   INTEGRATION    │            │  DIFFERENTIATION │
└──────────────────┘            └──────────────────┘
```

HIERARCHY CULTURE	MARKET CULTURE
Leader: Coordinator; Monitor; Organiser	**Leader:** Hard-Driver; Competitor; Producer
Effectiveness Criteria: Efficiency; Timeliness; Smooth Functioning	**Effectiveness Criteria:** Market Share; Goal Achievement; Beating Competitors
Organisation Theory Basis: Control fosters efficiency	**Organisation Theory Basis:** Competition fosters productivity

```
                    ┌─────────────────┐
                    │    STABILITY    │
                    │      AND        │
                    │    CONTROL      │
                    └─────────────────┘
```

Figure 9.4 Competing values and organisational cultures model. Source: Following Cameron and Quinn 1999.

- *Clan.* 'Some basic assumptions in a clan culture are that the environment can be managed best through teamwork and employee development, customers are best thought of as partners, the organization is in the business of developing a humane work environment, and the major task of management is to empower employees and facilitate their participation, commitment, and loyalty' (ibid, p. 37).
- *Adhocracy.* 'A major goal of an adhocracy is to foster adaptability, flexibility and creativity where uncertainty, ambiguity and/or information-overload are typical.

Effective leadership is visionary, innovative and risk-orientated. The emphasis is on being at the leading edge of new knowledge, products, and/or services. Readiness for change and meeting new challenges are important' (ibid, p. 38–39).

- *Market.* 'The major focus of markets is to conduct transactions with other constituencies to create competitive advantage. Profitability, bottomline results, strength in market niches, stretch targets, and secure customer bases are primary objectives for the organization. Not surprisingly, the core values that dominate market type organizations are competitiveness and productivity' (ibid, p. 35).
- *Hierarchy.* 'The organizational culture compatible with this form is characterised by a formalized and structured place to work. Procedures govern what people do. Effective leaders are good coordinators and organizers. Maintaining a smooth-running organization is important. The long-term concerns of the organization are stability, predictability, and efficiency. Formal rules and policy hold the organization together' (ibid, p. 34).

Despite Hofstede's (2002) assertion, values remain a pervading basis for identifying organisational cultures. Mentzer et al. (2001) note that commitment, cooperative norms, organisational compatibility, top management support, and trust are important cultural components of shared values in supply chains.

Dowty and Wallace (2010) adopt the grid/group (rules/relationships) typology of Douglas (1999). The resultant 2×2 matrix comprises: high grid, high group (hierarchist); high grid, low group (isolationist/fatalist); low grid, high group (egalitarian); low grid, low group (individualist). As this is a model of cultural bias/preferences, it is held to operate in a contingency manner in that 'organizational representatives may use one cultural bias to make decisions in interactions with one organization, while they may adopt a different cultural bias to interact with another organization' (Dowty and Wallace 2010, p. 58). This can occur because although each organisation has one dominant culture type (at any time), the others are also present in the organisation.

9.3.3 Fitting with Other Cultures

It is common for people to move between cultures (organisational and national) and, consequently, need to adapt to the cultural context into which they move – such as secondment of staff to projects overseas. The concerns include selection of appropriate staff (and families) and intercultural training for them. Once transferred to the new culture, there is a usual pattern of development – acculturation – that occurs over a considerable time (usually a few years).

Acculturation is different from cultural accommodation. Acculturation involves '…the changes that an individual experiences as a result of being in contact with other cultures and participating…' (Berry 1990, p. 460). It includes attitudes about the desire to maintain one's own identity while relating to other groups, different behaviour and life-styles, and stress due to living in a different society. The progression of a person's acculturation depends on their valuation of their home cultural identity relative to their relationships in the host society; the expected duration of the situation is also likely to impact. The outcome comprises four types of attitude towards acculturation: integration, assimilation, separation, and marginalisation (Navas et al. 2007).

Differences between cultures are a common source of misunderstandings, which may escalate into conflicts and disputes. This occurs between all levels of culture and categories – national, organisational, occupational, generational, etc. A consequence that is particularly relevant to construction is '…we have come to accept the conflict between engineering and management as "normal", leading members of each culture to devalue the concerns of the other rather than looking for integrative solutions that will benefit both' (Schein 1996, p. 18; see also Rooke et al. 2003 – regarding claims).

A further concern is how conflict and its management is affected by culture. Euwema and Van Emmerik (2007) advance a typology of conflict behaviour (forcing, avoiding, compromising, accommodating, and problem-solving) and behavioural preferences of persons from different cultures. Fellows and Liu (2008) discuss interactions between the selection of project personnel, the perceived hierarchies of objectives, the causes and sources of conflicts, and the preferences of persons for conflict management styles. They note evidence of changing preferences in the rapidly changing society of Hong Kong. People from Western, particularly North American, Anglo, etc. cultures, tend to adopt competitive, confrontational approaches to conflict management, while people from Eastern, Chinese, etc. cultures tend to adopt more compromising, even withdrawal, approaches; thus, cultural differences impact on incidences, progression, and outcomes of conflict episodes. Furthermore, conflict management behaviour is impacted by significant differences in how people from different societies perceive fairness (Fellows et al. 2011).

Cultural issues have been found to be important throughout business and negotiations – including perceptions of the status of others (achievement, ascription), approaches to power and its use, punctuality and use of time, and ethical considerations (Hofstede 2001; Trompenaars and Hampden-Turner 1997). Fellows et al. (2004) analyse ethics in briefing for construction projects and examine various dilemmas and issues which occur, including self-protectionism and compliance with wishes of the paymaster. Furthermore, the culturally bound perceptions of ethics and morals also impact on what is regarded as corrupt and other unacceptable means and ends, many of which are situationally dependent also (e.g. opportunism and forms of cheating). An important difficulty is that once an absolute stance is abandoned, the value judgements which are required to secure benefits based on relative evaluations are complicated (e.g. the greatest good for the greatest number in cost–benefit analyses).

In dealing with different cultures, two further factors are important – attitude towards members of the other culture; notably, adventurism, which regards the other culture as interesting, exciting, and a challenge (rather than seeing it as a threat) – and empathy towards people in the other culture. These two factors operate together to generate a person's cultural competence – '…the ability to think and act in interculturally appropriate ways' (Navas et al. 2007, p. 431). Cultural competence involves managing uncertainty which, especially for persons from a low context culture moving into a high context culture, involves ambiguity as well as variability (as in the assumptions underpinning everyday existence and consequently how people make sense of their world) (Schein 2004; Weick et al. 2005).

Thus, given the importance of communications, especially in construction, Redmond (2000) identifies six competences of intercultural communications: language, adaptation, social decentering, social integration, knowledge of the other culture, and communication effectiveness. Additionally, Tone et al. (2009), from their study of communications on construction projects in Samoa, stress, inter alia, that it is essential for managers to include explanatory reasons for decisions and actions in order to minimise the possibilities of misunderstandings.

9.3.4 Organisational Climate

A common difficulty is to differentiate organisational culture and organisational climate. Mullins (2002, p. 906) explains organisational climate as 'Relating to the prevailing atmosphere surrounding the organisation, to the level of morale, and to the strength of feelings or belonging, care and goodwill among members. Organisational climate is based on the perceptions of members towards the organisation'. Organisational climate operates between organisational culture and organisational behaviour and, consequently may change more quickly than organisational culture, but far less rapidly than organisational behaviour.

Victor and Cullen (1988) discuss organisational climate at two levels. The first level is aggregate perceptions of organisational conventions concerning forms of structure and procedures for rewards and control (perceptions of practices and procedures – Schneider 1975). The second level is aggregate perceptions of organisational norms concerning warmth towards and support for peers and subordinates (organisational values – Denison 1996; Ashforth 1985).

'Organizational Climate is a relatively enduring quality of the internal environment of an organization that (a) is experienced by its members, (b) influences their behaviour, and (c) can be described in terms of the values of a particular set of characteristics (or attributes) of the organization' (Tagiuri and Litwin 1968, p. 27). Thus, the climate of an organisation distinguishes it from other, similar organisations. As shared experience of members of an organisation, it reflects their perceptions about autonomy, trust, cohesion, fairness, recognition, support, and innovation and thus leads to the members of the organisation having shared knowledge and meanings. An organisation's climate is an important contributor to homogeneity amongst members.

Organisational climate, and its groundings in organisational culture and national culture, is important in establishing the organisation's identity, as strongly manifested in its (business) objectives and behaviour – both of the organisational entity and of its individual agents. The usual prevailing competitive business context tends to reinforce and perpetuate self-orientated (individualistic) behaviour, often increasingly short-term (Hutton 1996, 2002). There are, however, antithetical pressures on organisations and their agents in supply chains to produce profits in support of dividend imperatives, while behaving as collaborative partners with others in the supply chain.

The delicate art for managers in such supply chains is to secure appropriate balancing of interests and rewards, for which it is essential to carry the other actors along a path for common anticipated benefit. The mechanisms for such 'balancing acts' are founded in understanding and being sensitive to incorporating others' requirements with their own requirements. This requires thorough appreciation of the organisational context and how, in relation to the project, it impacts on the behaviour of the actors through influence on their motivation and commitment.

9.3.5 Project Atmosphere

In project-based industries, such as construction, failure to appreciate the deep-seated, enduring (but dynamic) nature of culture is common and has given rise to notions of formation and change of culture which can be effected in the short term – the rather mistaken concept of 'project culture'. Such a conceptualisation fosters the functionalist notion of control by management such that a change in culture can be effected rapidly and under managerial control (but see Ogbonna and Harris 1998). While very long duration, megaprojects may develop their own cultures, this is extremely unlikely to be possible for smaller projects of quite short durations. On most projects, the concept of a 'project culture' consists of an amalgam of the cultures of the participating organisations (and societies) as transmitted to the project by their more powerful agents (Liu and Fellows 2008) – therefore, this is termed, more appropriately, 'project atmosphere' (analogous to workplace atmosphere – e.g. Hovmark and Nordqvist 1996).

9.3.6 Behaviour Modification

Modifying the behaviour of employees is a common endeavour of motivational schemes – usually associated with positive, systematic reinforcement, via rewards, of behaviour which enhances productivity. Provision of the reward is contingent on both behavioural change and consequently enhanced performance. It is the link between the behaviour modification and the increase in output of the employees which is important and thus which occasions the reinforcement reward. The reinforcements are likely to vary between people and can operate in the negative direction too – as 'punishments' for unwanted behaviour modification.

The simplicity of the required associations in behaviour modification has led to its acceptance being limited, as well as to issues of ethics relating to employee choice. However, a significant aspect of behaviour modification is that the causal chain operates rapidly. A further potential, detrimental effect is that the effectiveness (e.g. financial rewards – extrinsic motivator – as reinforcement) may be only temporary. In using legislation to effect behaviour modification, sustaining the legislation, coupled with a programme of education, can lead to culture change by amending people's beliefs about their practices – important for industrial safety, for example.

What is of fundamental importance is not to confuse behaviour modification with changing culture! Behaviour modification can occur very rapidly (and be temporary) whereas culture change occurs only over a long period of time and requires continuous reinforcement for it to occur; once it has occurred it is enduring because the beliefs – the intrinsic motivator – of the people involved have altered.

9.4 Values and Value

Usually, value is manifested in monetary terms. Such interpretation of exchange value is essential in market business contexts but is also increasingly recognised as restrictive and inadequate to provide comprehensive measurement of performance. Stakeholder perspectives, which include recognition that an individual may fulfil diverse roles (organisational employee, consumer, environmentally aware person, etc.) and CSR metrics (including environmental impacts, ethics) are becoming widespread to lend credibility to the range of values (as things of worth to people) which are important. This extends the perspective of use value.

Thus, generic value – what something is perceived to be worth (to society or to an individual decision maker) – determines what is produced and how production occurs. For construction projects to be successful, it is essential to determine who the stakeholders are and what their objectives are for the project – i.e. what their values are and the structuring of those values – in order to examine compatibility and, within the parameters of the project, to secure acceptance of performance targets derived from the values to apply to the project. This extends the remit for briefing even beyond the strategic and project typology (e.g. Green and Simister 1999) into the realm of determination of the values of the major stakeholders and their compatibility, as investigated by Mills et al. (2006).

It is the values of the project stakeholders which determine the pressures on the supply chain members of what to provide. It is argued here that stakeholders at different vertical positions in the supply chain are likely to have different value structures. However, even though stakeholders at the same vertical level in the supply chain may have similar values, this, in itself, does not ensure collaborative, synergetic performance. It is vital not only to determine what the values are (their identities), but also how they operate in the context of the project. Indeed, different values of different stakeholders may be complementary for performance (e.g. project-functional orientation of designers and project-quality orientation of constructors), while similar values of different stakeholders may yield detrimental consequences (own, short-term, profit orientation). The critical issue is compatibility of the values and their practical manifestations within the project constraints.

'A transaction is the exchange of *values* between two parties. The things-of-value need not be limited to goods, services, and money; they include other resources such as time, energy and feelings' (Kotler 1972, p. 48). Rokeach (1973) considers that values are the deeply held, enduring beliefs of people (although this does contrast with Hofstede – see Figure 9.2); while value is the benefit resulting from an exchange and arises from people's preferences. Thus, Kotler is referring to exchanges of things (tangible and/or intangible) to which the transacting parties attach values, both in exchange and in use.

Values are 'desirable, transsituational goals, varying in importance, that serve as guiding principles in people's lives' (Schwartz and Bardi 2001, p. 269). Values are positive, because they are desirable, and generic, because they are transsituational and so are different from specific objectives, which they underpin. Schwartz and Bilsky (1987, p. 551) advance 'five features that are common to most ... definitions of values ... (a) concepts or beliefs that are, (b) about desirable end states or behaviours, and (c) that transcend specific situations, (d) guide selection or evaluation of behaviour and events, and (e) are ordered by relative importance'.

Values can be classified as ends (situations: outcomes – as in the functioning of a project in use) or means (instrumental values: processes – as in project realisations which consume less resource and produce less pollution).

It is usual to regard values as motivators of human behaviour (Schwartz and Bilsky 1987 suggest nine motivational domains of values; amended to ten motivational types of values in Schwartz and Sagiv 1995), along with needs (e.g. Maslow 1943; Alderfer 1972), and means (e.g. Vroom 1964). Values refer to what people believe to be important and thus are instrumental in generating goals and targets. Schwartz and Bilsky's (1987) motivational domains of values support the perspective of congruence between people's values and those expressed for tasks/projects having a positive effect on performance.

Schwartz and Sagiv (1995) and Schwartz and Bardi (2001) advance a model of motivational types of values which is derived from research into individual's values and consistency between them (Table 9.1). They discuss the organisation of the value types

Table 9.1 Higher order values, constituent motivational types of values, and goals.

Higher order value	Motivational types of values	Goals
Universalism	Broad-minded, wisdom, social justice, equality, world at peace, world of beauty, unity with nature, protecting the environment	Understanding. appreciation, tolerance, protection of the welfare of all people and of nature
Benevolence	Helpful, honest, forgiving, loyal, responsible	Preservation and enhancement of the welfare of people with whom one is in frequent personal contact
Conformity	Politeness, obedient, self-discipline, honouring parents and elders	Restraint of actions, inclinations and impulses likely to upset or harm others and violate social expectations and norms
Tradition	Humble, accept position in life, devout, respect for tradition, moderate	Respect, commitment, and acceptance of the customs and ideas that traditional culture or religion provide
Security	Family security, national security, social order, clean, reciprocation of favours	Safety, harmony, and stability of society, of relationships, and of self
Power	Social power, authority, wealth, preserving public image	Social status and prestige, control or dominance over people and resources
Achievement	Successful, capable, ambitious, influential	Personal success through demonstrating competence according to social standards
Hedonism	Pleasure, enjoying life	Pleasure and sensuous gratification for oneself
Stimulation	Daring, a varied life, an exciting life	Excitement, novelty, challenges in life
Self-direction	Creativity, freedom, independent, curious, choosing own goals	Independent thought and action, choosing, creating, exploring

Source: Derived from Schwartz and Bardi (2001).

into two dimensions: Openness to Change versus Conservation; Self-Enhancement versus Self-Transcendence. Honesty and other 'pro-social' values are important, while power values, including wealth, are far less important. Consensus over the level of importance of hedonism values is low. Notably, the research reveals that the identity of values and their interrelationships are quite uniform across cultures but there are differences between value hierarchies (relative importance of the values) of different occupational and national groups. This confirms the necessity for identification of the values of project stakeholders in the supply chain and, more especially, for developing frameworks to secure acceptably compatible manifestations of these values for project realisation – performance targets, etc. This is a primary task of the project (supply chain) manager to address right from the initiation of the project (project conception).

9.5   Ethics

Ethics concern human interactions – what people do, how the things are done, and with what impacts on other people, living things and the environment; as such, they are related to values very closely and constitute an important, integral component of culture. (In fact, ethics are the manifestations of moral values.) Furthermore, ethical concerns feature ever more widely in evaluations of projects and organisations. Codes of ethics (conduct) usually indicate the boundaries of what is acceptable behaviour and being acknowledged as ethical is recognised as valuable for marketing advancement. The 'bottom line', of course, is that legal systems are grounded in morals and ethics and denote the absolute limits of acceptable behaviour – notably, there are significant differences between cultures.

A reputation for honesty and good, moral behaviour attracts business and tends to lead to reduction of transaction costs through reducing promotion and scrutiny requirements. These are important cost and organisational issues for a project supply chain manager to address.

Hinman (1997) distinguishes morals as first-order beliefs and practices about what is good and what is bad which guide behaviour; and ethics as second-order, reflective consideration of moral beliefs and practices. Rosenthal and Rosnow (1991, p. 231) note 'ethics *refers* to the system of moral values by which the rights and wrongs of behaviour … are judged' [italics added].

Issues of definition and perspective on ethics have led to the development of four primary paradigms. In deontology (relating to duty or moral obligation), a universal moral code applies. In scepticism (relativism; subjectivism), ethical rules are arbitrary and relative to culture and to time; this is extended into ethical egoism where ethics become matters for the conscience of the individual. Thus, egoism concerns pursuit of self-interest and, as such, can be related to common business performance criteria (notably, profit maximisation). Teleology (the branch of philosophy relating to 'ends' or final causes) constitutes a utilitarian approach where ethics are dependent upon the anticipated consequences – prompting a cost–benefit perspective, perhaps invoking the judgemental criterion of 'the greatest good for the greatest number' which, itself, is likely to necessitate subjectively determined weightings. Objectivism asserts that there are definitions of right and wrong which are accepted generally (either universally or more locally) (Leary 1991, pp. 261–262).

Thus, given the diversity of ethical paradigms, there remains great scope for variability in determination of what is ethical – a 'tip' in one context/society may constitute a 'bribe' elsewhere. Due to the deep-seated nature of ethics and their moral foundation, the project supply chain manager should establish a code of ethical behaviour which is appropriate to the project location and to the stakeholders; this code should be documented and communicated to all members of the supply chain and adherence to it policed (with suitable sanctions for transgressions).

Ethics concerns how the actions of one person may impact on others and thus imposes a 'duty of care' not to harm others. This perspective generates questions of to whom such a duty is owed, together with concerns over whether it should be applied absolutely or relatively (deontologically or teleologically). Law and codes of conduct endeavour to denote boundaries of application (the 'neighbour' principle; the client). Clearly, national law employs wide boundaries and applies to all people in the country (the jurisdiction of the law); codes of conduct apply more restrictively and may be specific regarding behaviour towards specified others likely to be encountered in the course of activities (notably, the client of a construction consultant).

However, stepping beyond such rather arbitrarily drawn boundaries into the realm in which a professional is a person who possesses special knowledge, which itself concerns generic 'good', and uses that knowledge for the benefit of the immediate client and wider society, the boundaries of application vanish. Benefit from professional activities is the objective but, at the same time, leaving distributions of such benefits open to judgement due to the diversity of people affected by a professional's work.

The prospect of 'compartmentalisation' through the presence of artificial boundaries around behavioural requirements and perspectives – as governed by circumstances (domestic, employment, etc.) – prompts Fellows et al. (2004) to discuss the notion of 'personal shielding', in which a person amends his or her (ethical) behaviour to accord with the expressed or perceived ethics of another, usually an employing, organisation. Such shielding may feature in principal-agent circumstances, including those between a commissioning client and a design consultant as well as employer–employee relationships.

9.6   Organisational Citizenship Behaviour (OCB) and Corporate Social Responsibility (CSR)

OCB concerns the voluntary behaviour of employees towards the benefit of the organisation in excess of the requirements of both the contract of employment specifications and the norms of behaviour of similar employees (Organ 1988); the employees of the organisation 'go the extra mile' (for the organisation's benefit). For such behaviour to occur, employees must feel committed to the organisation, which results from their own disposition and their perception of how the organisation (and it's superiors) treats them. Thus, adopting the perspective that an organisation has a personality and behaviour separate from its members (Wayne et al. 1997), it is appropriate to examine the reciprocations in the relationship between employees and the organisation.

For a construction project, OCB can occur within a single supply chain organisation (a firm), within the project temporary multiorganisation (TMO) assembled to realise the project, or both. Although OCB is examined most often in relation to a single firm regarding relationships between the firm and persons within it, OCB's applicability may be extended, by analogy, to apply to behaviour at the interfirm level, as for a whole supply chain/network.

Organ (1988) employs dimensions of altruism (discretionary behaviour which assists others), conscientiousness (fulfilling role requirements in excess of the minimum), sportsmanship (accepting minor frustrations without complaint), courtesy (respecting the needs of others and behaving accordingly), and civic virtue (appropriate participation at work) to examine the presence of any OCB.

Van Dyne et al. (1994, p. 769) note that 'The global perception that an organization supported its members and valued their contributions was an important correlate of employee behaviour and affective states'. Eisenberger et al. (2001) argue that based on the norm of reciprocity, employees are motivated to compensate beneficial treatment by acting in ways that support the organisation. However, they note that employees may differ in their acceptance of the norm of reciprocity that underlies the exchange relationship.

Whether employee behaviour constitutes OCB is determined by causal analysis which, in many cases, is problematic. Contractual requirements should be express, provided a contract of employment exists. Norms are established by custom and practice and may change (rapidly) in response to conditions/situations. Thus, norms of employee behaviour can be amended by pressures (threats, power exercising, inducements) by employers/bosses, resulting in apparent rather than real OCB by employees/subordinates. Hence, in practice, especially during recessionary periods and other times of difficulty in securing (alternative) employment, it can be very difficult to identify OCB.

CSR is discretionary behaviour by the organisation, in excess of the requirements of law and the norms of the market(s). CSR is evidenced most commonly by ethical behaviour towards customers, society, and the natural environment. Sharp Paine (2003) documents a variety of case studies concerning potential CSR actions and demonstrates that such organisational behaviour is not only a question of actions, but that the timing and overt causes of the actions are germane.

Organisations may be tempted to use CSR actions as marketing and legal defence mechanisms in efforts to improve profitability; such motivations may be criticised from a 'pure CSR' perspective as there has been no value change in such organisations, but instead, CSR is employed as (part of) the means to the end. Instrumental values have altered, while situational values have not. This, in itself, does raise the question of whether it is the actions and their consequences that are important or the reasons for the actions (the means–ends dichotomy as in deontology–teleology). Perhaps, given the importance of environmental preservation and ethical behaviour towards others, it is appropriate for a project supply chain manager to adopt the pragmatic stance of examining behaviour and effects (the phenomenal level) in preference to the reasons underpinning the behaviour (the ideational level). However, as with all cultural aspects, if a long-term (enduring) change is desired, the people concerned must be convinced of the merits of the change, so that it will become part of their usual behaviour.

9.7 Teams and Teamwork

A team is two or more people who are collaborating in pursuit of a common objective(s) – goal congruence – which distinguishes a team from a group. However, individual actors are likely to have differing goals, whether self-determined or imposed by others (as in principal-agent relationships), which impacts what information is sought and how it is used – in the (motivated) reasoning to underpin decisions. The people constituting a team may be quite different from each other, notably in technical knowledge and abilities, and probably in sociopolitical skills too. What is important is their preparedness and ability to collaborate towards realisation of the goal(s) which, over enduring periods, is likely to require subjugation of individual desires and behaviour for the ultimate outcome. Such collaboration is dependent upon the team members recognising their interdependence in striving for success and then acting according to that recognition (see, e.g. Crainer 1996). This is likely to require people communicating and pooling their interpretations of events – the meanings they obtain from environmental cues through sensemaking.

The rhetoric of teams and teamwork has been widespread and strong in the construction industry for many years – including partnering, alliances, and joint venturing.

Synergetic performance is believed to result from teamwork. In many instances, such beliefs are coupled with competition within teams (for membership) and between teams as performance stimulants (for rewards) as well as some elements and degrees of conflict as further motivators for performance (see, e.g. Robbins 1984). However, competitive perspectives are culturally bound and, although seemingly apposite for Western participants and contexts, may be quite inappropriate elsewhere (e.g. Asia).

As teams and teamworking are dependent upon integration, communication, self-subjugation, and coordination (see, for example, Belbin 1981), their existence is rare – it is exactly the lack of these key constituents for which the construction industry is criticised (see, e.g. Latham 1994; Construction Industry Review Committee 2001). As the size of any team is subject to upper limits, it is appropriate to view project TMOs as collectives of collectives, in which each collective may be a group or a team, dependent upon its constituents and processes and contingent upon its environment. Furthermore, it is probable that if the TMO comprises groups, the TMO itself (as a meso-level collective) cannot be a team.

Here, personal dispositional variables impact the potential and probability of whether a human collective will behave as a group or as a team. Those dispositional variables are rooted in national cultures – notably, Individualism and Power Distance. The Femininity element of nurturing and the pursuit of longer-term perspectives reinforce tendencies towards teamwork by fostering integration and self-subjugation to a common good. The superimposed factors of organisational culture and climate filter and mediate the manifestations of the basic cultural traits.

Thus, it seems that, commonly, efficient and effective supply chain functioning is hampered by the absence of real teams and teamwork. While suitable supply chain structuring and systems may be conducive to the development of teams and teamworking, it is the forging of collaborative relationships between appropriate combinations of persons that is critical (see, e.g. Belbin 1981, who discusses characteristic requirements for successful teams).

9.8 Sensemaking

Sensemaking is a perspective which assists people to understand the continuous, ambiguous, complex, and equivocal dynamics of their lives (Weick et al. 2005) to achieve a coherent, plausible (not necessarily accurate) understanding of the meaning of their situation and to organise. Weick (1995) identifies seven properties of sensemaking: social setting, identity, retrospect, cues, ongoing development, plausibility, and enactment (SIR COPE). These properties are employed in sensemaking by reflectively and mindfully considering the situation, task, intent, concerns, and calibration (STICC) (Weick 2009). The process of sensemaking is impacted by the discourse that takes place and by the environment in which it occurs (Helms Mills et al. 2010) and is boundedly rational.

Traditionally, sensemaking is retrospective. However, increasingly, attention is shifting to 'prospective' sensemaking by '…casting ourselves figuratively into the future and acting as if events have already occurred and then making "retrospective" sense of those imagined events' (Corley and Gioia 2011, p. 25) – as in project design and project planning.

While several investigations of sensemaking have examined major, critical events (e.g. Weick et al. 2005) recently, attention is moving to (continuous) sensemaking relating to everyday (mundane) occurrences (e.g. Patriotta and Brown 2011). Thus, '... actors engage ongoing events from which they extract cues and make plausible sense retrospectively while enacting more or less order into those ongoing events' (Weick 2001, p. 463), hence '...organization emerges from organizing and sensemaking' (Brown et al. 2015, p. 267).

'An organized activity provides actors with a given set of cognitive categories and a typology of action options' (Tsoukas 2005, p. 124), i.e. a map. This occurs in an organisational context that provides or prescribes a 'governing episteme for ordering the world' (O'Leary and Chia 2007, p. 393), according to the organisational culture (Fellows and Liu 2016). This governing episteme (or logic) determines that only some cues are noticed and how these cues are interpreted through framing, language, and experiences and knowledge used (i.e. cognitive framing bias applies in which gaps are filled with information consistent with the frame – Hahn et al. 2014).

Frequently, sensemaking is triggered by a disruption to normal activity. The disruption triggers the process of sensemaking – creation, interpretation, and enactment – which cycles until order is restored (Weick 1995). Through enactment, people generate what they interpret (Weick 1995) which, if correct, may constitute a self-fulfilling prophesy or, if incorrect, a mistake.

Often, people follow routines without considering them consciously and cope with situations spontaneously (Sandberg and Tsoukas 2011) – this is absorbed coping. A disruption, especially if highly stressful, is likely to invoke a 'System 1' ('automatic cognition'; intuitive) (Kahneman 2011) reaction, due to training, experience, etc. However, in most instances, a heedful, mindful, reflective, and thoughtful response is needed – 'System 2' (ibid). Unfortunately, System 2 may be invoked only after the System 1 remedy is seen as a clear mistake (ibid, p. 31).

Sensemaking leads to new understandings of situations. Abolafia (2010, p. 357) identifies that '...every organization has a variety of plots that it draws from in making sense of its environment' and that decision makers employ operating models that are '...dominant perceptual filter[s] that shape{s} and bias{es} sensemaking' ([ ], { } added, ibid, p. 363).

Gioia and Chittepeddi (1991) discuss cycles of sensemaking and sensegiving in a strategic organisational change. Sensegiving becomes a process of aligning the views and practices of members with the (new) goals and values of the new leader (hierarchy) to produce a 'strong' organisational culture with perceived performance advantages.

Vlaar et al. (2008) discuss two further elements of securing understandings – sense-demanding and sense-breaking. In sense-demanding, persons seek more information so that they have adequate bases for decisions with acceptable levels of uncertainty (plausibility); sense-breaking involves challenging norms and current understandings (through reframing, etc.) to test them and thus open them to new interpretations – an important component of reflective practice and of decision making.

9.9 Motivated Reasoning

Motivated reasoning, as a behavioural perspective, has been applied in political science and in research processes to explain how persons take new data and information into account in developing their opinions and consequent actions (e.g. voting; accommodating evidence). The impact of new evidence depends upon the reasoning criterion which the decision maker uses. Broadly, the criteria relate to accuracy of use of data to determine the action/outcome or to the achievement of a particular type of outcome (e.g. maximisation of financial return) – a directional motivation (e.g. Kunda 1990).

Updating models are many and varied – from replacement, through (weighted) updating to ignoring. Motivated reasoning asserts that the updating, and hence the impact of the new data, depends on whether the decision criterion is accuracy, which should apply in professionalism, or some directional, distorting objective (possibly imposed by a principal in an agency situation). However, the evidence confirms that the biasing effect of a directional criterion is limited by plausibility and, if applicable, the potential requirement to justify the decision (and impact of the data). 'When people are motivated to be accurate, they expend more cognitive effort on issue-related reasoning, attend to relevant information more carefully, and process it more deeply, often using more complex rules' (Kunda 1990, p. 481). Thus, the accuracy criterion invokes cognitive, System 2 reasoning and more reflexive sensemaking and decisions.

However, 'Motivational bias does not simply sway judgements in a particular direction; it is strong enough to influence choice…Specifically, if the preferences of the person to whom one must justify are known, individuals tend to tailor their judgements or decisions in the direction of those preferences' (Boiney et al. 1997, p. 19). Hence, outcome actions are affected as the principal's desires are important influences on agents, even if limited by plausibility and 'professionalism'. The influence works in the opposite way for evidence contrary to the desired outcome or set of beliefs. Also, people reinterpret information after actions and decisions to favour the outcome – as in absorbed coping and the 'garbage can' decision model.

Vagueness in the environment fosters motivated reasoning (Piercey 2009). This is exacerbated where probabilities in forecasting can be assessed only verbally (uncertainties) and thus elastic (re-) definitions (comprising quantification, precision, and skew of the descriptors) are easy, rendering more biased, but perceived as more justifiable, predictions (ibid).

Information processing may be on-line (processing on perception as System 1), or memory processing (System 2 and cognitive sensemaking). Directional goals reinforce using on-line processing while accuracy goals reinforce memory processing (Redlawsk 2002). Thus, 'motivated reasoning predicts a process similar to anchoring and adjustment … but one in which the anchor is far stronger than the adjustment and the adjustment may in fact be in the wrong direction. Such processes would clearly violate … Bayesian updating' (ibid, pp. 1025–1026).

Motivated reasoning is insightful for the nature of goal setting and decision making on construction projects – both precontract and postcontract – and hence for the performance and outcomes which are achieved. Motivated reasoning is a bias vector that influences how actors arrive at understandings, reach decisions, and behave to generate performance and outcomes. The goals tend to be directional (e.g. business desires – revenue maximisation; profitability) and impact subsequent reasoning. In decision making, amending existing data (and beliefs) occurs in many ways. However, the nature of the goal(s), whether accuracy (as in professionalism) or directional (potentially, through agency), influences the processes and outcomes. Requirements to justify a decision, if only to self, mitigate the effects of directional bias. Conversely, nonquantification (of probabilities, data) may enhance such bias, if only by easing justification through elastic defining (Piercey 2009; Fellows and Liu 2018).

9.10 (Strategic) Alliances

A business alliance is 'an ongoing, formal business relationship between two or more independent organizations to achieve common goals' (Sheth and Parvatiyar 1992, p. 72). ul-Haq (2003) suggests that there are four principal types:

- A formal co-operative venture
- A joint venture
- Joint ownership
- A strategic investment in a partner

Whatever the formal arrangements are which bring businesses into close contact for individual transactions at one extreme or for enduring alliances at the other, 'Usually the corporate culture of the most powerful or economically successful company dominates' (Furnham 1997, p. 570). Hence, for integration to occur successfully, whether through take-over, alliancing, merger, or forming a subsidiary joint venture organisation, not only goal congruence but also compatibility of organisational cultures is critical.

There are 'two basic organizational modes of alliance: equity joint ventures (EJVs) and non-equity joint ventures (NEJVs)' (Glaister et al. 1998, p. 169). This classification is supported by Pangakar and Klein (2001) who adopt the classification of equity alliances or contractual relationships.

In construction, informal alliances are common. Every project may be viewed as a joint venture due to the dependence of the output on interrelationships and interactions between participants (interdependencies). Informal alliances constitute a hybrid in which the contract binds the participants while the effectiveness of the project team is determined by the quality of interpersonal relationships.

Sheth and Parvatiyar (1992) employ a two-dimensional analysis – purpose (strategic/operational) and parties (competitors/noncompetitors) – to examine forms, properties, and characteristics of business alliances. They note that the strategic purposes of alliances are growth opportunity, strategic intent, protection against external threats and diversification. The operational purposes, on the other hand, are resource efficiency, increased asset utilisation, enhanced core competence and a closed performance gap. A contextual factor is that an alliance form may be stipulated as the legally required structure for nondomestic organisations to operate in the location – most commonly in less developed economies. Horizontal alliances may be made with existing competitors, potential competitors, indirect competitors, and (potential) new entrants, while vertical alliances occur with customers, potential customers, suppliers, and potential suppliers.

Contractual alliances provide much greater entry and exit flexibility for participants and at much lower cost than equity alliances. However, such apparent advantage results in reluctance of the participants to make significant alliance-specific investment (Pangakar and Klein 2001). In equity alliances, the alignment of objectives and performance incentives acts to deter participants from free-riding and from other forms of opportunistic behaviour (ibid.).

Uncertainty and trust are the two primary constructs which affect formal alliance relationships and their institutional arrangements (Sheth and Parvatiyar 1992). Bachmann (2001) examines trust and power as means for social control within business relationships. Bachmann notes that '… today, trust based on individual actors' integrity can only fulfil a supplementary function, compared with trust produced by institutional arrangements' (p. 348). Strong institutional arrangements are demonstrated to foster the development of trust, while otherwise, business actors resort to power to safeguard their interests.

Because of the cyclical and volatile nature of the demand for property development and (to a lesser extent) construction sectors, risks are perceived to be high, given the typical returns. Hence, work allocation has a universally strong focus on cost minimisation, to the potential, virtual exclusion of other considerations. Arguably, however, design and construction work should be awarded to the parties who can provide the most suitable and reliable assurance of performance for the work (package) in question, in the context of also being compatible with parties engaged already, and others likely to be engaged over the course of the project's realisation (see also: Baiden et al. 2006).

Given the importance of relationships and behaviour to the operation and performance (success) of joint ventures, it is clear that culture has a fundamental impact, especially when considering compatibilities amongst participants. Those concerns are reinforced by Das and Teng (1999, p. 56), who note that 'Because of incompatible organizational routines and cultures, partner firms often do not work together efficiently'.

Studies have often used measures of cultural distance, investment risk, and market potential to explain which mode of entry to employ in new markets. Such decisions reflect how the firms respond to the externalities which they perceive in the target location.

> '… firms choose a higher control form in response to conditions of high external (market and political) uncertainty … [and] … in countries that have greater market potential… firms … need to get established early in emerging markets … regardless of the market potential and/or country risk, firms resort to sharing of risks and managerial resources'.
>
> (Agarwal 1994, p. 74).

Brouthers and Brouthers (2001) investigate the relationship between cultural distance and entry mode and find that investment risk moderates this relationship.

Shenkar (2001) recognises the impact of the theory of familiarity in that firms are less likely to invest in markets which they perceive to be culturally distant. Often, 'follow-my-leader' strategies are adopted by oligopolistic organisations as a method for reducing risks. Organisations which are second or later entrants to new markets are likely to adopt similar forms of entry to those adopted by initial entrants.

Kogut and Singh (1988) find that both the cultural distance from the home country and the score for Uncertainty Avoidance are correlated with preference for a joint venture form of entry to a new country market.

Measurement of cultural distance is itself an issue for debate. Normally, cultural distance has been measured through use of indices (as in Kogut and Singh 1988, who employed Hofstede's (1980) initial four dimensions of national culture). However, this approach to measuring cultural distance involves assumptions which may be inappropriate. The problems include the fact that cultural distance is not symmetrical; home-country culture is embedded in the firm, host-country culture is embedded both in the alliance partner(s) and in the local, operating environment (Shenkar 2001).

Cultural distance, as indices of measurements of dimensions of culture, is challenged through concerns over the relative sizes of in-group versus between-group variances. Furthermore, cultures are dynamic temporally and vary within national borders. Not all cultural facets are of equal importance nor do they necessarily operate in the same direction. Intracultural variations (national and organisational) may exceed intercultural variations (Au 2000). Hofstede (1989) confirms that differences between cultures vary in significance and that differences in Uncertainty Avoidance are potentially the most problematic for international business alliances.

For supply chains, the need to examine cultural differences on appropriate dimensions, rather than via a single, aggregate index of cultural distance, is demonstrated by Cadden et al. (2013) who use six dimensions of organisational culture (results, employee, open, loose, norm, market) to examine compatibilities of suppliers and buyers. The best performing supply chains have similar collaborative cultures (shared values, beliefs, and behaviours) while the poor performing supply chains have similar adversarial cultures (job-focused, process-driven, and inflexible).

A factor which is not addressed in the cultural distance indices is the volatility of the host country – hardly surprising as business perspectives on international volatility focus on political, economic, and financial variables and thus may be treated as additional complications for analyses of cultural differences (at both national and organisational levels). Agarwal (1994, p. 67) notes that '...problems associated with bounded rationality and opportunistic behaviour are aggravated under conditions of high volatility', which would be exacerbated by increasing rates of changes on cultural dimensions that are apparent in many countries. Hence, how to 'fit' into another culture can be of great pragmatic relevance.

Adopting the paradigm that all construction projects are realised through (informal) joint venturing, and that culture varies to a large degree both between organisations and within societies, the issues relating to international alliances also apply to domestic construction projects.

9.11 Supply Chain Participants and Behaviour

A vital concern is the transferring of findings from one society, organisation, or other social group to another directly with no suitable amendments (or allowance) for cultural differences (e.g. as found by car manufacturers operating in Japan and the USA – Womack et al. 1990). However, determining the differences is itself problematic. Hofstede (1994b, p. 81) warns that 'A management technique or philosophy that is appropriate in one national culture is not necessarily appropriate in another'.

Thus, studies of Asian supply chains (deKoster and Shinohara 2006 – Western European and Japanese; Ryu et al. 2006 – Korean) conclude that collectivist cultural orientation (Asian) is more likely to yield long-term relationships but slower processes (due to collective responsibility and consensual decision making).

Figure 9.1 provides a schematic, systems representation of a construction project. Not only are there extensive differences between the value perspectives of the various individual actors performing the functions identified, there will also be differing value perspectives amongst team members that constitute project actor firms. These values may be classified as business values, technical values, and personal values.

Business values, and most particularly the performance criteria/objectives (and parameters) derived from them, have much global commonality; however, because the performance focus is self-orientated (ultimately, at least), these values are likely to give rise to conflicts between members of the supply chain. Technical values concern the specialist activity and how it is carried out; thus, for each specialist, the values are bespoke and tend to be complementary with the technical values of other participants. Personal values are the most overtly culturally determined and variable.

Construction projects are realised through TMOs which have highly diverse members, many of whom have involvement or roles that are transient, and subject to highly varying degrees of integration. Usually, the TMOs operate not as teams but as flexible, multigoal coalitions based upon fluid power structures. 'Adversarialism and opportunism are rife at all stages, as low barriers to entry maintain the high degree of fragmentation and low levels of profitability and investment within these markets' (Ireland 2004, p. 273). However, '… construction companies are effectively the "integrator" for a myriad of construction supply chains… are faced with the challenge of obtaining a regular workload that is sufficiently profitable …' (ibid).

'In general terms, it can be argued that supply chains must exist as structural properties of power … the physical resources that are necessary to construct a supply chain will exist in various states of contestation … based on the horizontal competition between those who compete to own and control a particular supply chain resource … also … on the vertical power struggle over the appropriation of value between buyers and suppliers at each point in the chain… possession of these power attributes will be demonstrated by the relative capacity of the owners of particular resources to appropriate value for themselves' (Cox 1999, p. 173). Power, according to Emerson (1962, p. 32), '… is a property of the social relation …' and '… resides implicitly in the other's dependency' and so is context dependent.

In construction, value measurements and perceptions are based upon comparisons of anticipations/forecasts/expectations of performance with performance realisations. A consequence is that cognitive dissonance (the mental conflict which arises when assumptions are contradicted by new information – here, when the performance realised falls significantly short of the forecasts given) (Festinger 1957) is likely to occur and constitute a component of the value perceived (for both the supplier and the consumer). Such comparisons, and their consequences for value perceptions, constitute a significant area of risk for project participants in addition to the tangible risks. Additionally, the vast number of interdependent, component transactions, coupled with diversity amongst participants, leads to complexity and, consequently, boundary management issues and risks (see, e.g. Fellows and Liu 2012).

Thus, comparisons of actual performance with forecasts and targets occurs throughout project realisation, as essential parts of performance control by managers, as well as on completion of the project supply process. Differences in performance realisations from these forecasts are, almost invariably, attributed to the realisation processes and ignore the presence of (possibly considerable) variability in the forecasts themselves. Given that attention tends to focus on performance realisation shortfalls and a human tendency to blame others readily, it is all too easy for people to become frustrated and demotivated and, especially clients, generally dissatisfied with the performance achieved. The desire to avoid responsibility and consequences (liquidated damages, for example) encourages project actors to shift blame onto others, especially those in weaker positions, as well as to pursue other elements of opportunistic behaviour – notably claims (see Rooke et al. 2003, 2004).

Lawrence and Lorsch (1967) investigate the dichotomy of differentiation and integration within organisational processes and determine that appropriate degrees of both are required for effectiveness and efficiency – analogous to clan organisational culture. However, Tavistock (1966), and virtually all subsequent reports on construction industry organisation, criticises its performance and cites causes rooted in fragmentation, poor communications, low levels of coordination and lack of trust and adversarialism; indeed, the industry remains characterised by 'mutual mistrust and disrespect' amongst participants.

It is usual for fragmentation to be identified as a major cause of the problems on construction projects. Fragmentation arises through two forces – a strong force for differentiation (specialisation, division of labour, etc.) but a weak force for integration. The result is proliferation of separate organisations which operate largely independently in pursuit of their own interests. The effects are compounded through the operation of the common procurement methods which '… have focussed on organisations' individual … capability rather than their collective ability to integrate and work together effectively' (Baiden et al. 2006, p. 14). This is reflected in the zero-sum games typical of construction projects.

Braunschneidel et al. (2010, p. 899) find that '…a direct link between organizational culture and organizational performance exists and is mitigated by the integration practices an organization adopts….organization culture…has a direct effect on the internal and external integration practices a firm employs'. Using the competing values framework of Cameron and Quinn (1999), Braunschneidel et al. (2010) detail that hierarchy culture score relates to delivery performance negatively and seems to hamper both internal and external integration; however, adhocracy has a positive relationship with performance and with external integration. They note that integration practices mediate the relationships of both hierarchy and adhocracy culture scores with performance.

Cao et al. (2015) use a similar 'competing values' model of organisational culture to investigate associations between organisational cultural types (development: short- or long-term orientation; group: cooperation and team spirit; rational: reward systems; and hierarchical: centralised or distributed control). They found that '…development, group and rational culture are positively related to SCI [supply chain integration], but hierarchical culture is negatively related to SCI' ([..] added; Cao et al. 2015, p. 36).

The SCI studies are noteworthy as internal integration is viewed widely as a precursor to external integration (Stevens 1989) and external integration is regarded as a factor that leads to improved performance (see, e.g. Peters and Waterman 1982; Latham 1994; Egan 1998; Construction Industry Review Committee 2001).

Although a vast amount of rhetoric concerns teams and teamwork in realisations of construction projects (see, e.g. Latham 1994), in practice precious little teamwork can be found. Nicolini (2002) notes five categorical factors which are critical to success and superior performance of crossfunctional teams – task design, group composition, organisational context, internal processes, and group psychosocial traits. These factors are important contributors to 'project chemistry', which is a range of antecedent variables necessary for project management success. Dainty et al. (2005) assert that project affinity, emotional attachments to the project (objectives/purpose) outcome, enhances how people work, especially their OCB, thereby fostering performance. Both constructs are culturally bound and, to their degrees of presence, enhance performance via team formation and commitment of personnel.

Given the pressures on businesses to secure competitive (financial) returns continuously, it is unsurprising that the organisational members of supply chains/networks, and their representatives on projects, succumb to opportunistic behaviour aimed at appropriation of value for self. Many systems and procedures in common usage encourage such behaviour either overtly as in 'lowest competitive bid wins' or implicitly through tight regulation as under many conditions of contract. In particular, it is such combinations which appear highly detrimental to the well-being of the industry – well-being is manifested in levels of return for all participants commensurate with the risks assumed, a cooperative and collaborative context, and levels of trust which require minimum surveillance and enforcement to secure high and continuously improving levels of performance.

Elmuthi and Kathawala (2001) note that the main risks and problems which strategic alliances and, following the joint venture paradigm adopted here, construction project supply chain/network members, face are:

- 'Clash of cultures' and 'incompatible personal chemistry'
- Lack of trust
- Lack of clear goals and objectives
- Lack of coordination between management teams
- Differences in operating procedures and attitudes among partners
- Relational risk (due to self-interest focus)

There is a cultural dimension to each of the items on the list. Differences, of course, need not yield negative outcomes. Indeed, positive effects of differences are emphasised in development of effective teams (as noted, above). Thus, it is not the differences themselves which are detrimental, but the ways in which they are managed – or, perhaps more appropriately, *not managed*.

While differences between cultures may appear quite obvious even through casual observation, how to assess the differences and then how to deal with them remains open to question. One solution, firmly rooted in the functionalist paradigm, is to measure 'cultural distance'. Kogut and Singh (1988) use an index of cultural distance to analyse organisational fit for firms entering overseas markets because '...differences in national culture have been shown to result in different organizational and administrative procedures and employee expectations...' (p. 414) on the premise that lower levels of organisational fit lead to increasingly high post acquisition costs and thus formal joint ventures are more appropriate.

In a case study of an international joint venture in Hong Kong, Liu and Fellows (2008, p. 268) find that 'The organisational cultures of the parent companies are consistent with their own national culture characteristics but ... the joint venture organisational culture is highly influenced by the dominant national culture of the management team' and reinforces the conceptualisation of 'project culture' as an amalgam of the cultures of participant organisations, i.e. project atmosphere.

Social capital is the '… goodwill that is engendered by the fabric of social relations and that can be mobilized to facilitate action' (Adler and Kwon 2002). As such, social capital constitutes a powerful intangible resource with particular importance for the formation and working of an informal system of relationships amongst individuals and organisations. Social capital comprises two primary components – bridging and bonding.

'Bridging social capital examines the external linkages of individuals and groups that help to define their relationships … bonding social capital focuses on the internal relationships of a focal actor and specifically examines the linkages and corresponding relationships among individuals and groups within a focal group or organization' (Edelman et al. 2004, pp. 60–61). The two components are important in determining membership of a group or team and how the teams integrate and relate to each other and their members. However, Edelman et al. warn that '… loss of objectivity is a function of actors becoming deeply embedded in an existing network. This can lead to the exclusion of new actors or ideas that are potentially beneficial' (p. 61).

Newell et al. (2004) examine the use of social capital for acquisition and sharing of knowledge amongst members of a project realisation collective. Social capital is important as the knowledge in question is personal/tacit knowledge and thus the possessors of knowledge must become aware of its existence and then be willing and able to communicate it to relevant other parties. That exchange process is important in realisations of projects, especially for those at the forefront of knowledge and involving innovation.

Newell et al. (2004, p. 54) note that the '… project team must develop "strong" relationships internally if the information and knowledge derived from … external networks is to be integrated'. However, '… individual members did appear to be using their social capital, but more for their own personal good than for the public good of the project… as the project became more insecure, the individual team members increased their networking with their functional departments but very much to secure their own personal goals'. Hence, trust, motivation, and commitment are vital ingredients of bridging and bonding behaviour.

Goal congruence, an essential component of teamwork, arises from setting, communicating, and accepting of goal(s). Furthermore, the goals themselves may constitute performance incentives, depending on their content and their level. Goal content concerns the subject of the goal (e.g. time performance) and operates within the context of the total project – notably, the project function as in 'project affinity' as discussed by Dainty et al. (2005). Goal level concerns the quantity of the goal and thus its incentivisation may enhance or detract from the motivation of the goal content. Locke et al. (1988, p. 32) summarise the situation as 'commitment declines as the goal becomes more difficult and/or the person's perceived chances of reaching it decline'; hence the motivation of increasingly difficult to achieve goals is an inverted U-shape. Furthermore, they recognise that the effectiveness of different styles adopted for setting goals depends on the culture(s) (values) of the participants – perhaps Power-Distance in particular.

Cultures evolve in path-dependent directions (along trajectories/pathways that are, largely, determined by history and environmental variables), punctuated by periods of stability and by rapid, step-type changes, 'The evolution of culture is shaped by agency and power, but cannot be created by fiat' (Weeks and Gulunic 2003). However, '… despite agreement that cultural evolution occurs …, espoused approaches to culture interventions are more commonly revolutionary in nature …' (Harris and Ogbonna 2002). When faced with change, most people exhibit a strong preference for the familiar and so tend to resist; if change does occur, a strong tendency to revert to prior norms remains.

Perspectives on changes in cultures span two extremes. 'Functionalists' believe that organisational culture can be directly controlled by management and so are instrumental in promoting the cultural basis for determining organisational performance. The alternative perspective regards culture as a context within which action must be taken and thus necessitates compatibility of action with the cultural environment. However, a third category, falling between these two extremes, is the perspective that culture is malleable and thus may be adapted – albeit that adaptations are likely to be difficult, replete with ethical problems, and require effort over long periods.

Even the most carefully devised and conducted change initiatives are likely to have unanticipated consequences – including ritualisation of change, cultural erosion, hijacking of the process, and uncontrolled and uncoordinated effects (Harris and Ogbonna 2002).

Hult et al. (2007) compare supply chains that pursue a culture of competitiveness with those that pursue a knowledge development culture; they conclude that firms that endeavour to satisfy the market (competitiveness) are '…likely to reap positive advantages in stable markets' (p. 1047) but do less well in turbulent markets (such as construction or property development). Conversely, firms that pursue knowledge development have better performance in turbulent markets. Firms/supply chains that meld the two cultures seem likely to secure inimitability and so establish considerable competitive advantage (according to the resource-based view – see, e.g. Barney 1991).

9.12 Conclusion

Culture impacts our lives dynamically and constantly. This generates questions of what culture is, how it operates, how it changes, and what the impacts are, and whether culture is a cause or a consequence. There is a considerable and increasing diversity of paradigms, models, etc., of culture at all levels and significant differences are involved – this is of particular note regarding emic and etic studies, different philosophical approaches (positivist, constructivist, etc.), and at the interface between culture and personality (as depicted in Figure 9.3). Thus, there are many different methods of researching and understanding culture, the main drivers of which are philosophical stances and the purpose of the study. Given the variety of philosophical approaches and resultant paradigms for studying human behaviour and its underpinning beliefs and values, what is valid in one paradigm may well be invalid in another.

Cultures may be analysed by the nature of relationships (e.g. tight – loose) – how closely and dependently people regard themselves with respect to others (e.g. in-group vs out-group), context and the environment, and time – whether highly interdependent or independent. This is important for construction supply chains with their tradition of individualism (opportunistic behaviour, adversarialism, etc.) but, contrasting with this, advocacy of teamwork and cooperation and now, with legislative support, 'partnering'. Indeed, legislation and industry procedures often present a paradox for construction supply chains through at one-and-the-same-time advocating/requiring both 'partnering' and competition (in selecting 'partners')!

The issue of boundaries is fundamental. However, pragmatics impact, especially in relation to 'national' culture. In most cases, the cultural profiles must be treated as indicators only, particularly for nations comprising diverse (ethnic) groups, recalling that within group variability is likely to be large; an analogous reasoning applies to major, multinational organisations, etc. In analysing and understanding cultures, it is important to avoid both the ecological and atomistic fallacies (ecological – inferences made about an individual using data about a group; atomistic – inferences made about a group using data about an individual). The mere existence of a definite boundary (country, company, profession) does not guarantee that it contains a single, tightly defined, coherent culture.

Awareness, understanding, and accommodation of culturally based differences are important for successful performance of a project supply chain because success is judged by culturally determined performance metrics – usually derived from the goals of the project paymaster. Positivist/functionalist persons regard culture as a tool which may be employed to effect changes to advance performance against often predetermined and sometimes distantly determined criteria and targets. This approach tends to confuse effecting cultural change (long-term, permanent) with behaviour modification (short-term and reversed easily); here it is important to recall that people are, generally, risk averse as a basic trait (Kahneman 2011; Kahneman and Tversky 1979) (but, around which, some are more risk seeking than others – see Fiolet et al. 2016) and consequently endeavour to return to the *status quo*. Others regard culture as a medium in which adaptation must occur (social constructivism, interpretivism) but in which 'creep' (evolutionary change) takes place in response to changing conditions, usually steadily but with occasional perturbations, yielding long-term change.

It is inevitable that all construction projects and their realisation supply chains include a variety of cultures – organisational, national, or both. Given this environment, successful project supply chain managers must be interculturally competent, especially to get the best from the contributing stakeholders/actors (individuals and organisations), many of whom change rapidly and frequently throughout the life of the project. Intercultural competence requires the managers to think and act in ways which are appropriate to the cultures involved and with empathy for the various cultures, i.e. to see the project through the eyes of the different stakeholders and to appreciate the (performance) requirements which they place on the project (from their own points of view). Such perceptions occur through sensemaking by the supply chain managers who, in the pressured world of construction project business, are likely to be subject to various behavioural pressures – notably through motivated reasoning – in pursuit of project performance.

Cultural understanding and sensitivity necessitate a high level of open mindedness, flexibility, and tolerance of others which is likely to be vested in only the more adventurous persons; those who regard new situations as desirable challenges, rather than threats (i.e. persons with low Uncertainty Avoidance). These attributes will be manifested in their conflict management style, which requires attention to their own and others' objectives, coupled with the ability to evaluate immediate and longer-term outcomes.

It is important that managers of supply chain/network TMOs are aware of the myriad issues and difficulties concerning culture – identification, understanding, accommodation, adaptability, etc. This requires sensitivity to others as well as self-awareness. A common problem is the delusion of control – nowhere is this so evident and so important as in the selecting of appropriate combinations of people, organisations, and processes to foster an environment conducive to successful performance, whatever this is determined to be for the combination of stakeholders in a construction project. Project performance results from the amalgam of the members of the supply chain, not just their 'technical' abilities, but their ability and preparedness to cooperate; only then can synergy result (see, e.g. Baiden et al. 2006).

It is increasingly clear that hierarchical organisational cultures are detrimental to both internal and external integration of supply chains and, seemingly consequentially, to good performance. Furthermore, adversarialism is also detrimental to performance.

While many devote great energy to devising systems and procedures to facilitate control, much control is illusory (Kahneman and Lovallo 1993). Good systems, including contracts, provide frameworks, and only frameworks, within which projects can be realised (see, for example, Pryke et al. 2018). Selection of participant organisations can render collaborative working more (or less) likely. Such facets of 'project hardware' cannot guarantee performance or satisfaction and success, which may stem from good performance; these can be secured only through the 'project software', the people involved with the project and how they relate to the others, the organisations, and the project itself.

This chapter has addressed the behavioural aspects of the 'project software', what underpins the behaviour exhibited and what the consequences are likely to be. People – clients, superiors, colleagues, subordinates, and self – are fickle, have differing perspectives, goals, and pressures; to ignore this is to ignore a fundamental reality. Despite the voices of critics, the overwhelming weight of evidence does portray people as knowledgeable and skilled specialists who want to do a good job (Deming 2000). What stands in their way, *our way*, is lack of integration.

The appropriate maxim seems to be 'We're all in this TOGETHER'.

References

Abolafia, M.Y. (2010). Narrative construction as sensemaking: how a central bank thinks. *Organization Studies* 31 (3): 349–367.

Adler, P.S. and Kwon, S.W. (2002). Social capital: prospects for a new concept. *Academy of Management Review* 27 (1): 17–40.

Agarwal, S. (1994). Socio-cultural distance and the choice of joint ventures: a contingency perspective. *Journal of International Marketing* 2 (2): 62–80.

Alderfer, C.P. (1972). *Existence, Relatedness and Growth: Human Needs in Organizational Settings*. New York: Free Press.

Aldrich, H. (1979). *Organizations and Environments*. Englewood Cliffs, NJ: Prentice-Hall.

Ashforth, B.E. (1985). Climate formation: issues and extensions. *Academy of Management Review* 10 (4): 837–847.

Au, K.Y. (2000). Inter-cultural variation as another construct of international management: a study based on secondary data of 42 countries. *Journal of International Management* 6 (3): 217–238.

Bachmann, R. (2001). Trust, power, and control in trans-organizational relations. *Organization Studies* 22 (2): 337–367.

Baiden, B.K., Price, A.D.F., and Dainty, A.R.J. (2006). The extent of team integration within construction projects. *International Journal of Project Management* 24 (1): 13–23.

Barney, J.B. (1991). Firm resources and sustained competitive advantage. *Journal of Management* 17 (3): 99–120.

Belbin, R.M. (1981). *Management Teams: Why they Succeed or Fail*. Oxford: Butterworth-Heinemann.

Benedict, R. (1946). *The Chrysanthemum and the Sword: Patterns of Japanese Culture*. Boston, MA: Houghton Mifflin.

Berry, J.W. (1990). Psychology of acculturation. In: *Cross-Cultural Perspectives: Nebraska Symposium on Motivation* (ed. J. Berman), 457–488. Lincoln: University of Nebraska Press.

Boiney, L.G., Kennedy, J., and Nye, P. (1997). Instrumental bias in motivated reasoning: more when more is needed. *Organizational Behavior and Human Decision Processes* 72 (1): 1–24.

Braunschneidel, M.J., Suresh, N.C., and Boisnier, A.D. (2010). Investigating the impact of organizational culture on supply chain integration. *Human Resource Management* 49 (5): 883–911.

Brouthers, K.D. and Brouthers, L.E. (2001). Explaining the national cultural distance paradox. *Journal of International Business Studies* 32 (1): 177–189.

Brown, A.D., Colville, I., and Pye, A. (2015). Making sense of sensemaking. *Organization Studies* 36 (2): 265–277.

Burns, T. and Stalker, G.M. (1961, 1968). *The Management of Innovation*, 2e. Tavistock.

Cadden, T., Marshall, D., and Cao, G. (2013). Opposites attract: organisational culture and supply chain performance. *Supply Chain Management: An International Journal* 18 (1): 86–103.

Cameron, K.S. and Quinn, R.E. (1999). *Diagnosing and Changing Organizational Culture*. Massachusetts: Addison-Wesley Longman.

Cao, Z., Huo, B., Li, Y., and Zhao, X. (2015). The impact of organizational culture on supply chain integration: a contingency and configuration approach. *Supply Chain Management: An International Journal* 20 (1): 24–41.

Chen, C.C., Meindl, J.R., and Hunt, R.G. (1997). Testing the effects of vertical and horizontal collectivism: a study of allocation preferences in China. *Journal of Cross-Cultural Psychology* 28 (1): 44–70.

Construction Industry Review Committee (2001). *Construct for Excellence ('The Tang Report')*. Hong Kong: Government of the Hong Kong Special Administrative Region.

Corley, K.G. and Gioia, D.A. (2011). Building theory about theory building: what constitutes a theoretical contribution? *Academy of Management Review* 66 (1): 12–32.

Cox, A. (1999). Power, value and supply chain management. *Supply Chain Management* 4 (4): 167–175.

Crainer, S. (1996). *Key Management Ideas: Thinkers that Changed the Management World*. London: Pitman Publishing.

Dainty, A.R.J., Bryman, A., Price, A.D.F. et al. (2005). Project affinity: the role of emotional attachments in construction projects. *Construction Management and Economics* 23 (3): 241–244.

Das, T.K. and Teng, B.-S. (1999). Managing risks in strategic alliances. *The Academy of Management Executive* 13 (4): 50–62.

deKoster, M.B.M., and Shinohara, M. (2006) Supply Chain Culture Clashes in Europe: Pitfalls in Japanese Service Operations. Erasmus Research Institute of Management Report Series: Research in Management. Rotterdam: Erasmus School of Economics, Erasmus University.

Deming, W.E. (2000). *Out of the Crisis*. Cambridge, MA: MIT Press.

Denison, D. (1996). What is the difference between organizational culture and organizational climate? A native's point of view on a decade or paradigm wars. *Academy of Management Review* 21 (3): 619–654.

Denison, D.R. (2009), Organizational Culture and Leadership Surveys. The Denison Model. Denison Consulting. http://bit.ly/rW1Snd (accessed 7 October 2009).

Douglas, M. (1999). Four cultures: the evolution of a parsimonious model. *GeoJournal* 47 (3): 411–415.

Dowty, R.A. and Wallace, W.A. (2010). Implications of organizational culture for supply chain disruption and restoration. *International Journal of Production Economics* 126 (1): 57–65.

Eccles, R. (1981). The quasifirm in the construction industry. *Journal of Economic Behavior and Organization* 2 (4): 335–357.

Edelman, L.F., Bresnen, M., Newell, S. et al. (2004). The benefits and pitfalls of social capital: empirical evidence from two organizations in the United Kingdom. *British Journal of Management* 15 (S1): 59–69.

Egan, J. (1998) *Rethinking Construction*. Report from the Construction Task Force, Department of the Environment, Transport and the Regions. London: HMSO.

Eisenberger, R., Armeli, S., Rexwinkel, B. et al. (2001). Reciprocation of perceived organizational support. *Journal of Applied Psychology* 86 (1): 42–51.

Elmuthi, D. and Kathawala, Y. (2001). An overview of strategic alliances. *Management Decision* 39 (3): 205–217.

Emerson, R.M. (1962). Power-dependence relations. *American Sociological Review* 27 (1): 31–41.

Euwema, M.C. and Van Emmerik, I.J.H. (2007). Intercultural competences and conglomerated conflict behaviors in intercultural conflicts. *International Journal of Intercultural Relations* 31 (4): 427–441.

Fellows, R.F. (2006). Understanding approaches to culture. *Construction Information Quarterly* 8 (4): 159–166.

Fellows, R.F., Liu, A.M.M. (2008) A culture-based approach to the management of conflict on multi-national construction projects: participants and performance. In Proceedings CIB W112 International Conference on Multi-National Construction Projects, 'Securing high Performance through Cultural Awareness and Dispute Avoidance', Shanghai, China, 21–23 November.

Fellows, R.F. and Liu, A.M.M. (2012). Managing organisational interfaces in engineering construction projects: addressing fragmentation and boundary issues across multiple interfaces. *Construction Management and Economics* 30 (8): 653–671.

Fellows, R.F. and Liu, A.M.M. (2013). Use and misuse of the concept of culture. *Construction Management and Economics* 31 (5): 401–422.

Fellows, R.F. and Liu, A.M.M. (2016). Sensemaking in the cross-cultural contexts of projects, festschrift for Peter Morris. *International Journal of Project Management* 34 (2): 246–257.

Fellows, R.F. and Liu, A.M.M. (2018). Where do I go from here? Motivated reasoning in construction decisions. *Construction Management and Economics,* 36 (11): 623–634.

Fellows, R.F., Liu, A.M.M., and Storey, C. (2004). Ethics in construction project briefing. *Journal of Science and Engineering Ethics* 10 (2): 289–302.

Fellows, R.F., Liu, A.M.M. and Zhang, S.B. (2011) What is fair? Perceptions of justice. In Proceedings, International Construction and Business Management Symposium, Kuala Lumpur, Malaysia, 21–22 September.

Festinger, L. (1957). *A Theory of Cognitive Dissonance.* Stamford: Stamford University Press.

Fiolet, J.-C., Haas, C., and Hipel, K. (2016). Risk-chasing behaviours in on-site construction decisions. *Construction Management and Economics* 34 (12): 845–858.

Furnham, A. (1997). *The Psychology of Behaviour at Work: The Individual in the Organization.* Hove: Psychology Press.

Gioia, D.A. and Chittepeddi, K. (1991). Sensemaking and sensegiving in strategic change initiation. *Strategic Management Journal* 12 (6): 443–448.

Glaister, K.W., Husan, R., and Buckley, P.J. (1998). UK international joint ventures with the triad: evidence for the 1990s. *British Journal of Management* 9 (3): 169–180.

Gomez, C., Kirkman, B.L., and Shapiro, D.L. (2000). The impact of collectivism and in-group/out-group membership on the generosity of team members. *Academy of Management Journal* 43 (6): 1097–1100.

Graen, G.B. (2006). In the eye of the beholder: cross cultural lessons in leadership from project GLOBE: a response viewed from the third culture bonding (TCB) model of cross-cultural leadership. *Academy of Management Perspectives* 20 (1): 95–101.

Green, S.D. and Simister, S.J. (1999). Modelling client business processes as an aid to strategic briefing. *Construction Management and Economics* 17 (1): 63–76.

Griffith, A. and Sidwell, A.C. (1995). *Constructability in Building and Engineering Projects (Building & Surveying).* Basingstoke: Macmillan.

Hahn, T., Preuss, L., Pinkse, J., and Figge, F. (2014). Cognitive frames in corporate sustainability: mangerial sensemaking with paradoxical and business case frames. *Academy of Management Review* 39 (4): 463–448.

Hall, E.T. (1976). *Beyond Culture.* Garden City, NY: Anchor Press/Doubleday.

Hall, E.T. and Hall, M.R. (1990). *Understanding Cultural Differences.* Yarmouth, ME: Cultural Press.

Handy, C.B. (1985). *Understanding Organisations,* 3e. Harmondsworth: Penguin.

Harris, L.C. and Ogbonna, E. (2002). The unintended consequences of cultural interventions: a study of unexpected outcomes. *British Journal of Management* 13 (1): 31–49.

Hatch, M.J. (1993). The dynamics of organisational culture. *Academy of Management Review* 18 (4): 657–693.

Helms Mills, J., Thurlow, A., and Mills, A.J. (2010). Making sense of sensemaking: the critical sensemaking approach. *Qualitative Research in Organizations and Management* 5 (2): 182–195.

Herzberg, F., Mausner, B., and Bloch Snyderman, B. (1967). *The Motivation to Work,* 2e. New York: Wiley.

Hinman, L.M. (1997). *Ethics: A Pluralistic Approach to Moral Theory.* Orlando: Harcourt Brace Jovanovich.

Hofstede, G.H. (1980). *Culture's Consequences: International Differences in Work-Related Values.* Beverley Hills, CA: Sage Publications.

Hofstede, G.H. (1983). The cultural relativity of organizational practices and theories. *Journal of International Business Studies* 14 (Fall): 75–89.

Hofstede, G.H. (1989). Organizing for cultural diversity. *European Management Journal* 7 (4): 390–397.

Hofstede, G.H. (1994a). The business of international business is culture. *International Business Review* 3 (1): 1–14.

Hofstede, G.H. (1994b). *Cultures and Organizations: Software of the Mind*. London: Harper Collins.

Hofstede, G.H. (2001). *Culture's Consequences: Comparing Values, Behaviors, Institutions, and Organizations across Nations*, 2e. Thousand Oaks, CA: Sage.

Hofstede, G.H. (2002). Dimensions do not exist: a reply to Brendan McSweeney. *Human Relations* 55 (11): 1–7.

Hofstede, G.H. (2006). What did GLOBE really measure? Researchers' minds versus respondents' minds. *Journal of International Business Studies* 37: 882–896.

Hofstede, G., Hofstede, G.J., and Minkov, M. (2010). *Culture and Organizations: Software of the Mind: Intercultural Cooperation and its Importance for Survival*, 3e. New York: McGraw-Hill.

House, R., Javidan, M., and Dorfman, P. (2001). Project GLOBE: an introduction. *Applied Psychology: An International Review* 50 (4): 489–505.

Hovmark, S. and Nordqvist, S. (1996). Project organization: change in the work atmosphere for engineers. *International Journal of Industrial Ergonomics* 17 (5): 389–398.

Hult, G.T.M., Ketchen, D.J. Jr., and Arrfelt, M. (2007). Strategic supply chain management: improving performance through a culture of competitiveness and knowledge development. *Strategic Management Journal* 28 (10): 1035–1052.

Hutton, W. (1996). *The State We're In*, 2e. London: Vintage.

Hutton, W. (2002). *The World We're In*. London: Abacus.

Ireland, P. (2004). Managing appropriately in construction power regimes: understanding the impact of regularity in the project environment. *Supply Chain Management* 9 (5): 372–382.

Javidan, M. and House, R. (2001). Cultural acumen for the global manager: lessons from project GLOBE. *Organizational Dynamics* 29 (4): 289–305.

Javidan, M., Dorfman, P.W., de Luque, M.S., and House, R. (2006a). In the eye of the beholder: cross cultural lessons in leadership from project GLOBE. *Academy of Management Perspectives* 20 (1): 67–90.

Javidan, M., Dorfman, P.W., de Luque, M.S., and House, R. (2006b). A failure of scholarship: response to George Graen's critique of GLOBE. *Academy of Management Perspectives* 20 (4): 102–114.

Jin, X.-H., Doloi, H., and Goa, S.-Y. (2007). Relationship-based determinants of building project performance in China. *Construction Management and Economics* 25 (3): 297–304.

Kahneman, D. (2011). *Thinking, Fast and Slow*. London: Penguin Books – Allen Lane.

Kahneman, D. and Lovallo, D. (1993). Timid choices and bold forecasts: a cognitive perspective on risk taking. *Management Science* 39 (1): 17–31.

Kahneman, D. and Tversky, A. (1979). Prospect theory: an analysis of decision under risk. *Econometrica* 47 (2): 263–291.

Kogut, B. and Singh, H. (1988). The effect of national culture on the choice of entry mode. *Journal of International Business Studies* 19 (3): 411–433.

Kotler, P. (1972). A generic concept of marketing. *Journal of Marketing* 36 (2): 46–54.

Kroeber, A.L. and Kluckhohn, C. (1952). Culture: a critical review of concepts and definitions. In: *Papers of the Peabody Museum of American Archaeology and Ethnology*, vol. 47. Cambridge, MA: Harvard University Press.

Kunda, Z. (1990). The case for motivated reasoning. *Psychological Bulletin* 180 (3): 480–498.

Latham, S.M. (1994). *Constructing the Team*. London: HMSO.

Lawrence, P.R. and Lorsch, J.W. (1967). *Organization and Environment: Managing Differentiation and Integration*. Boston: Division of Research, Graduate School of Business Administration, Harvard University.

Lazzarini, S., Chaddad, F., and Cook, M. (2001). Integrating supply chain and network analyses: the study of netchains. *Journal on Chain and Network Science* 1 (1): 7–22.

Leary, M.R. (1991). *Introduction to Behavioral Research Methods*. Belmont, CA: Wadworth.

Liu, A.M.M. and Fellows, R.F. (2008). Organisational culture of joint venture projects: a case study of an international JV construction project in Hong Kong. *International of Human Resources Development and Management*. 8 (3): 259–270.

Locke, E.A., Latham, G.P., and Erez, M. (1988). The determinants of goal commitment. *The Academy of Management Review* 13 (1): 23–39.

Markus, H.R. and Kitayama, S. (1991). Culture and the self: implications for cognition, emotion, and motivation. *Psychological Review* 98 (2): 29–43.

Maslow, A.H. (1943). A theory of human motivation. *Psychological Review* 50 (4): 370–396.

Mentzer, J.T., DeWitt, W., Keebler, J.S. et al. (2001). Defining supply chain management. *Journal of Business Logistics* 22: 1–25.

Mills, G., Austin, S., and Thomson, D. (2006) Values and value: two perspectives on understanding stakeholders. In Proceedings, Joint International Conference on Construction Culture, Innovation and Management (CCIM), The British University in Dubai, November.

Minkov, M. and Blagoev, V. (2012). What do project GLOBE's cultural dimensions reflect? An empirical perspective. *Asia Pacific Business Review* 18 (1): 27–43.

Mullins, L.J. (2002). *Management and Organisational Behaviour*, 6e. Harlow: Prentice Hall.

Navas, M., Rojas, A.J., García, M., and Pumares, P. (2007). Acculturation strategies and attitudes according to the relative acculturation extended model (RAEM): the perspectives of natives versus immigrants. *International Journal of Intercultural Relations* 31 (1): 67–86.

Newell, S., Tansley, C., and Huang, J. (2004). Social capital and knowledge integration in an ERP project team: the importance of bridging AND bonding. *British Journal of Management* 15 (S1): 43–57.

Nicolini, D. (2002). In search of 'project chemistry'. *Construction Management and Economics* 20 (2): 167–177.

Ogbonna, E. and Harris, L.C. (1998). Managing organizational culture: compliance or genuine change? *British Journal of Management* 9 (4): 273–288.

O'Leary, M. and Chia, R. (2007). Epistemes and structures of sensemaking in organizational life. *Journal of Management Inquiry* 16 (4): 392–406.

Organ, D.W. (1988). *Organizational Citizenship Behavior: The Good Soldier Syndrome*. Lexington MA: Lexington Books.

Ouchi, W.G. and Wilkins, A.L. (1985). Organizational culture. *Annual Review of Sociology* 11: 457–483.

Pangakar, N. and Klein, S. (2001). The impacts of Alliance purpose and partner similarity on lliance governance. *British Journal of Management* 12 (4): 341–353.

Patriotta, G. and Brown, A.D. (2011). Sensemaking, metaphors and performance evaluation. *Scandinavian Journal of Management* 27 (1): 34–43.

Peters, T.J. and Waterman, R.H. (1982). *In Search of Excellence: Lessons from America's Best-Run Companies*. New York: Harper and Row.

Piercey, D.M. (2009). Motivated reasoning and verbal vs. numerical probability assessment: evidence from an accounting context. *Organizational Behavior and Human Decision Processes* 108 (2): 330–341.

Pryke, S.D., Badi, S.M., Almadhoob, H. et al. (2018). Self-organizing networks in complex infrastructure projects. *Project Management Journal* 49 (2): 18–41.

Redlawsk, D.P. (2002). Hot cognition or cool consideration? Testing the effects of motivated reasoning on political decision making. *The Journal of Politics* 64 (4): 1021–1044.

Redmond, M.V. (2000). Cultural distance as a mediating factor between stress and intercultural communication competence. *International Journal of Intercultural Relations* 24 (1): 151–159.

Robbins, S.P. (1984). *Essentials of Organizational Behaviour*. Englewood Cliffs, NJ: Prentice-Hall.

Rokeach, M. (1973). *The Nature of Human Values*. New York: Free Press.

Rooke, J., Seymore, D.E., and Fellows, R.F. (2003). The claims culture: a taxonomy of attitudes in the industry. *Construction Management and Economics* 21 (2): 167–174.

Rooke, J., Seymore, D.E., and Fellows, R.F. (2004). Planning for claims: an ethnography of industry culture. *Construction Management and Economics* 22 (6): 655–662.

Rosenthal, R. and Rosnow, R.L. (1991). *Essentials of Behavioral Research: Methods and Data Analysis*, 2e. Boston, Mass: McGraw-Hill.

Ryu, S., Han, J.-H., and Frank, J. (2006). Does culture matter? Collectivism, long-term orientation, and supply chain management in Korea. *International Journal of Internet and Enterprise Management* 4 (2): 162–179.

Sandberg, J. and Tsoukas, H. (2011). Grasping the logic of practice: theorizing through practical reality. *Academy of Management Review* 36 (2): 338–360.

Schein, E.H. (1984). Coming to an awareness of organizational culture. *Sloan Management Review* 25 (Winter): 3–16.

Schein, E.H. (1990). Organizational culture. *American Psychologist* 45: 109–119.

Schein, E.H. (1996). Three cultures of management: the key to organizational learning. *Sloan Management Review* (Fall): 9–20.

Schein, E.H. (2004). *Organizational Culture and Leadership*, 3e. San Francisco: Jossey-Bass.

Schneider, B. (1975). Organizational climate: an essay. *Personnel Psychology* 28: 447–479.

Schneider, W.E. (2000). Why good management ideas fail: the neglected power of organizational culture. *Strategy and Leadership* 28 (1): 24–29.

Schwartz, S.H. and Bardi, A. (2001). Value hierarchies across cultures: taking a similarities perspective. *Journal of Cross-Cultural Psychology* 32 (3): 268–290.

Schwartz, S.H. and Bilsky, W. (1987). Toward a psychological structure of human values. *Journal of Personality and Social Psychology* 53 (3): 550–562.

Schwartz, S.H. and Sagiv, L. (1995). Identifying culture-specifics in the content and structure of values. *Journal of Cross-Cultural Psychology* 26 (1): 92–116.

Sharp Paine, L. (2003). *Value Shift: Why Companies Must Merge Social and Financial Imperatives to Achieve Superior Performance*. New York: McGraw-Hill.

Shenkar, O. (2001). Cultural distance revisited: towards a more rigorous conceptualization and measurement of cultural differences. *Journal of International Business Studies* 32 (3): 519–535.

Sheth, J.N. and Parvatiyar, A. (1992). Towards a theory of business Alliance formation. *Scandinavian International Business Review* 1 (3): 71–87.

Smirchich, L. (1983). Concepts of culture and organizational analysis. *Administrative Science Quarterly* 28 (3): 339–358.

Stevens, J. (1989). Integrating the supply chain. *International Journal of Distribution and Materials Management* 19 (8): 3–8.

Tagiuri, R. and Litwin, G.H. (eds.) (1968). *Organizational Climate*. Graduate School of Business Administration, Harvard University.

Tavistock Institute of Human Relations (1966). *Interdependence and Uncertainty: A Study of the Building Industry*. London: Tavistock Publications.

The Chinese Culture Connection (a team of 24 researchers) (1987). Chinese values and the search for culture-free dimensions of culture. *Journal of Cross-Cultural Psychology* 18 (2): 143–164.

Thompson, J.D. (1967). *Organizations in Action: Social Science Bases of Administrative Theory*. New York: McGraw-Hill.

Thorelli, H.B. (1986). Networks: between markets and hierarchies. *Strategic Management Journal* 7 (1): 37–51.

Tone, K., Skitmore, R.M., and Wong, J.K.W. (2009). An investigation of the impact of cross-cultural communication on the management of construction projects in Samoa. *Construction Management and Economics* 27 (4): 343–361.

Triandis, H.C. (1989). The self and social behavior in differing cultural contexts. *Psychological Review* 96 (3): 506–520.

Triandis, H.C. (1990). Cross-cultural studies of individualism and collectivism. In: *Cross-Cultural Perspectives, Nebraska Symposium on Motivation 1989* (ed. J.J. Berman), 41–133. University of Nebraska Press: Lincoln.

Trompenaars, F. and Hampden-Turner, C. (1997). *Riding the Waves of Culture*, 2e. London: Nicholas Brealey.

Tsoukas, H. (2005). *Complex Knowledge: Studies in Organizational Epistemology*. Oxford: Oxford University Press.

ul-Haq, R. (2003). *Executive Briefing – Strategic Alliances*. Birmingham: the Birmingham Business School.

Values Survey Module, 2008. VSM 08 available at https://geerthofstede.com/research-and-vsm/vsm-08/ (accessed August 2019).

Van Dyne, L., Graham, J.W., and Dienesch, R.M. (1994). Organizational citizenship behaviour: construct redefinition, measurement and validation. *Academy of Management Journal* 4 (4): 765–802.

Victor, B. and Cullen, J.B. (1988). The organizational bases of ethical work climate. *Administrative Science Quarterly* 33 (1): 101–125.

Vlaar, P.W.L., van Fenema, P.C., and Tiwari, V. (2008). Cocreating understanding and value in distributed work: how members of onsite and offshore vendor teams give, make, demand, and break sense. *MIS Quarterly* 32 (2): 227–255.

Vroom, V. (1964). *Work and Motivation*. New York: Wiley.

Wagner, J.A.I.I.I. and Moch, M.K. (1986). Individualism-collectivism: concept and measure. *Group and Organization Studies* 11 (3): 280–304.

Wayne, S.J., Shore, L.M., and Linden, R.C. (1997). Perceived organizational support and leader-member exchanges: a social exchange analysis. *Academy of Management Journal* 40 (1): 82–111.

Weeks, J. and Gulunic, C. (2003). A theory of the cultural evolution of the firm: the intraorganizational ecology of memes. *Organization Science* 24 (8): 1309–1352.

Weick, K.E. (1995). *Sensemaking in Organizations*. Thousand Oaks, CA: Sage.

Weick, K.E. (2001). *Making Sense of the Organization*. Malden MA: Blackwell.

Weick, K.E. (2009). *Making Sense of the Organization: The Impermanent Organization*. Chichester: Wiley.

Weick, K.E., Sutcliffe, K.M., and Obstfeld, D. (2005). Organizing and the process of sensemaking. *Organization Science* 16 (4): 409–421.

Williams, A., Dobson, P., and Walters, M. (1989). *Changing Culture: New Organizational Approaches*. London: Institute of Personnel Management.

Witkin, H.A. (1976). Cognitive style in academic performance and in teacher-student relations (eds. S. Messick and Associates). In: *The Individuality in Learning*, 38–72. San Francisco, CA: Jossey-Bass.

Witkin, H.A. (1979). Socialization, culture and ecology in the development of group and sex difference in cognitive style. *Human Development* 22 (5): 358–372.

Womack, J.P., Jones, D.T., and Roos, D. (1990). *The Machine that Changed the World*. New York: Free Press.

Yamazaki, Y. (2005). Learning styles and typologies of cultural differences: a theoretical and empirical comparison. *International Journal of Intercultural Relations* 29 (5): 521–548.

Part II

Chapters that Principally, but not Exclusively, Deal with Case Study Material

10

Managing Megaproject Supply Chains: Life After Heathrow Terminal 5

Dr Juliano Denicol

10.1 Motivation for the Research

A megaproject is an extreme case of the temporariness phenomenon faced by the construction industry, where a number of firms will work together during an extended (but still limited) period of time to deliver a final one-off asset to the client (Priemus and Van Wee 2013). It is an endeavour conducted throughout several years, commonly 5 to 10 years in the construction phase, and it is not uncommon for one or two decades to elapse from initial inception to final handover. If the front-end stages of inception, planning, and approval are considered, a large proportion of projects will end up with at least one decade of life-cycle until final delivery (Artto et al. 2016).

The UK broke the paradigm and started the shift from old practices to a new approach through the Heathrow Terminal 5 breakthrough project (Potts 2009; Davies et al. 2009). The lessons learned were adapted and incorporated in the London Olympics 2012 (Davies and Mackenzie 2014; Brady and Davies 2014), and Crossrail – incorporating, what was claimed at the time to be, the world's first innovation strategy (Davies et al. 2014; Dodgson et al. 2015). In this way, the UK construction industry is 'passing the baton' and implementing best practices and innovations to subsequent megaprojects: Thames Tideway Tunnel, a 25 km sewer tunnel under the river Thames in London, and High Speed Two, a high-speed railway connecting London to the north of the UK. In this context, there is a necessity and an opportunity to analyse these five megaprojects and their ecosystems in order to understand how their supply chains were structured.

Scholars are calling for megaproject studies in a number of refereed journals: *Project Management Journal* (Söderlund et al. 2015), *International Journal of Project Management* (Flyvbjerg 2015), and *Journal of Management in Engineering* (Wang et al. 2015). The UK is consolidating an expertise and methodology to deliver megaprojects that is attracting the attention of many countries, from developed countries such as the USA and European neighbours to developing nations such as Brazil, India, and China. Therefore, this chapter will theoretically frame the megaproject phenomenon towards supply chain management literature in order to illuminate the reasons why those projects are

Successful Construction Supply Chain Management: Concepts and Case Studies, Second Edition.
Edited by Stephen Pryke.
© 2020 John Wiley & Sons Ltd. Published 2020 by John Wiley & Sons Ltd.

critical in a global scale, the importance of their supply chains, and the opportunities for interorganisational coordination.

10.2 Construction Supply Chain Management

The construction industry historically has gathered and adapted management practices from other industries. The automobile industry, for example, influenced lean construction (Koskela 2000). Reports published by Latham (1994), Egan (1998), and the Strategic Forum for Construction (2002), which analysed the characteristics of the construction industry in the UK, have stimulated the adaption and translation of supply chain management concepts from well-developed sectors, such as retail, defence, and oil and gas, into construction. Arguably, Latham (1994) had a procurement approach for organising the supply chain, and from the perspective of Egan (1998), the construction industry needs main contractors to approach their supply chain through long-term partnerships, emphasising other factors related to performance instead of focusing on competition by the lowest price. In this context, Lambert et al. (1998) and O'Brien et al. (2008), respectively for the manufacturing and construction industries, pointed out that efficient supply chain management enables significant cost reduction through the increased integration between supply chain actors, ranging from distant 'n tier' suppliers to end users. Thus, the adoption of supply chain management practices can improve the overall performance of the construction industry, since the companies start to analyse the business holistically and make decisions through a 'systems lens', consolidating demands at an enterprise level, instead of oriented solely towards individual projects (Pryke 2009; Tommelein et al. 2009; Pryke 2017).

The fragmentation of construction supply chains generates a lot of waste and reduces productivity, since the chain is quite vertical containing several Tier 1 suppliers (Lambert et al. 1998). In terms of organisational structure, Pero and Rossi (2014) present a structure for the construction supply chain where there are four main categories of players, namely:

(i) Clients, which are ultimately owners and make their profit running operations exploiting the physical asset.
(ii) General contractors, which are organisations responsible for the delivery of the entire physical asset to clients. Usually, these are companies that rely on systems integration capabilities, through supply chain coordination (or project networks, depending of the maturity of the company, if the integration is realised at project or enterprise level).
(iii) First tier suppliers, which provide systems for the general contractors, often finished systems in the form of components ready to assemble on site.
(iv) Second tier suppliers, which provide subsystems and intermediate components for the production process of first tier suppliers.

This representation has parallels with other organisational structures present in the literature and with the understanding of the majority of construction supply chain scholars. The main variance among these representations is that some authors deconstruct the supply chain into more detail adding more levels of players, the lowest tier being responsible for the delivery of just raw materials or small components for the supplier in the above tier. The tier number depends on the focal perspective from which the supply chain structure was designed. Pero and Rossi (2014) emphasise the role of the general contractor, and represent Tier 1 suppliers as the ones below the general contractor, while Davies and Hobday (2005) present a structure where Tier 1 contractors are the general contractors or systems integrators.

Cox and Ireland (2006) present a different conceptualisation of construction supply chains, introducing the concept of gate-keeper, which relates to the issue of fragmentation, where each tier of the supply chain can only understand and manage its immediate lower tier supplier. This logic argues that the industry lacks a player with a holistic coordination lens, which can be translated into (programme or portfolio) management. Cox and Ireland present an evolution of construction supply chain categorisation, which is segmented according to three major supply chains (materials, labour, and equipment) that are integrated by two supply chain actors in different phases of the construction project, the professional service firm and the construction engineering firm. The professional service firm, in the shape of architects and other design consultants, acts as systems integrator for information, adding value reconfiguring and transforming the available information. Conversely, the construction engineering firm is the more traditional systems integrator (as seen in high volume industries, such as manufacturing) relying on their capability of fitting all the subsystems together to form the entire physical system desired by the client (Prencipe et al. 2003; Davies 2017).

The framework proposed by Cox and Ireland (2006) is different from the average description since it divides what other authors generically call construction supply chain into supply chain categories with similar characteristics. Arguably, it is a more efficient framework considering that construction is an engineer-to-order (ETO) industry as a whole. However, there are many specialised supply chains that provide systems and subsystems for construction systems integrators, and those organisations have different production strategies, from make-to-stock (MTS) to assemble-to-order (ATO). The production strategy is important and has direct implications and effects on management once it reflects a set of decisions regarding demand and supply, inventory levels, commercialisation of the end product, and operation of the production process. It is still a very rare scenario where just companies with high maturity recognise that the construction supply chain comprises a set of diverse supply chains. This recognition is the first step in establishing an extended understanding enabling further analysis of the context and ultimately leading to creative supply chain management solutions.

Koskela and Ballard (2006) presented a discussion regarding the different perspectives and interfaces of two project management schools – economics-based and production-based. The authors argue that the economics approach conceptualises the project as an information system where organisations perform transactions amongst themselves to deliver the final product. The focus of this approach is to design and manage an organisational structure and the subsequent contracts involved in the transactions. As an optimisation mechanism, the target is to reduce uncertainty in order to have a more reliable information system (or project). On the other hand, the production approach conceptualises the project as a production system, where there is transformation, flow, and value. The focus of this approach is to design and manage the production system. As an optimisation mechanism, the target is to eliminate waste continuously in the production system to increase the value of the overall system.

This is the duality faced by the construction industry. On the one hand, the industry understands and follows the production logic, in the context of its intrinsic nature. It focuses on the physical asset being constructed and allocates resources to optimise sequencing and enable integration of systems and subsystems. On the other hand, the project-based/ETO characteristic of the industry tends to move the focus from the economics approach observed in mass production industries. The practice in construction is to establish organisational structures and production strategies on a project-by-project basis, a fact that avoids consolidation of demands at the enterprise level and the development (and improvement) of relationships (not transactions) in a long-term perspective. This discussion about conflicting economics vs production conceptualisations reflects the dilemma in managing supply chains in the construction industry.

The economics and production approaches are not mutually exclusive and do not work in silos; they are intertwined and complement each other. Before any production can start (development and delivery activities), there are activities and strategies that need to be designed at the enterprise level to support the production activities. Contracts are not production mechanisms, they are artefacts that establish and enable the rules of the game that connect the supply chain actors. Therefore, the economics approach needs to come first to design and establish an information system that will enable production to be carried out effectively later. The production-based theorists might argue that they cover the economics approach when the design of production-systems is being conducted. Once again, the problem is one of terminology; both schools are referring to different topics. The design emphasised by the production-based approach is connected to the production system (design and construction), which is connected to the project level, since the production in construction follows a logic project-by-project. The design emphasised by the economics-based approach is concerned with organisational structures, functions, and information flows of the firm, which is located at the enterprise level. There is, therefore, a need to design a macro system at the enterprise level that enables the production function at the project level.

This perspective can be expanded through the temporary–permanent lens, with projects being temporary and the enterprise permanent. The project level is associated with construction endeavours that happen at the boundaries of the firm through temporary supply chains, assembled on a project-by-project basis. The enterprise level is permanent and composed of a centralised structure that works as an enabler to all projects. Ideally, this centralised structure would be able to consolidate requirements from all projects (or at least according to a specific segmentation) and apply portfolio management techniques to achieve economies of scale and standardisation. However, the challenge of construction companies is frequency of demand, which is translated into a lack of visibility about future projects and a consequential lack of certainty over workload. Construction enterprises tend to postpone investments without confirmation of demand or at least visibility of the upcoming pipeline of projects. The visibility of future investments by public and private sectors could provide them with confidence to invest in capability building based on the expectation of successful bids. The business development department of a project-based firm is constantly mapping the market, establishing relationships, and creating opportunities for future business. When there is confidence in a specific project, the business development department informs the bidding department; there is then an interface to pass the baton and align the strategy to secure that project, merging the current enterprise strategy and the future corporate plans (if known) with the future strategy to be deployed at the project.

Following a comprehensive research project dealing with a large dataset about major projects (Morris and Hough 1987), Morris concluded that the problems causing major disruption in the project environment are not related to the iron triangle (i.e. the pursuit of cost, time, and quality standards) during the execution, and can be tracked to the preconstruction phase – to the front-end of projects where strategic decisions are taken. Considering these findings, Morris expanded his research a few years later (Morris 1994), criticising the *status quo* of the literature and practice about the executional and operational bias of project management. Morris's research presented project management practices and perspectives and coined the term 'the management of projects'. Morris sought to distinguish between 'project management' – the executional practice with tools to optimise the production, stimulated by published bodies of knowledge and operationalised through guides and handbooks – and the 'management of projects' – a more systemic approach focused on the whole life-cycle of the project, emphasising the front-end stage where stakeholders are shaping the project and defining strategies into the various project phases. This focus on the management of projects caught the attention of the project management community who sought answers to critical questions previously neglected:

- What are the factors to be influenced in the front-end?
- What is the impact of these factors in the project life-cycle?
- Who are the stakeholders at this stage?
- To what extent should the stakeholders get involved to shape the project (Morris 2013)?

Project management or the management of projects aside, the unit of analysis, in terms of research, is still the project instead of the firm. This is in contrast to supply chain management, where the unit of analysis is a group or cluster of firms and the relationships between them, over time. The longer time scales associated with megaproject provides the opportunity to move beyond project management and the management of projects to understand how strategy is developed, needs are met, and value created through interorganisational and intraorganisational relationships within the supply chain serving the project.

10.2.1   Temporary vs Permanent Supply Chains (ETO vs MTS)

Competitive advantage is different for MTS (high-volume) and ETO (project-based) industries; the former has an emphasis on cost reduction and economies of scale, the latter has an emphasis on agile solutions and innovations, which should be available quickly to achieve the objectives in a temporary environment (Gosling et al. 2012). The construction industry is characterised as an ETO sector, which has high fragmentation, a low level of communication between the supply chain actors, one-of-a-kind projects, and temporary supply chains oriented by projects (Van der Vaart et al. 1996). Hicks et al. (2000) argued that supply chain management practices from high volume industries cannot be simply translated to project-based environments, considering that the context is not the same and there is not frequency of demand. Therefore, the context of the ETO supply chain should be carefully analysed before the application of traditional high-volume strategies like just-in-time, supplier base reduction and long-term relationships (Souza 2015).

In the construction industry, customisation is a *qualified* factor and not a *winner* factor, as seen in aerospace and shipbuilding (Amaro et al. 1999). In a context with high levels of customisation and without 'pipeline visibility' (certainty about demand from future projects), it is not possible to forecast the demand and apply a structured strategic process for supplier relationship management. The construction industry has neither constant contact with suppliers (e.g. daily, weekly) nor an extensive buyer power like original equipment manufacturers (such as GM, Ford, Fiat, Volkswagen). The challenge for the construction industry is to establish standard processes and procedures to maintain stability in the development of nonrepetitive one-off singular facilities (Ballard and Howell 1998).

The main problem for companies running their business through MTS or ATO production strategies relates to the large volume of intermediary or final products in inventory, which obviously impacts on cash flow. On the other hand, the main constraint for companies driven by make-to-order (MTO) and ETO production strategies is the lack of order confirmation that impacts the production planning and sequencing, ultimately reflecting in barriers to the management of supply chains. In addition, for the last two strategies there is the challenge of customer engagement and retention, once the client commitment happens early in the process and the lead-time between order placement and product delivery is high (Powel et al. 2014).

ETO and project-based production strategies are fundamentally similar but supported by different literatures. ETO is more closely associated with manufacturing and management science research; project-based production is more connected to the construction industry, media projects, and IT solutions, with the literature supported by organisational science and project management. It is possible to find in the literature different terminologies for the same, or similar, concepts. Three other expressions were identified: one-of-a-kind, MTO, and built-to-order. To a large extent, built-to-order is the least related to project-based settings, while MTO and ETO are often applied to project-based contexts, with just a few studies defining clearly the boundaries and differences between them.

Considering that the level of customers' requirements and customisation has increased over the last decades and that the world is moving towards project-based production through a phenomenon called 'projectification', there is a need for new and more robust mass customisation strategies to achieve 'customer delight'. Given this worldwide necessity, Gosling et al. (2011) proposed a framework expanding the usual four production strategies (ETO, MTO, ATO, MTS) and conceptualised a stage called research-to-order (RTO), which is a production strategy even more advanced than ETO, increasing the freedom for customisation and removing any level of standardisation in the procured item. In addition, this framework provides a clear representation of the decoupling point, which is a concept widely used for production planning and control. The decoupling point is associated with the moment where production ceases to be standardised and waits for the customer order with its customisation requirements (Olhager 2003). This is the point where production stops being pushed by the manufacturers (mass production) and starts to be pulled by the customers (customised production).

Figure 10.1 summarises the contrast between ETO supply chains (project-based: temporary) and MTS supply chains (high volume industries: permanent). On the

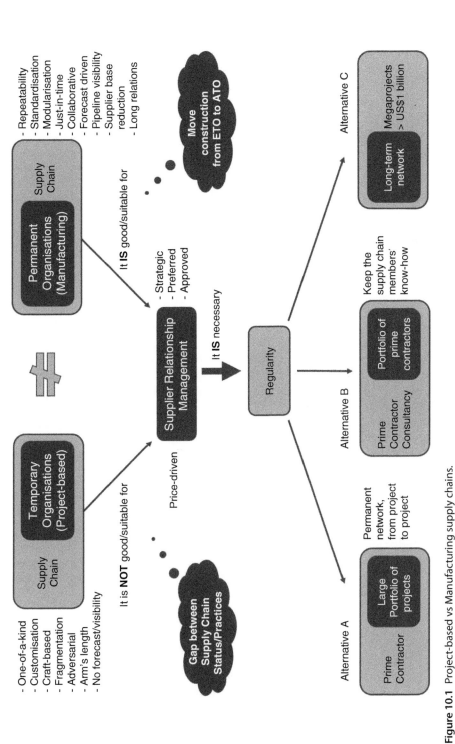

Figure 10.1 Project-based vs Manufacturing supply chains.

one hand, supply chains oriented by projects are usually reconstructed for every new project, which gives them the temporary characteristic and constrains improvements over time. Those supply chains present the following characteristics: one-of-a-kind, customisation, craft-based production, fragmentation of the supply base, adversarial relationships with business partners, arm's length as the predominant purchasing model, and high uncertainty of future projects which causes a lack of pipeline visibility constraining forecasts.

On the other hand, in contrast, manufacturing supply chains are permanent and stable, presenting the following characteristics: repeatability, standardisation, modularisation, just-in-time production, collaborative relationships with supply chain partners, pipeline visibility that allows forecast-driven demand, supplier base reduction over time, and long-term relationships with the supply chain. Arguably, these two supply chain types are in clear contrast with each other and it is possible to observe that the permanent supply chains are suitable for the establishment of long-term relationships with suppliers, while the temporary supply chain fails to achieve this goal due to the lack of regular contact between the organisations.

Frequency of demand and 'regularity' were identified as key characteristics for effective supply chain management implementation in ETO environments (Cox and Ireland 2006; Holti et al. 2000). An important critique of the previous reports which suggest the adoption of supply chain management to boost construction and foster innovations in the UK was that the context and examples provided were biased (Latham 1994; Egan 1998; Holti et al. 2000). These reports presented advice on the adoption of supply chain management in construction and building upon the consideration of the context of large clients. This generalisation is not helpful at all for the evolution of construction as an industrial sector, given that large clients are fundamentally different from the myriad of small and medium companies that comprise the supply chain. Large clients can (and should) apply supply chain management practices, capitalising on their portfolio of projects to integrate suppliers and exploit standardisation of processes and economies of scale. In other words, supply chain management is not for all organisations in construction.

Given this scenario, some questions emerged, namely:

- How to bridge the gap between the manufacturing and construction suppliers?
- How to create a business model that allows the creation of a constant supply and consequently a long-term relationship, in order to increase the productivity and competitiveness?
- How to expand towards integration with suppliers through performance measurement and supplier development?

In the light of these issues, three scenarios are proposed where it is possible to establish a proactive and effective construction supply chain management:

- Establish a prime contractor role (Holti et al. 2000) with a large portfolio of projects, each project running in a different stage, where the supplier base is replicated and reconfigured across projects, creating a permanent network.
- Utilise a prime contractor consultancy, where the focal consultancy acts to consolidate the demands from several prime contractors and maintain the supply chain know-how at a macro level.

- Explore megaprojects – large 'system of systems' projects that have an extended horizon available and power to proactively influence the supply chain.

If the client has a permanent business, running a portfolio of projects, delivering on time and budget produces financial savings. Furthermore, if there is a company/consultancy that can sell to a client the ability/assurance of delivery of a megaproject on time and budget, that company can negotiate a very attractive profit margin, considering that it will be saving a lot of money and creating value for the client, given that 98% of megaprojects tend to be delivered over budget and/or are delayed (McKinsey Global Institute 2013).

10.3   Why Are Megaprojects So Important?

Megaproject is a recent term to describe complex endeavours, with long timescales and total predicted expenditure of in excess of 1 billion US dollars. The use of the term, and its application, grew exponentially after Bent Flyvbjerg's book in 2003 (Flyvbjerg et al. 2003). Naturally, these projects always existed and are well documented in the literature under different terminologies over the last decades: major projects (Morris and Hough 1987), grand-scale projects (Shapira and Berndt 1997), large-engineering projects (Miller and Lessard 2000), mega capital projects (Rolstadås et al. 2011), large-scale projects (Merrow 2011), global projects (Scott et al. 2011) amongst others. Denicol et al. (2018a) presents a systematic literature review of the management of megaprojects drawing upon a comprehensive list of synonyms used to describe megaprojects over the years. Flyvbjerg (2014) points out that the term mega is usually associated with 1 billion projects but its correct relation would be millions, while gigaproject would be the correct term for 1 billion plus projects. However, for commercial and marketing reasons, the term gigaproject was never widely used by the market, which still utilises megaproject when referring to 1 billion projects. Galloway et al. (2012) has written one of the few books with gigaproject in the title (when comparing with titles on megaprojects). However, she proposes a complementary perspective rather than replacing one term for the other, where gigaprojects are projects with a cost in excess of 10 billion US dollars, leaving megaprojects in the classification between 1 and 10 billion US dollars. A recent classification of large-scale projects emerged when McKinsey launched the term ultra-large projects to represent projects with capital investment in excess of 5 billion US dollars (McKinsey 2017). The report suggests an analogy between marathons and ultra-marathons to support the argument that ultra-large projects are another category of projects with different characteristics from megaprojects. Analogies seem to be frequently used to convey messages about the complexity of megaprojects. Flyvbjerg (2014) presents an analogy from his colleague Patrick O'Connell advocating that project managers need to have a pilot's licence instead of regular driver's licence to deliver megaprojects.

Flyvbjerg (2014) suggests that the next frontier is teraprojects, where the cost would be above one trillion US dollars. However, to achieve this figure the author considers as a project the entire economic package launched by superpowers to incentivise their economy or the annual defence budget of a country. The scale of megaprojects is clearly increasing given the challenges faced by contemporary societies. However, there is a necessity for clearer boundaries and a need to relate measures to projects, programmes, and portfolios. Considering the number of projects available to establish a benchmark, one would expect that megaprojects would have evolved within its delivery form over the last decades; however, we are still facing quite the opposite. Major projects are rapidly growing in complexity, specification requirements, and budget, but the learning capacity across projects seems to be very limited since the overrun pattern and practices remain almost unchanged – a phenomenon called 'Productivity Paradox' (Flyvbjerg 2017). The success of megaproject deliveries is related to how much effort the managers concentrate in the front-end of the projects, understanding the particular risks of each stage of the project and its impacts on the whole system, as well as creating strategies to minimise and overcome them (KPMG 2013).

In the USA, Boston Big Dig was originally planned to open in 1998 but was delivered in 2007, nine years late with an outturn cost of 14.6 billion US dollars, more than 500% over budget (Greiman 2013). More recently, in the UK, the Wembley Stadium project was concluded four years late and 80% over budget. This project damaged the reputation of the UK construction industry, resulting in several suppliers seeking redress through the courts for contractual disputes; in several cases Tier 2 contractors were forced into insolvency. The effects of the Wembley Stadium project were evident when the UK government began to establish contact with the industry in order to build the London 2012 Olympic Park. Diverse actions were needed to engage the industry and earn back the confidence of the supply chain. Before Heathrow Terminal 5, the predominant mindset of the industry was to transfer as much risk as possible to the supply chain, creating an assurance shield for the client with all the legal guarantees, leading to extensive legal disputes in court, one of the principal causes for failures in megaprojects (Hart 2015). The extensively adopted lump-sum contract is at the root cause of under performances and legal disputes (Stinchcombe and Heimer 1985). In addition to Wembley Stadium, another two major projects are famous for having suffered severe contractual problems with serious consequences: the Channel Tunnel and the London Underground extension of the Jubilee line (Davies et al. 2017).

A great risk for megaprojects is what Wachs (1989) and Flyvbjerg et al. (2003) call 'strategic misrepresentation' and 'optimism bias', terms that explain the behaviour of the stakeholders at early stages of the project in order to obtain the go-ahead signal. In the feasibility and conceptual stages of the project there has been a systematic pattern of under estimation of the costs and difficulties for the project in many arenas, from planning applications and associated licence approval, to availability of materials and equipment, in order to provide an attractive business case with which to convince investors and other stakeholders to support the project. This behaviour has been statistically demonstrated by Flyvbjerg et al. (2003) who argued that it is not mathematically possible that such a large proportion of megaprojects suffered from the same mistake at the same stage, project after project, ignoring not only the technical capacity of the decision makers, but also the lessons that could have been learned from previous projects. The relevance of these issues becomes even more important if we think about post-war military and NASA projects in the USA, passing through the cold war and the space race, until the last three decades and the need for new infrastructure in the developing world and upgrades in developed countries. Gil (2017) presents evidence from practitioners stating that considerable optimism is necessary for any kind of project to receive a go-ahead decision, otherwise nothing would be constructed. However, the practitioners in this study suggested that they believed in optimism but not in falsification of documents and data to get approvals.

It can be argued that the benefits of these projects are not accurately quantifiable, especially in the very long term; this was exemplified by the Channel Tunnel and the Sydney Opera House. The latter project was highly criticised at the time of its construction and has suffered with the extensive cost and time overrun patterns. However, it is estimated that this single building contributes 775 million US dollars per annum to the Australian economy, through the boosting of tourism in Sydney (Deloitte Access Economics 2013). The measure of success of a megaproject depends when the evaluation is made. During the construction phase of the Channel Tunnel and the Sydney Opera House both were considered failures, but when the analysis is applied to the whole life of the facility, a very different picture emerges (Deloitte Access Economics 2013).

10.4 Megaproject Supply Chain Management

After the widely recognised managerial success of Heathrow Terminal 5 (Gil 2009), supply chain management practices were carried forward to the London 2012 Olympics (Mead and Gruneberg 2013), and were applied to Crossrail, the Thames Tideway Tunnel, and High Speed Two. Therefore, it is crucial to understand how these megaprojects designed their supply chain architecture.

There is a need for main contractors to outsource a large part of the production to specialist suppliers distributed throughout the supply chain tiers. This fragmentation is, arguably, a function of increasing specialisation in technically complex projects. The argument is to maintain in-house only the organisational core competencies and to maximise the outsourcing of other functions, given that specialist suppliers are likely to conduct these activities more effectively than the main contractor. The main core competence of Tier 1 contractors is the ability to oversee the entire construction process and coordinate the supply chain at project level, which is commonly called systems integration. Arguably, the innovation and value, which can be considered the hardware value, are hidden in the lower tiers of the supply chain. However, these suppliers would never be able to integrate all the pieces together due to their highly specialised business nature. Therefore, there is a large market for the software value, the systems integration capability, where companies are not producing components and products per se, but are acting as the head of the supply chain to deliver the project through integration. These companies provide the glue – the management capability to organise interdependent parts of the project. In megaprojects, this is the gap where the client can act jointly with main contractors (and/or delivery partners) in the decision-making process to design the supply chain structure and manage it over time to foster innovative solutions.

The industry is shifting its position from a risk-averse client that dumps the risk in the supply chain, towards the necessity of a more robust client organisation acting as an intelligent and strong client. The intelligent client is responsible for designing a supply chain structure to facilitate and stimulate bottom-up innovations. Suppliers of systems and subsystems are well placed to provide suggestions for problem solving and innovation. Considering the logic that suppliers will not innovate and share their best ideas without the appropriate incentives, there is a need for an organisation acting as the supply chain architect, orchestrator, manager, or systems integrator, which would be responsible for designing contracts that are appealing to the supply chain, as well as structuring the processes to absorb the innovations that will emerge from the incentivised suppliers. From this perspective, infrastructure clients in the UK are now emphasising their own decision-making processes at the front-end of projects. This is a reflection of a market evolution which increased their maturity, where they now understand the need to have a bigger voice and bear the risk during the process of developing the strategies that will drive the subsequent phases of the megaproject. These strategic decisions need to be conducted internally or be driven by the client when developed in collaboration with development and delivery partners. The intelligent client is the controlling mind of the megaproject.

In the case of Heathrow Terminal 5, the owner and client organisation (British Airport Authority – BAA) had an organisational memory of involvement in previous capital projects throughout the years (Potts 2009). This corporate knowledge, along with other factors and worldwide benchmarking research in airports and megaprojects, informed the strategic position of the client organisation in adopting a different supply chain management approach. Instead of following the common practice of dumping the risk in the supply chain and acting as the 'law enforcer of contracts' client, BAA decided to build upon its expertise in capital projects, internalise the risk, build a client-driven supply chain, and act as the supply chain manager or systems integrator, incorporating the principles of an intelligent client (Davies et al. 2009). BAA established contracts with key supply chain players and adopted a contracting approach that guaranteed the profit margin of suppliers with the adoption of open book production costs benchmarked through an independent cost consultant. This approach removed the risk and fear from the supply chain actors of not making profits in the project, opening the path to focus on improvements in the production process that would lead to cost reduction and innovation.

Asset owners that run an operation and have market knowledge are in a strong position to take the 'make or buy' decision (Winch 2014; Winch and Leiringer 2016). These companies usually conduct small expansions and improvement programmes internally. However, megaprojects demand another level of involvement and commitment. Considering this fact, megaprojects are usually conducted through different arrangements outside the boundaries and corporate structure of the firm. In other words, clients seek to separate their operational business from new major capital projects, building upon the understanding that these endeavours need dedicated structures such as an independent leadership team and bespoke governance structure to cope with the scale and pace of expenditure (Croft et al. 2016). BAA had to make a strategic decision considering its own project management and systems integration capabilities, and it was only possible because the company had a strong team with previous experience in capital projects and knowledge about aviation sector projects (Brady and Davies 2011). However, recent interviews with senior managers revealed that, given the scale and complexity of Terminal 5 compared with previous aviation projects from BAA, it was necessary to build a different and more sophisticated suite of processes and procedures to address the project. The executives were clear that the corporate memory of previous projects had influence only until a certain point in BAA's decision of bearing all the risk, all the time. Effective supply chain management relies upon a strong, experienced client organisation (Pryke 2009) or, alternatively, the appointment of a supply chain management agent such as a contractor or consultant (Holti et al. 2000)

Denicol et al. (2018b) present a comprehensive discussion that clarifies the roles of owners, sponsors, clients, and operators in megaprojects, which usually are different entities but sometimes might combine two or more terms. This is a terminology problem originating from the management literature, where different words are used as synonyms, influenced by the necessity to use new terminologies and concepts to aid publication in high impact-factor journals. The four terms previously mentioned are used interchangeably. However, owner and operator, and sponsor and client are the most commonly used synonyms.

Building upon Heathrow Terminal 5, the evolution of the megaproject ecosystem in the UK indicates that it is necessary to establish an increased interface between the owner/sponsor and the companies appointed to deliver the megaproject on their behalf. Denicol et al. (2017) present a perspective of project-based companies that work alongside clients (shifting the mind-set of working 'for clients') helping them to deliver major programmes according to their requirements. The research analysed CH2M, a company recognised worldwide for its programme management capabilities (Engineering News Record 2015), unpacking its involvement in several megaprojects. Increasingly, owners are moving from a role solely of sponsor towards a more proactive participation in decisions during the whole life-cycle of the project, becoming an 'intelligent client'. The movement towards a close interaction by clients is connected to historic failures of large and complex projects to achieve the requirements of the owner.

In the case of the London 2012 Olympics, the temporary client organisation (the Olympic Delivery Authority) recognised that as a brand new (pop-up) client it would not have the capacity and capability to build an organisation in time to manage all the Tier 1 contractors responsible for delivering parts of the Olympic park. Hence, the Olympic Delivery Authority recruited a consortium of construction companies (CH2M, Laing O'Rourke and Mace – CLM) to perform the function of a delivery partner, with the specific remit and focus on the production dimension, coordinating all Tier 1 contractors and integrating their work packages at a programme level. Recognising the need for a stronger involvement in the project, the temporary client organisation (the Olympic Delivery Authority) was responsible not only for establishing a closer interaction with the delivery partner, but also for focusing on and managing everything external that could affect the production progress. In contrast to previous failure models, the promoters of the London 2012 Olympics had a proactive engagement throughout the project stages rather than acting as the traditional hands-off sponsor with interest only in the final asset and limited involvement in the project.

In the case of Crossrail, The Department for Transport (DfT) and Transport for London (TfL) (the sponsors) created a temporary delivery vehicle, Crossrail Limited (the client), which was empowered to deliver the project on behalf of the sponsors. Crossrail Limited also recognised that as a pop-up client organisation it was necessary to complement its capabilities by working with project-based companies. To this end, the client appointed two consortiums of three companies each to increase its capability of managing the Tier 1 contractors and delivering the £14.8 billion megaproject currently unfinished and overbudget. The two consortiums were: The Programme Partner (CH2M, AECOM, and Nichols) and the Programme Delivery Partner (Bechtel, Halcrow, and Systra). Initially, the client structure responsible to manage the Tier 1 contractors was composed of three different entities: Crossrail Limited, the Programme Delivery Partner, and the Programme Partner. After a period of tensions, lack of clarity, and duplication of work between the three entities, Crossrail Limited senior leadership decided to remove the corporate boundaries and integrate the seven companies into one single client team. In this case client and delivery partners worked as an integrated team where everyone was recognised as a Crossrail Limited employee (even if one of the other six companies was paying the salary). Comparing Crossrail with the London 2012 Olympics, although the Olympic Delivery Authority and CLM worked very closely with clear and open communication channels, there was a much stronger separation between the two entities, with clearer distinction of roles. Regardless of the integration model, both structures were driven by the same managerial problem, where both temporary client organisations (the Olympic Delivery Authority and Crossrail Limited) recognised the necessity of a systemic approach to optimise the management of Tier 1 contractors, consolidating and exploiting the similarities at programme level.

In the case of the Thames Tideway Tunnel, Bazalgette Tunnel Limited, also known as Tideway (the client), hired CH2M to be the programme manager (Delivery Partner) to increase its client capabilities and assist in the management of the Tier 1 contractors. The contractors are organised in three joint ventures and divided in geographical areas (West, Central, and East). The West joint venture is composed of BAM Nuttall Ltd, Morgan Sindall plc, and Balfour Beatty Group Limited. The Central joint venture is formed by Ferrovial Agroman UK Ltd and Laing O'Rourke Construction. The East joint venture comprises Costain Ltd, Vinci Construction Grands Projets and Bachy Soletanche. Recognising the technical challenges of integration along the Tunnel route, Tideway awarded a systems integration contract to Amey that was to work with the joint ventures to achieve technical consistency and prepare the asset for operation. An interesting feature of this project was the alliance agreement between Tideway, Thames Water, the three joint ventures, and the systems integrator. This agreement had the objective of incentivising a horizontal collaboration between all parties, in addition to the vertical contractual incentives between the client and the organisations delivering the project. Comparing the Thames Tideway Tunnel with Crossrail, it is possible to see an evolution of the industry and a bigger appetite for risk once bigger and fewer contracts were awarded on Tideway's three geographical areas. This is a reflection and learning from Crossrail's experience, where the client divided the project into several contracts and increased its challenges to manage a large number of interfaces and integrate them all. One factor that incentivised Crossrail to divide the project into several contracts was the economic recession period in which the procurement happened, resulting in a market with reduced appetite for risks and big contracts.

High Speed Two is currently the largest infrastructure project in Europe and by far the most challenging in terms of complexity, scale, and timescales. This railway will connect London with the north of England and is divided in three phases: (i) Phase 1 – London to Birmingham; (ii) Phase 2a (Birmingham to Crewe); and (iii) Phase 2b – Crewe to Manchester and Birmingham to Leeds. The DfT is the sponsor of the £55.7 billion megaproject and has created a client organisation called High Speed Two Limited to deliver the project. High Speed Two Limited has separated the development and delivery stages and appointed a Development Partner and an Engineering Delivery Partner to increase the capacity and capability of the client organisation. Similarly to Crossrail, the partners appointed work integrated and collocated with the client (High Speed Two Limited) to procure and manage an extensive supply chain that provides services and products during development and delivery. Regarding the delivery stage of Phase 1, the client divided the route from London to Birmingham into three geographical areas (South, Central, and North) and appointed Tier 1 contractors for each one of them. This is similar to Tideway's approach in the rationale of dividing the project geographically and also of awarding bigger contracts than Crossrail. High Speed Two Limited has awarded three contracts to deliver the enabling works, one for each geographical area, and seven contracts to deliver the main works civils (two contracts for South, three for Central and two for North). Each contract is delivered by a joint venture of two or more Tier 1 contractors, who have the freedom to design their own supply chains from Tier 2 suppliers onwards. High Speed Two Limited is responsible for managing the interfaces between contracts and to act as the ultimate systems integrator.

Recently in the UK, the involvement of the sponsor and its delivery vehicle (the client) has been particularly strong in the early stages of megaprojects, given the impact decisions taken at the front-end are likely to have on subsequent phases of the project. It is important to consider that this involvement is continuous and also encompasses the construction phase with much more emphasis on reporting and control than at the front-end, however. A particular feature of Heathrow Terminal 5 was its opening failure with massive media coverage of the baggage system problems. This example sheds light on the importance of the early engagement of the owner and operator with the development and delivery teams, to ensure consistency in the requirements throughout the project and co-create strategies to transition the asset into operation. Recently Morris has argued that the role of the owner is essential and considered it as the most influential supply chain player in the front-end of projects, shaping complex decisions such as the degree of technology to be adopted and the governance structure that would frame the functioning of complex interfaces with external stakeholders and the project supply chain (Morris 2014). Another layer of complexity relies on the permanent and temporary nature of the owner and its extent of influence in the project. If the owner is a permanent organisation with a wealth of experience in the sector, there is a constant challenge to consolidate and align the corporate and project strategies. The owner should provide guidance about complex decisions at a very early stage of the journey. Analogously, the owner can be seen as the captain of the ship – the one that has the power and whose decisions have potential to change completely the direction of the endeavour. Considering the relevance of this supply chain actor, it is surprising that so little literature deals with clients and owners (when compared with literature about main contractors) regarding interfaces with other stakeholders and the possible impacts on the project. Pryke (2012) does, however, adopt a network perspective to understand the role of four construction clients – two of which actively manage their supply chains and two of which take more passive roles as clients.

The UK megaproject clients are actively engaged in the front-end, assessing the risks and establishing strategies to impact the supply chain such as: (i) development of integrated project teams to manage the interfaces with the supply chain players; (ii) risk analyses to design the supply chain architecture; (iii) performance control and innovation management at the execution phase through a structured reporting system; (iv) creation of a more flexible organisational structure to stimulate and absorb the innovations from all supply chain actors, in advance or during construction; (v) establishment of robust change management processes, specifying the levels of responsibilities; and (vi) directives for the use of technologies such as Building Information Modelling (BIM) and virtual reality for supplier performance control, material tracking, and just-in-time deliveries at the construction site.

10.5 Conclusion

The adversarial nature of the construction industry is well documented in the literature, mostly driven by inherent competing priorities amongst organisations, enhanced by the fragmentation of the sector. Supply chains increasingly consist of highly specialised actors. Therefore, in public megaprojects there is a need for a closer interaction between the public entity (or their representative) and the private organisations contracted to deliver the project. These entities will form a temporal and evolutionary supply chain, with organisations leaving and entering the project coalition considering the changes of scope throughout the phases of the project. It is particularly important to have an intelligent public client organisation driving the project, functioning as the controlling mind, designing the supply chain architecture and orchestrating the interfaces, relationships, and behaviours amongst the supply chain actors at different points in time.

The constant review and evaluation of the supply chain form (intra- and interorganisational structures) is an important capability for public sector clients. The capability of infrastructure clients to evaluate the dynamics of their internal processes and structures (intraorganisational relationships) is the first step of a mature organisation in the direction of the appropriate design of a megaproject supply chain. The understanding of what is happening inside the boundaries of the client organisation (sometimes the one project firm) is essential to design the interfaces of the firm with the supply chain it intends to assemble. How would you know what the complementary capability is that you need to hire without a clear understanding of what is already there? Organisations that do not have a clear understanding of what is happening inside their own boundaries are likely to be in a 'lock-in' position after unintentionally awarding nonattractive contracts to the supply chain (and to themselves). It is extremely challenging, and more expensive, to work backwards and reengineer organisational behaviours in a scenario where contracts were let without the strategic thinking of how the vertical and horizontal relationships would work. The commercial relationship is driven by the contract and all organisations will have experts analysing the contract and its propositions put forward by a client. Evidently, lower tiers of the supply chain tend to have less resources to invest in strong legal capability to analyse contracts, whereas Tier 1 contractors (systems integrators) would be in a stronger position in terms of legal advice to influence the relationship.

The vertical relationship to be designed is the one between the client, the contractor, and the wider supply chain. It is often not fully considered precontract and, in the cases where it is, there is a focus on the single interface between the client and Tier 1 contractor. The design of the relationships should go deeper and be expanded to how that contractor is planning to manage its supply chain (Tiers 2, 3, and 4) during the project, exploring topics such as prequalification, selection, performance measurement, fair payment, and others. This alignment is essential and supported by numerous industrial reports and academic publications informing that Tier 1 contractors are usually integrators of components and systems, while the innovation and value added remain at the lower tiers of the fragmented construction industry.

Considering the external environment, it is equally important for the supply chain actors to be clear as to which operations are being managed by whom and how. Naturally, there are challenges with the concept of supply chain management across a single business entity through interorganisational process integration. To start with, considering a focal company and its relationship with Tier 1 suppliers, there are differences and tensions regarding power relationships, levels of interest for the development or execution of a given task (or product), industrial and marketing core competences (and therefore secrets), and multilateral agreements with competitors, among others. Therefore, it is clear that if in a binary relationship there are challenges about processes and terminologies (the same word means different things for different people), it is even harder to achieve business process integration when amplifying the spectrum for a supply chain with several organisations involved (or affected). Metrics and indicators should be established to evaluate supply chain relationships, enabling analyses to identify the weaknesses of the system and continuous improvement through the implementation of action plans.

Arguably, the exploitation of large programmes and portfolios to implement supply chain management in construction relates to big clients with relatively constant demand and a pipeline of projects that can be forecasted in terms of utilisation of resources. For those clients, which are empowered by a favoured power relationship with the supply chain through its buyer power and supply chain knowledge, it is beneficial to develop an in-house management department to interact with supply chain partners. From the lower tier suppliers perspective there is a high level of interest in establishing a relationship with big clients, mainly supported by two facts, namely: (i) the possibility of accessing that given market, which is usually a considerable part of their current production, operations, or distribution; and (ii) the incentive of integration in a long-term relationship through a portfolio environment that might provide high visibility of current and future projects. This can be translated into certainty of demand with positive implications for production planning and control. In this scenario, suppliers are confident to invest their resources in developing productive relationships towards continuous improvement for value creation. Considering the arguments presented it is possible to state categorically that supply chain management practices for project-by-project environments are unlikely to achieve success in managing supply chains and are difficult to sustain in a long-term business perspective.

The construction industry suffers with temporary supply chains oriented towards projects and the organisational culture biased towards short-term solutions. This inhibits the implementation of supply chain management practices, depending on the context. Critically, following the concept of regularity, it is possible to establish a parallel between supply chain management in mass production and construction environments involving the necessity of a continuous flow of operations to develop the supply chain actors. Continuous improvement is closely related to regularity (or frequency) of relationships, and therefore there should be an active portfolio of projects where the buyer can integrate the supplier and improve business process considering a learning curve over time. The operations scenario from high volume and process industries can be translated to construction in the form of programmes and portfolios, where there is a management perspective relying on consolidation of demands at an enterprise level instead of project level. In summary, supply chain management relies on the establishment and development of long-term relationships; this necessarily implies a constant and close interaction between buyer and supplier.

The challenge for the construction industry is how to transfer the set of supply chain management routines successfully applied in mass production and high-volume industries to a project-based environment with production oriented towards small batches and high uncertainty in relation to future demand. Drawing upon the manufacturing experience, clients and main contractors could follow Toyota's model and embed small teams from the focal organisation managing the supply chain into the suppliers' companies to work alongside their teams. This initiative would work for short periods of time as a small project or consultancy, aiming to guarantee the focal company's requirements (clients and main contractors), as well as to provide cross-fertilisation and access to new knowledge to develop improvements and innovation through co-creation.

Drawing upon the discussion of this chapter, seven avenues for future construction supply chain management research are suggested:

1. *Enterprise and project supply chain management* – In a project-based environment, supply chain management and integration are not appropriate for all projects. It is necessary to have frequency of demand, which can be translated into programmes or a portfolio of construction projects (or sites) to manage activity of the supply chain over time, relying on a list of homologated suppliers (strategic, preferred, approved). The connection between the enterprise and project levels is yet to be improved by main contractor organisations, which could unlock business-wide value by acting holistically to integrate and exploit multiple temporary supply chains at the enterprise level.

2. *Client-driven supply chain management in construction* – The research on supply chain management has been focused on main contractors (Tier 1 systems integrators) and there is limited research available considering the perspective and involvement of the client, which is surprising and intriguing given that this is a highly influential player in the process. Property developers are clients and sometimes also the Tier 1 integrators. However, they are transient clients since they will sell the units to end customers, the ultimate owners (commercial companies or residents), immediately after the construction is finished. It would be interesting to see a comparative study of supply chain management practices between property developers and infrastructure client organisations (pop-up clients and permanent owners) (Pryke 2012).

3. *Focus on the structure not on the technology* (focus on software not hardware) – The construction industry is very resistant to change. The business model is based upon a craft industry with all the materials being brought to site and the building produced on site. How can we change or challenge this business model? How about the Chinese who are trying to manufacture their buildings in factories and just carrying out final assembly on site? First it was a hotel with 30 floors in 15 days, then another hotel with 15 floors in six days, and lastly a building with 57 floors in 19 days. The industry might reflect on a situation where construction moves from an ETO to an ATO production strategy, focusing on the maximisation of standardisation, modularisation, and off-site construction. The increasing adoption of BIM and other technologies will help and foster innovation in the near future. However, technology is not a deterministic answer or solution – the industry needs to focus on the supply chain structure, the umbrella under which technologies are included. If an efficient operation is automatised, efficiency will maximise: if an inefficient operation is automatised, however, it will be faster, but it will still be an inefficient operation.

4. *Supply chain management metrics and indicators* – Surprisingly there is little literature regarding the interface between supply chain and metrics and indicators, which

can be used to provide a diagnosis of interorganisational relationships, enabling the development of action plans for improvements and extended integration and control.

5. *The design of supply chain architectures with the focus on collaboration over time* – Clients should identify which actors are the ones likely to have long-term relationships in a megaproject and work backwards to design a favourable structure that would incentivise them to engage in collaborative initiatives with the right behaviour during the project. This strategic decision implies a shift in the mind-set of infrastructure clients in two areas:

   a) Recognising its role as the party to which the risk will return if something goes wrong during the process and therefore the necessity for an upfront proactive posture to internalise and manage risk, rather than transferring it to other supply chain actors.

   b) Moving decisions from cost-driven, arm's length, and project-by-project basis to a programme management perspective, where commonalities of multibillion programmes are consolidated at the client level and exploited.

6. *Use of data analytics to go beyond Tier 1 suppliers* – With the advancement of artificial intelligence and machine learning techniques, in the future infrastructure clients might be able to combine a range of datasets to extract business intelligence regarding the wider network of companies that might comprise the supply chain. This information could be useful to design supply chains informed by the capabilities and capacity of the players in the market. Data-driven technology will enable the necessary visibility to inform strategic decisions and avoid awarding contracts to companies likely to fail from the outset through lack of capacity.

7. *Explore and translate supply chain management practices from other project-based sectors* – The construction industry needs to try to connect practices from other industries with project-based production strategy, which have the potential to translate best practices such as offshore, shipbuilding, and aerospace. Although these industries are similar through the use of small batches and a fixed layout with dynamic production activities happening around the final product, it is important to notice that the context of each industry is very relevant and there are peculiarities that do not allow a simple observation of the practice and adoption in other industries. The process of adoption by the construction industry of management techniques from manufacturing, such as Total Quality Management and Lean Production, requires some modification or a process of translation to render these techniques effective in the construction context. This would be similar to the cultural translation conducted by Toyota when it transferred the Japanese Lean Production philosophy to manufacturing plants in the USA.

References

Amaro, G., Hendry, L., and Kingsman, B. (1999). Competitive advantage, customisation and a new taxonomy for non make-to-stock companies. *International Journal of Operations and Production Management* 19 (4): 349–371. https://doi.org/10.1108/01443579910254213.

Artto, K., Ahola, T., and Vartiainen, V. (2016). From front end of projects to the back end of operations: managing projects from value creation throughout the system lifecycle. *International Journal of Project Management* 34 (2): 258–270. https://doi.org/10.1016/j .ijproman.2015.05.003.

Ballard, G. and Howell, G. (1998) What kind of production is construction? In Proceedings, 6th Annual Conference of the International Group for Lean Construction, Guaruja, Brazil.

Brady, T. and Davies, A. (2011). Learning to deliver a mega-project: the case of Heathrow terminal 5. In: *Procuring Complex Performance: Studies of Innovation in Product-Service Management* (eds. N. Caldwell and M. Howard), 174–198. Oxford, UK: Routledge. ISBN: 9780415800051.

Brady, T. and Davies, A. (2014). Managing structural and dynamic complexity: a tale of two projects. *Project Management Journal* 45 (4): 21–38. https://doi.org/10.1002/pmj.21434.

Cox, A. and Ireland, P. (2006). Relationship management theories and tools in project procurement. In: *The Management of Complex Projects: A Relationship Approach* (eds. S. Pryke and H. Smyth), 251–281. Oxford, UK: Blackwell Publishing. ISBN: 9781405124317.

Croft, C., Buck, M., and Adams, S. (2016). *Lessons Learned from Structuring and Governance Arrangements: Perspectives at the Construction Stage of Crossrail*. Crossrail Learning Legacy Available at: https://learninglegacy.crossrail.co.uk/documents/lessons-learned-from-structuring-and-governance-arrangements-perspectives-at-the-construction-stage-of-crossrail.

Davies, A. (2017). The power of systems integration: lessons from London 2012. In: *The Oxford Handbook of Megaproject Management*. (in Press) (ed. B. Flyvbjerg). Oxford: Oxford University Press.

Davies, A. and Hobday, M. (2005). *The Business of Projects: Managing Innovation in Complex Products and Systems*. Cambridge: Cambridge University Press. ISBN: 9780521843287.

Davies, A. and Mackenzie, D. (2014). Project complexity and systems integration: constructing the London 2012 Olympics and paralympics games. *International Journal of Project Management* 32 (5): 773–790. https://doi.org/10.1016/j.ijproman.2013.10.004.

Davies, A., Gann, D., and Douglas, T. (2009). Innovation in megaprojects: systems integration at Heathrow terminal 5. *California Management Review* 51 (2): 101–125. https://doi.org/10.2307/41166482.

Davies, A., MacAulay, S., DeBarro, T., and Thurston, M. (2014). Making innovation happen in a megaproject: London's crossrail suburban railway system. *Project Management Journal* 45 (6): 25–37. https://doi.org/10.1002/pmj.21461.

Davies, A., Dodgson, M., and Gann, D.M. (2017). Innovation and flexibility in megaprojects: a new delivery model. In: *The Oxford Handbook of Megaproject Management* (ed. B. Flyvbjerg). Oxford: Oxford University Press.

Deloitte Access Economics. (2013) How do you value an icon? The Sydney Opera House: economic, cultural and digital value. https://www2.deloitte.com/au/en/pages/sydney-opera-house/articles/value-an-icon-sydney-opera-house.html (accessed August 2019)

Denicol, J., Davies, A., Brady, T. and Thurston, M. (2017) Building and leveraging capabilities to deliver megaprojects: the case of CH2M. In Proceedings of the 15th Engineering Project Organization Conference with 5th International Megaprojects Workshop, Stanford Sierra Camp, USA.

Denicol, J., Krystallis, I. and Davies, A. (2018a) Understanding the causes and cures of poor performance in megaprojects: a systematic literature review and research agenda. In Proceedings of the 14th International Research Network on Organizing by Projects (IRNOP), Melbourne, Australia.

Denicol, J., Pryke, S. and Davies, A. (2018b) Exploring innovative organisational structures to deliver megaprojects: the role of owners, sponsors and clients in the Project System Organisation (PSO). In Proceedings of the 34th European Group of Organizational Studies (EGOS), Tallinn, Estonia.

Dodgson, M., Gann, D., MacAulay, S., and Davies, A. (2015). Innovation strategy in new transportation systems: the case of crossrail. *Transportation Research Part A* 77: 261–275. https://doi.org/10.1016/j.tra.2015.04.019.

Egan, J. (1998). *Rethinking Construction*. London: Department of Environment, Transport and the Regions.

Engineering News Record. (2015) The 2015 Top 50 Program Management Firms. http://www.enr.com/toplists/2015_Top_50_Program_Management_Firms (accessed August 2019).

Flyvbjerg, B. (2014). What you should know about megaprojects and why: an overview. *Project Management Journal* 45 (2): 6–19. https://doi.org/10.1002/pmj.21409.

Flyvbjerg, B. (2015). Call for papers-special issue on classics in megaproject management. *International Journal of Project Management* 33: 1–2. https://doi.org/10.1016/j.ijproman.2014.08.002.

Flyvbjerg, B. (2017). *The Oxford Handbook of Megaproject Management*. Oxford: Oxford University Press. ISBN: 9780198732242.

Flyvbjerg, B., Bruzelius, N., and Rothengatter, W. (2003). *Megaprojects and Risk: An Anatomy of Ambition*. Cambridge: Cambridge University Press. ISBN: 0521804205.

Galloway, P., Nielsen, K.R., and Dignum, J.L. (2012). *Managing Giga Projects: Managing Giga Projects: Advice from those who've Been There, Done that*. Reston, Virginia: ASCE Press. ISBN: 978-0-7844-7693-2.

Gil, N. (2009). Developing cooperative project-client relationships: how much to expect from relational contracts. *California Management Review* 51 (2): 144–169. https://doi.org/10.2307/41166484.

Gil, N. (2017). A collective-action perspective on the planning of megaprojects. In: *The Oxford Handbook of Megaproject Management* (ed. B. Flyvbjerg). Oxford: Oxford University Press.

Gosling, J., Hewlett, B., and Naim, M. (2011). A framework for categorising engineer-to-order construction projects. In: *Proceedings 27th Annual ARCOM Conference*, 5–7 September 2011, Bristol, UK, Association of Researchers in Construction Management (eds. C. Egbu and E.C.W. Lou), 995–1004.

Gosling, J., Naim, M.M., and Towill, D.R. (2012). A supply chain flexibility framework for engineer-to-order systems. *Production Planning and Control* 24 (7): 552–556. https://doi.org/10.1080/09537287.2012.659843.

Greiman, V.A. (2013). *Megaproject Management: Lessons on Risk and Project Management from the Big Dig*. Hoboken New Jersey: Willey. ISBN: 9781118115473.

Hart, L. (2015). *Procuring Successful Mega-Projects: How to Establish Major Government Contracts Without Ending up in Court*. Surrey, UK: Gower Publishing. ISBN: 978-1472455086.

Hicks, C., McGovern, T., and Earl, C.F. (2000). Supply chain management: a strategic issue in engineer to order manufacturing. *International Journal of Production Economics* 65 (2): 179–190.

Holti, R., Nicolini, D., and Smalley, M. (2000). *The Handbook of Supply Chain Management: The Essentials*. London: CIRIA and Tavistock Institute.

Koskela, L. (2000) An exploration towards a production theory and its application to construction. PhD. Dissertation, VTT Publications 408, Espoo, Finland, 296.

Koskela, L.J. and Ballard, G. (2006). Should project management be based on theories of economics or production? *Building Research and Information* 34 (2): 154–163. https://doi.org/10.1080/09613210500491480.

KPMG (2013). Megaprojects. *Insight* 4: 1–76. https://home.kpmg/content/dam/kpmg/pdf/2013/02/insight-megaprojects.pdf.

Lambert, D., Cooper, M., and Pagh, J. (1998). Supply chain management: implementation issues and research opportunities. *International Journal of Logistics Management* 9 (2): 1–20. https://doi.org/10.1108/09574099810805807.

Latham, M. (1994). *Constructing the Team*. London: HMSO.

McKinsey. (2017) The art of project leadership: delivering the world's largest projects. McKinsey and Company. https://www.mckinsey.com/industries/capital-projects-and-infrastructure/our-insights/the-art-of-project-leadership-delivering-the-worlds-largest-projects (accessed August 2019)

McKinsey Global Institute. (2013) Infrastructure productivity: how to save $1 trillion a year. McKinsey and Company. https://www.mckinsey.com/industries/capital-projects-and-infrastructure/our-insights/infrastructure-productivity (accessed August 2019)

Mead, J. and Gruneberg, S. (2013). *Programme Procurement in Construction: Learning from London 2012*. Oxford: Wiley. ISBN: 978-0-470-67473-4.

Merrow, E.W. (2011). *Industrial Megaprojects: Concepts, Strategies, and Practices for Success*. Hoboken, New Jersey: Wiley. ISBN: 9780470938829.

Miller, R. and Lessard, D.R. (2000). *The Strategic Management of Large Engineering Projects: Shaping Institutions, Risks, and Governance*. Cambridge, MA: The MIT Press. ISBN: 9780262122368.

Morris, P.W. (1994). *The Management of Projects*. London, UK: Thomas Telford. ISBN: 9780727716934.

Morris, P.W. (2013). *Reconstructing Project Management*, 978. London, UK: Wiley. ISBN: 0 470-65907-6.

Morris, P.W.G. (2014). Reflections on the discipline: development and research. In: *Thinking about the Management of Projects: A Celebration* (ed. A. Edkins), 54–57. London: The Bartlett School of Construction and Project Management.

Morris, P.W.G. and Hough, G.H. (1987). *The Anatomy of Major Projects: A Study of the Reality of Project Management*. Winchester: Wiley. ISBN: 9780471915515.

O'Brien, W.J., Formoso, C.T., Vrijhoef, R., and London, K.A. (2008). *Construction Supply Chain Management Handbook*. Boca Raton: CRC Press. ISBN: 978-1-4200-4745-5.

Olhager, J. (2003). Strategic positioning of the order penetration point. *International Journal of Production Economics* 85 (3): 319–329. https://doi.org/10.1016/S0925-5273(03)00119-1.

Pero, M. and Rossi, T. (2014). RFID technology for increasing visibility in ETO supply chains: a case study. *Production Planning and Control: The Management of Operations* 25 (11): 892–901. https://doi.org/10.1080/09537287.2013.774257.

Potts, K. (2009). From Heathrow express to Heathrow terminal 5: BAA's development of supply chain management. In: *Construction Supply Chain Management: Concepts and Case Studies* (ed. S. Pryke), 160–181. Oxford, UK.: Wiley https://doi.org/10.1002/9781444320916.ch8.

Prencipe, A., Davies, A., and Hobday, M. (2003). *The Business of Systems Integration.* Oxford: Oxford University Press. ISBN: 9780199263233.

Priemus, H. and Van Wee, B. (2013). *International Handbook on Mega-Projects.* Cheltenham: Edward Elgar. ISBN: 9781781002292.

Pryke, S.D. (2009). *Supply Chain Management: Concepts and Case Studies.* Oxford, UK: Wiley https://doi.org/10.1002/9781444320916.

Pryke, S.D. (2012). *Social Network Analysis in Construction.* Wiley.

Pryke, S.D. (2017). *Managing Networks in Project-Based Organisations.* Oxford, UK: Wiley. ISBN: 978-1-118-92992-6.

Rolstadås, A., Hetland, W., Jergeas, G.F., and Westney, R. (2011). *Risk Navigation Strategies for Major Capital Projects: Beyond the Myth of Predictability.* London: Springer.

Scott, W.R., Levitt, R.E., and Orr, R.J. (2011). *Global Projects: Institutional and Political Challenges.* Cambridge: Cambridge University Press. ISBN: 978-1107004924.

Shapira, Z. and Berndt, D.J. (1997). Managing grand-scale construction projects: a risk-taking perspective. In: *Research in Organizational Behavior*, vol. 19, 303–360. ISBN: 0762301791.

Söderlund, J., Sankaran, S., and Biesenthal, C. (2015). Call for papers – megaprojects – symbolic and sublime: an organizational theory perspective. *Project Management Journal.*

Souza, D. V. S. (2015) A conceptual framework and best practices for designing and improving construction supply chains, PhD thesis, University of Salford.

Stinchcombe, A.L. and Heimer, C. (1985). *Organizational Theory and Project Management: Administering Uncertainty in Norwegian Offshore Oil.* Oslo: Norwegian University Press.

Strategic Forum for Construction (2002). *Accelerating Change.* London: Strategic Forum for Construction.

Tommelein, I.D., Ballard, G., and Kaminsky, P. (2009). Supply chain management for lean project delivery. In: *Construction Supply Chain Management Handbook* (eds. W.J. O'Brien, C.T. Formoso, R. Vrijhoef and K.A. London), 105–126. Boca Raton: CRC Press. ISBN: 978-1-4200-4745-5.

Van der Vaart, J.T., De Vries, J., and Wijngaard, J. (1996). Complexity and uncertainty of materials procurement in assembly situations. *International Journal of Production Economics* 46–47: 137–152. https://doi.org/10.1016/0925-5273(95)00193-X.

Wachs, M. (1989). When planners lie with numbers. *Journal of the American Planning Association* 55 (4): 476–479.

Wang, X., Chong, H.Y., and Kwak, Y.H. (2015). Call for papers – supply chain management in megaprojects. *Journal of Management in Engineering.*

Winch, G.M. (2014). Three domains of project organising. *International Journal of Project Management* 32 (5): 721–731. https://doi.org/10.1016/j.ijproman.2013.10.012.

Winch, G.M. and Leiringer, R. (2016). Owner project capabilities for infrastructure development: a review and development of the "strong owner" concept. *International Journal of Project Management* 34 (2): 271–281. https://doi.org/10.1016/j.ijproman.2015.02.002.

11

Anglian Water @one Alliance: A New Approach to Supply Chain Management

Grant Mills, Dale Evans, and Chris Candlish

11.1   Introduction

Within the construction industry, supply chain management is frequently seen as a functional role applied to a project by either a client or contractor. Some see the client as most important in this process (Latham 1994; Dainty et al. 2001; Briscoe et al. 2004; Potts 2009; Rimmer 2009), while others see it as the role of the contractor (King and Pitt 2009). These roles are key in supply chain management. The client defines the transitional outcomes that supply chain management must achieve. They specify the product and service requirements, create governance structures, and establish the procurement method that ensure whole life cost and facilitate early involvement, opportunities management, and shared risk allocation. They must incentivise multiorganisational collaboration across task and team boundaries and promote innovation.

The contractor will have a significant stake in the development of capabilities across the collaborating partners and drive innovation and improvement to technical systems through collaborative planning, supporting integrated team decision making. While client and contractor(s) will take up these key roles in alliance supply chain management, particularly important is the way that specialist suppliers can be empowered through alliancing. When they are, they will take a more active integration role in the delivery of the programme or project, and can reduce transaction costs, bring innovative technology, make year-by-year asset management improvements, and redesign their service offering through mutually beneficial relationships. These relationships will not see the subcontractor or supplier subservient to the Tier 1 contractor and the contractor being subservient to the client. Some may develop for themselves capabilities to integrate and manage other suppliers to innovate, improve and share risks in order to gain prominence with the client, understand opportunities, and enhance their market position. Other suppliers, with a more specialised and standardised product, may for example see long-term opportunities to advance manufacturing and develop innovative production processes to enhance delivery to time, cost, and quality.

There is a need for a new approach to supply chain management that improves interorganisational supply-chain integration and delivers mutual benefits for all parties (Dainty et al. 2001). Perhaps the best known interorganisational strategy for achieving

Successful Construction Supply Chain Management: Concepts and Case Studies, Second Edition.
Edited by Stephen Pryke.
© 2020 John Wiley & Sons Ltd. Published 2020 by John Wiley & Sons Ltd.

delivery model integration is alliances and so this chapter claims a new approach to alliance supply chain management. However, in reality what this chapter is doing is documenting the practices identified from working alongside an advanced alliance that has made a long-term commitment to supply chain management.

Once the literature on alliancing and supply chain management have been reviewed, the best practices of Anglian Water @one Alliance are described. This chapter illustrates how they have set up an alliance-based supply chain and then looks at the advanced approaches it has used to integrate design, construction, and operation across projects. It then describes how it has increased value through interorganisational learning, capability development and innovation. It shows how collaborative planning and production, early supplier engagement, and work cluster design has been adopted from the manufacturing sector, and how strong alliancing relationships have been managed through the development of an efficient structure and process.

11.2 Supply Chain Management

There is very little literature on the application of supply chain management in a construction alliance situation. One notable exception is Rimmer (2009), who identified supply chain relationships lasting many decades at Slough Estates during the 1990s and early 2000s. As such, little is known about alliancing strategies to integrate across the whole supply chain to achieve organisational efficiency and responsiveness to customer needs. Much of the discussion has remained on whether supply chain management should be client-centric (Latham 1994; Egan 1998) or contractor-centric (King and Pitt 2009), and what are the abilities of these to establish mutually beneficial win-win outcomes or not (Cox 2004; Cox et al. 2007; King and Pitt 2009).

At a tactical level, supply chain management approaches such as clustering and team integration (Nicolini et al. 2001; Baiden et al. 2006), design chains (Austin et al. 2001), partnership (Barlow and Jashapara 1998), and alliancing (Walker and Lloyd-Walker 2016) are well understood. But at a more strategic level, supply chain management remains hampered by traditional attitudes and behaviours, such as quasi-competitive controls, mistrust, and conflict (Bresnen and Marshall 1999; Moore and Dainty 2001) and by project, stage, and bidding silos. This fails to capture the advantages of more collaborative alliance-based supply chain approaches (Construction Industry Institute 2003), and falls short of interconnecting business systems to achieve interorganisational learning, innovation, and efficiency (Borys and Jemison 1989; Christopher and Ryals 1999; Tan 2001; Muckstadt et al. 2001; Chopra and Meindl 2004; and Wolter and Veloso 2007). Alliance supply chain environments must therefore develop a more sophisticated approach to scaling and positioning organisations in the supply chain (Thakkar et al. 2011; Department for Business Innovation and Skills 2013), and to modifying the proximity to the client (Harland 1996) to empower integration, innovation, and to achieve higher all-round performance. A key part of supply chain management outside of the construction sector has been early supplier engagement, and access to supplier design and manufacturing knowledge (Dowlatshahi 1998). This involves, for example:

- Establishing long-term strategic relationships through early procurement (to analyse make-buy decisions, negotiate price and terms, orders, lead times, stock, and quality)

- Improving product innovation and standardisation (to agree material standards, quality control, defects, efficiencies, capabilities, and research and innovation)
- Increasing the speed of project design and production information (to conceptualise, define functions, performance and material requirements, plan work packages, identify standard components, identify costs, minimise waste, and establish the market for supply chain partners), and
- Continuous manufacturing process improvement (production processes, schedules and runs, to set goals for inventory, set-up, efficiency, handling, and turnover)

(Dowlatshahi 1998).

Early contractor involvement requires changes in culture, teamwork, and contracting practice (Song et al. 2009; Rahman and Alhassan 2012). It benefits from relational contracts (Mosey 2009; Van Valkenburg et al 2008) that reduce the conflicting interests and barriers to innovation that may exist in the supply chain (Miller et al. 2002). It allows for shared risk management and appropriate risk allocation and mitigation (Mosey 2009; Rahman and Kumaraswamy 2004). It also provides invaluable design input on for example buildability (Gil et al. 2002) and more accurate cost control and reduced contract administration (Song et al. 2009). Alliance supply chain management presents an opportunity for owners, alliance partners, and construction firms to engage suppliers early to work together to achieve cost and time savings, better product quality and more efficient production processes.

11.3   Alliance Supply Chain Management

Various definitions of supply chain management have focused on the broad interorganisational process of delivery as a 'network' (Pryke 2009, 2017) to produce, handle and/or distribute resources, or a collaboration to 'leverage strategic positioning and to improve operating efficiency' (Bowersox et al. 2013). Others have defined an upstream and downstream network (or flow) of customer–supplier relationships to share materials, parts, other information, ideas, and sometimes people (Christopher and Ryals 1999). Austin et al. (2001) saw the flow of goods going one way in a supply chain (from supplier to client), while information flowed in the other direction (from client to supplier). This chapter investigates these definitions in the context of an alliance, whereby leading, sourcing, specification, planning, procuring, monitoring, and controlling activities achieved an effective and efficient flow (Cooper et al. 1997) because they were done collaboratively.

Literature on alliancing has not frequently focused specifically on the arrangements of supply chains, but rather on the interorganisational mechanisms and temporary supply networks (Galbraith 2002; Lawrence and Lorsch 1967; Walker and Lloyd-Walker 2016; Hietajärvi et al. 2017). Barlow and Jashapara (1998) described the strategic 'involvement climate' that provides mutually beneficial alliancing relationships, while others (e.g. Nicolini et al. 2001) investigated tactical strategies to supply chain integration. However, despite these advances, specialist suppliers continue to take a subordinate position in the supply chain of some alliances, creating suboptimal integration. Alliance supply chain management must bring partners together to achieve the common goals of improving productivity and waste minimisation. They must share problems and work together to: align outcome focus, find better client solutions, manage risk effectively, improve profit margins, increase development of mutual understanding, and enhance reputations (Hall 2001). To do this, however, some suppliers will need to develop new innovation and integration capabilities because they may be limited by involvement in a specific project stage or a narrow functional or technical capability. Greater understanding of the process by which alliances evolve and engage project suppliers over time, and between projects, must build on tactical approaches, such as 'work clusters' (Holti et al. 2000; Nicolini et al. 2001) to achieve supply chain integration.

11.4 Anglian Water Alliance Case Study

11.4.1 Strategic Approach to Alliance Supply Chain Management

Anglian water has a long history of alliance supply chain management. The Anglian Water @one alliance reports that, since 2004, it has halved carbon emissions, made time efficiencies of 40%, consistently outperformed Anglian Water's business plan targets, quintupled the length of alliance partner contracts (now 15 years), and established an integrated supplier network through supplier frameworks and category management. Key has been establishing a strong strategic direction, clear transitional goals, fair commercial arrangements, and robust selection and governance frameworks. The @one alliance itself is engaged by the client at the earliest stages of need identification and undertakes the delivery of the outcome in a highly integrated manner with Anglian Water.

During the period from 2015 to 2030 the @one Alliance is scheduled to design and build some 800 projects and aims to outperform the £1.2 billion target value. In order to ensure value for money on each project, the current baseline cost for delivering the required output is assessed. Anglian Water uses baseline performance to set targets for whole-life cost, capital cost, carbon, and time. As projects are delivered, the partners recover their direct costs, but can only generate a return by delivering solutions for less than the baseline. Outperformance, or conversely overspends, contribute to a communal programme pool that is periodically shared between the alliance partners, including Anglian Water. As a result, everyone is in the alliance together, and all alliance partners generate a return only by collectively and continuously outperforming the baseline. Partners are mutually dependent on each other to deliver work below baseline to make a profit. Consistent use of a contract form has allowed transparent risk allocation, so attention can be placed on the management of the risk rather than purely allocation.

The @one alliance integrated procurement team are strategically aligned with the Anglian Water procurement team to deliver 80% of the programme of works in collaboration with its supply chain partners who deliver the remaining 20% of the expenditure. They directly manage and develop over the long term the capabilities and behaviours of the supply chain partners, develop efficient and collaborative engagement strategies, and promote early engagement to identify effective and innovative solutions that deliver the required client and customer outcomes. Collaborative working groups allow supply chain partners to work closely together, share best practice, and allocate work on a best for task basis.

The Cambridge Water Recycling Centre provides an ideal project that shows the strength of the alliancing approach to supply chain management. This was a £22 million investment in the installation of a pumping station and eight new treatment tanks.

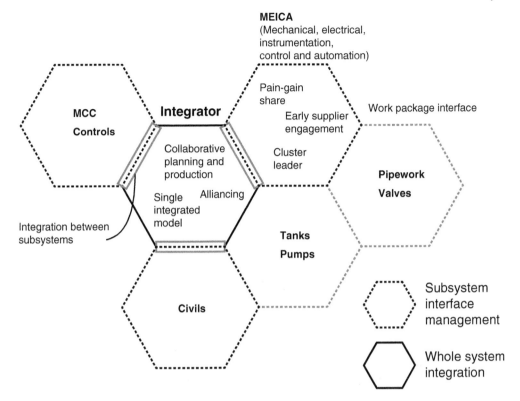

Figure 11.1 Integration between alliance partners/work packages. Source: Adapted from Nicolini et al. (2001).

Work commenced on site in May 2015 and was completed within a tight 10-month construction deadline. The project cost came in £9 million below the budget. Chosen as a pathfinder project for innovation, the Alliance used several initiatives to work closely with supply chain partners (as shown in Figure 11.1) on the development of new technology and delivery methods in the planning and construction phases. Eighty-two percent of suppliers on the Cambridge project were in a long-term relationship with Anglian Water (beyond three years) and 60% were in stable framework or alliancing relationships.

The Cambridge project exhibited a high intensity of innovation relating to the use of modularised and standard products. There was innovation in:

- *The use of new materials* – Recycled plastic (weholite) for tanks, overground pipework, and fibre reinforced concrete.
- *New techniques* – Soil stabilisation techniques, innovative platform design and 'profibus' control technology.
- New management system (level sensor integration).

These were all deemed as contributing to the improvement in project functions.

11.4.2 Alliance Supply Chain Work Clusters

From the start of the project the boundaries of several organisations, subsystems, and activities were redrawn in order to prevent the emergence of interface issues, to increase buildability, facilitate the exchange of information, to support the appropriate allocation of risks, and to exploit all the knowledge, expertise, and innovation potentially carried by the members of the supply chain. As show in Figure 11.1 the project was organised around clusters and work package, the most complex of which required subsystem interface management (e.g. MCC, MEICA, and Civils). Other suppliers provided customised and standard package products (e.g. plant rooms, tanks, pumps, pipework, valves, preconstruction, and construction materials).

The following two examples provide an illustration of the approach taken to organising supply chain relationships into clusters. The first, a mechanical and electrical cluster, resolved the complexity of interface conflicts and handover issues between pumps, cables, and motors. Working together as a joint venture, these systems were integrated to reduce cost, time, and increase quality (e.g. fewer defects). The second was a tank cluster, whereby one of the suppliers led a combined work package to incentivise integration and resolution interface issues as early as possible (e.g. reducing risk of tank leaks) and reducing fragmented responsibilities between separate suppliers tasked with excavation and foundations, and base and precast wall construction. High supplier productivity was incentivised.

The @one alliance project manager redrew the boundaries of groups of tasks as work clusters in order to prevent the emergence of interface issues, mitigate rework, reduce waste, and reduce time and cost. The analysis of interorganisational product and service interfaces was the basis for working together, and illustrated who was best placed to establish quality criteria for integration. The project manager reviewed risks and opportunities and enabled strategic relationships to be built with risk allocated to the party best able to manage it. This constructed innovative work package boundaries that avoided interface conflicts and enhanced quality. Incentives were given to integration leaders, who due to their position in the supply chain and the time of involvement, were best placed to double-check everything themselves and make sure it was happening correctly. Clusters were organised to learn lessons from previous scheme failures. Suppliers were given scope to establish control over cost, programme, and quality in completing the work of the cluster. This caused suppliers to work together to specify value (e.g. a durable and reliable water system) instead of being nonspecific (e.g. make good plumbing).

11.4.3   Alliance Supply Chain Early Involvement and Collaboration

Long-term relationships and established framework and contractual terms facilitated the preferred supplier being available at an early stage. Suppliers were selected to be involved in project strategy development which is decided on a project-by-project basis. This involved suppliers engaged in early design and option exploration and paid by way of professional services contracts prior to full contract engagement for delivery. All parties worked together to improve productivity, and reduce time and cost. A series of tools, approaches, and processes were used to share information. Collaborative planning structured and facilitated the process by which the owner and alliance partners created a dialogue with subcontractors to develop an integrated programme and a technically integrated product.

Figure 11.2 shows the evolution of the project team in the Cambridge project. It shows the extent of integrator, advisor, and supplier involvement and the duration of the relationship ordered by the organisations contract value at each level of involvement.

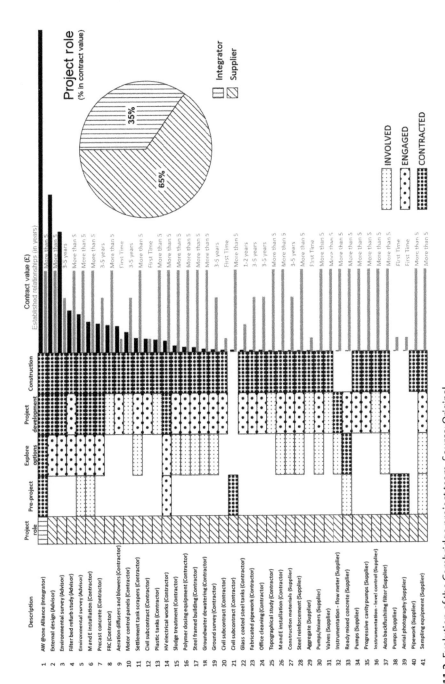

Figure 11.2 Evolution of the Cambridge project team. Source: Original.

It shows that 65% of the project value went directly to suppliers. There was early specialist supplier involvement preproject in collaborative planning and to explore options. This level of engagement happened because the majority of suppliers are involved in long-term @one alliance frameworks.

During the course of the Cambridge project collaborative planning and production mapped work flows (strategic through to tactical phases, areas, and milestone). These were supported by 'look-aheads', contract early warnings and waste reduction methods, in addition to stand up daily and weekly meetings which coordinated the sequence of information, labour, and materials. Teams were encouraged collectively to understand and remove the blockers, develop and reuse standard ways of doing things, and measure and improve performance. This mutual dependence built a shared commitment, trust, and transparency.

11.5 Evaluation of the Value of Alliance Supply Chain Management

This section takes the literature on alliances and supply chain management, and insights gained from the study, and uses them to evaluate and characterise the effectiveness of alliance supply chain management.

11.5.1 Strategic Approach to Alliance Supply Chain Management

The case study has shown the importance of thinking about the supply chain in terms of interconnected business systems. It showed how learning, capability innovation, and efficiencies were harnessed across interorganisational and stage boundaries to integrate systems. This mutually enhanced ongoing and long-term business relationships (e.g. Borys and Jemison 1989; Wolter and Veloso 2007). For example, one supplier in the alliance supply chain said:

> ... the framework is brilliant because you deal with the same people all the time ... this one must have been the eighth or tenth ... scheme of a decent size, so all the lessons learnt from all of those were carried forward

11.5.2 Alliance Supply Chain Management Provides an Effective Environment for the Early Engagement of Specialist Suppliers

The study showed that the alliance promoted early engagement, utilised new technologies, and suppliers were involved in both product development, innovation and education, and measuring customer satisfaction (Barratt 2004). The success is shared here by a supplier's view of the project.

> 'If it wasn't for [our early engagement] it wouldn't have happened'. '... we took almost a third off of the project ... from budgets to final completion ... it was a combination of probably five or six suppliers that were involved in that process, heavily involved in that process'.

11.5.3 Alliance Supply Chain Management Can Create a Win-Win-Win Reciprocal Relationship

The case study highlights that alliances with an integrated approach to supply chain engagement provide an ideal environment in which all parties can benefit. The win-win-win relationships previously proposed by King and Pitt (2009) were evident in the management of the alliance supply chain relationship. Table 11.1 further illustrates some of the interconnected relationships between client, contractor, and supplier and how alliance supply chain management could be applied to align benefits.

What this has shown is that alliance partners understood the size and position of the organisation in relation to the supply chain and this enabled them to achieve higher all-round performance (Thakkar et al. 2011; Department for Business Innovation and Skills 2013). The case study has illustrated the innovation impacts of giving suppliers greater proximity to the client (Harland 1996) and empowering them to drive integration, innovation, and achieve efficiency. Particularly important was paying for the supplier's knowledge in design, which facilitated integration in construction and operation. The added prominence of suppliers involved alongside contractors and clients in design, promoted efficient ways of working in construction and better operational solutions.

11.5.4 Alliance Supply Chain Management Can Drive Team Innovation and Create New Service Relationships

The alliance has consistently outperformed Anglian Water's business plan targets and as a result has been able to increase significantly the length of supplier alliance contracts.

Table 11.1 Benefits traded between client, contractor, and supplier.

Supply chain position	Benefits of alliance supply chain management
Client	Suppliers who understand transitional outcomes
	Innovation to achieve integration and product and service efficiencies
	Clear allocation of risks between parties
Alliance partners	Long-term and stable fee to integrate T2 and T3 suppliers
	Profit share of innovation and efficiency improvements
	Mutually beneficial exchange of technical and operational capabilities (existing and new)
	Clear allocation of risks between parties
Suppliers and manufacturers	Long-term and stable fee with funded early involvement
	Opportunities to integrate and collaborate with alliance partners and suppliers to develop innovative new product and services that can market relationships
	Mutually beneficial exchange of technical and operational and capabilities (existing and new)
	Clear allocation of risks between parties
	Efficiencies from the development of production processes and standardisation
	Visible demand provides assurance in workforce and resource planning. Less need for agency staff

The decision to bring the @one alliance partners together in an integrated alliance with integrated delivery teams to gain value from their complementary competencies and resources (Boon-itt and Wong 2011) was a central principle of the alliance. Data has illustrated that alliances provide an ideal environment in which partners can agree who is best placed to do what activities and to lead innovation. What this has shown is that when alliances are integrated and suppliers are in a trusted collaborative relationship, supply chain management can deliver significant benefits: improving productivity and waste minimisation, and improving client solutions, profits, and reputation (Hall 2001).

11.5.5 Long-Term Approaches to Alliance Supply Chain Management Can Drive Strategic Business Benefits

The Anglian Water @one alliance has since 2004 halved carbon emissions, made time efficiencies of more than 40%, and improved health and safety. This alliance achieved sufficient supply chain maturity to realise the benefits of whole architectural- and modular-system innovation (Slaughter 1998, 2000), economies of repetition and recombination (Davies and Brady 2000; Grabher 2002; Davies et al. 2009), standardisation (Mattsson 1973; Page and Siemplenski 1983), and so reduced transaction costs (Spekman et al. 1998; Wolter and Veloso 2007).

11.5.6 Alliance Supply Chain Management that Uses Advanced Production Systems Can Deliver Tactical Benefits

There was clear evidence of the use of work clusters (Nicolini et al. 2001) by the alliance at both the project and organisational level. Collaborative planning and production facilitated the dynamic delivery of 'integrity of processes', 'interdependence of activities', 'leadership' and 'incomplete clustering' leading to increased opportunities, reduced risks at the interfaces, improved sequencing, economies of scale, higher quality, and reduction of waste (Nicolini et al. 2001). Interface issues were identified from the beginning and discussed throughout a routine governance process. The use of this approach on the case study prevented the emergence of interface issues, increased buildability, facilitated the exchange of information, supported the appropriate allocation of risks, and exploited all the knowledge and expertise of the supply chain. As a result, a considerable number of innovations and efficiency savings resulted, and the business objectives of the client, alliance partners, and specialist suppliers were achieved.

11.6 Conclusions

This chapter has described a new approach to alliance supply chain management that has been developed from working alongside an advanced alliance that has made a long-term commitment to supply chain management. The literature on alliancing and supply chain management was first reviewed, then the best practices of Anglian Water @one Alliance were described.

The chapter shows how a long-term commitment to alliance supply chain management was applied to integrate design, construction, and operation across projects to increase the value created by all collaborating partners. It has also shown how collaborative planning and production, early supplier engagement, and work cluster design has been adopted from the manufacturing sector, and how early partner engagement and strong long-term alliancing relationships have been managed to align the supply chain with the same goals.

The implications of this work show that alliancing can significantly enhance supply chain management. It provides an effective environment for early specialist supplier engagement and can result in strong win-win-win reciprocal relationships. Alliance supply chain management can drive team innovation and create new service relationships that advance production systems and ultimately achieve improved business value for all partners.

References

Austin, S., Baldwin, A., Hammond, J. et al. (2001). *Design Chains: A Handbook for Integrated Collaborative Design*. Loughborough: Thomas Telford.

Baiden, B.K., Price, A.D.F., and Dainty, A.R.J. (2006). The extent of team integration within construction projects. *International Journal of Project Management* 24 (1): 13–23.

Barlow, J. and Jashapara, A. (1998). Organisational learning and inter-firm partnering in the construction sector. *The Learning Organisation* 5 (2): 86–98.

Barratt, M. (2004). Understanding the meaning of collaboration in the supply chain. *Supply Chain Management: An International Journal* 9 (1): 30–42.

Boon-itt, S. and Wong, C.W. (2011). The moderating effects of technological and demand uncertainties on the relationship between supply chain integration and customer delivery performance. *International Journal of Physical Distribution and Logistics Management* 41 (3): 253–276.

Borys, B. and Jemison, B.D. (1989). Hybrid arrangements as strategic alliances: theoretical issues in organizational combinations. *The Academy of Management Review* 14 (2): 234–249.

Bowersox, D., Closs, D., and Cooper, B.M. (2013). *Supply Chain Logistics Management*, 4e. McGraw-Hill Education.

Bresnen, M. and Marshall, N. (1999). Achieving customer satisfaction? Client–contractor collaboration in the UK construction industry. In: *Proceedings of the CIB W65/W55 Joint Triennial Symposium* (eds. P. Bowen and R. Hindle). Delft, Netherlands: CIB.

Briscoe, G., Dainty, A., Millett, S., and Neale, R. (2004). Client led strategies for construction supply chain management. *Construction Management and Economics* 22 (2): 193–201.

Chopra, S. and Meindl, P. (2004). *Supply Chain Management: Strategy, Planning, and Operation*, 2e. New Jersey: Person and Prentice Hall.

Christopher, M. and Ryals, L. (1999). Supply chain strategy: its impact on shareholder value. *The International Journal of Logistics Management* 10 (1): 1–10.

Construction Industry Institute (2003) Improving Capital Projects Supply Chain Performance. https://www.construction-institute.org/resources/knowledgebase/knowledge-areas/procurement-contracts/topics/rt-172/pubs/rr172-11 (accessed August 2019).

Cooper, M., Lambert, D., and Pagh, J. (1997). Supply chain management: more than a new name for logistics. *The International Journal of Logistics Management* 8 (1): 1–14.

Cox, A. (2004). *Win-Win? The Paradox of Value and Interests in Business Relationships*. Stratford-upon-Avon: Earlsgare Press.

Cox, A., Ireland, P., and Townsend, M. (2007). *Managing in Construction Supply Chain and Markets*. London: Thomas Telford.

Dainty, A.R.J., Millet, S.J., and Briscoe, G.H. (2001). New perspectives on construction supply chain integration. *Supply Chain Management: An International Journal* 6 (4): 163–173.

Davies, A. and Brady, T. (2000). Organisational capabilities and learning in complex product systems: towards repeatable solutions. *Research Policy* 29 (7–8): 931–953.

Davies, A., Gann, D.M., and Douglas, T. (2009). Innovation in megaprojects: systems integration at London Heathrow Terminal 5. *California Management Review* 51 (2): 101–125.

Department for Business Innovation and Skills (2013). *UK Construction: An Economic Analysis of the Sector*. London: The Crown Publishing Group.

Dowlatshahi, S. (1998). Implementing early supplier involvement: a conceptual framework. *International Journal of Operations and Production Management* 18 (2): 143–167.

Egan, J. (1998). *Rethinking Construction*. London: Strategic Forum for Construction.

Galbraith, J.R. (2002). Organizing to deliver solutions. *Organizational Dynamics* 31: 194–207.

Gil, N., Tommelein, I.D., Kirkendall, R.L., and Ballard, G. (2002). Leveraging specialty contractor knowledge in design build organizations. *Engineering Construction and Architectural Management* 8 (5/6): 355–367.

Grabher, G. (2002). The project ecology of advertising: tasks, talents and teams. *Regional Studies* 36 (3): 245–262.

Hall, M. (2001). Root cause analysis: a tool for closer supply chain integration in construction. In: *Proceedings of 17th Annual ARCOM Conference*, 929–938. University of Salford, , 5–7 September 2001.

Harland, C.M. (1996). Supply chain management: relationships, chains and networks. *British Journal of Management* 7 (1): 63–80.

Hietajärvi, A.-M., Aaltonen, K., and Haapasalo, H. (2017). Managing integration in infrastructure alliance projects: dynamics of integration mechanisms. *International Journal of Managing Projects in Business* 10 (1): 5–31.

Holti, R., Nicolini, D., and Smalley, M. (2000). *The Handbook of Supply Chain Management*. London: Construction Industry Research and Information Association.

King, A.P. and Pitt, M.C. (2009). Supply chain management: a Main contractor's perspective. In: *Construction Supply Chain Management: Concepts and Case Studies* (ed. S.D. Pryke). Oxford: Wiley Blackwell.

Latham, M. (1994). *Constructing the Team: Final Report: Joint Review of Procurement and Contractual Arrangements in the United Kingdom Construction Industry*. London: HMSO.

Lawrence, P.R. and Lorsch, J.W. (1967). Differentiation and integration in complex organizations. *Administrative Science Quarterly* 12 (1): 1–47.

Mattsson, L.-G. (1973). Systems selling as a strategy on industrial markets. *Industrial Marketing Management* 3 (2): 107–120.

Miller, C.J.M., Packham, G.A., and Thomas, B.C. (2002). Harmonization between main contractors and sub contractors: a Prequisite for lean construction? *Journal of Construction Research* 3 (1): 67–82.

Moore, D.R. and Dainty, A.R.J. (2001). Intra-team boundaries as inhibitors of performance improvement in UK design and build projects: a call for change. *Construction Management and Economics* 19 (6): 559–562.

Mosey, D. (2009). *Early Contractor Involvement in Building Procurement: Contracts, Partnering and Project Management.* Wiley-Blackwell.

Muckstadt, J., Murray, D., Rappold, J., and Collins, D. (2001). Guidelines for collaborative supply chain system design and operation. *Information Systems Frontiers* 3: 427–453.

Nicolini, D., Holti, R., and Smalley, M. (2001). Integrating project activities: the theory and practice of managing the supply chain through clusters. *Construction Management and Economics* 19 (1): 37–47.

Page, A.L. and Siemplenski, M. (1983). Product systems marketing. *Industrial Marketing Management* 12 (2): 89–99.

Potts, K. (2009). From Heathrow Terminal 5: BAA's development of supply chain management. In: *Construction Supply Chain Management: Concepts and Case Studies* (ed. S.D. Pryke). Oxford: Wiley Blackwell.

Pryke, S.D. (ed.) (2009). *Construction Supply Chain Management: Concepts and Case Studies.* Oxford: Wiley Blackwell.

Pryke, S.D. (2017). *Managing Networks in Project-Based Organisations.* Chichester, West Sussex: Wiley.

Rahman, M. and Alhassan, A. (2012). A contractor's perception on early contractor involvement. *Built Environment Project and Asset Management.* 2 (2): 217–233.

Rahman, M.M. and Kumaraswamy, M.M. (2004). Contracting relationship trends and transitions. *Journal of Management in Engineering* 20 (4): 147–161.

Rimmer, B. (2009). Slough estates in the 1990s – client driven SCM. In: *Construction Supply Chain Management: Concepts and Case Studies* (ed. S.D. Pryke). Oxford: Wiley Blackwell.

Slaughter, E. (1998). Models of construction innovation. *Journal of Construction Engineering and Management* 124 (3): 226–231.

Slaughter, S. (2000). Implementation of construction innovations. *Building Research and Information* 28 (1): 2–17.

Song, L., Mohamed, Y., and AbouRizk, S.M. (2009). Early contractor involvement in design and its impact on construction schedule performance. *Journal of Management in Engineering* 25 (1): 12–20.

Spekman, R., Kamauff, J., and Myhr, N. (1998). An empirical investigation into supply chain management: a perspective on partnerships. *Supply Chain Management: An International Journal* 3 (2): 53–67.

Tan, K.C. (2001). A framework of supply chain management literature. *European Journal of Purchasing and Supply Management* 7 (1): 39–48.

Thakkar, J., Kanda, A., and Deshmukh, S.G. (2011). Supply chain issues in SMEs: select insights from cases of Indian origin. *Production Planning and Control* 24 (1): 47–71.

Van Valkenburg, M., Lenferink S., Nijsten, R. and Arts, J. (2008) Early Contractor Involvement: A New Strategy for 'Buying the Best' in Infrastructure Development in the Netherlands, 323–356

Walker, D. and Lloyd-Walker, B. (2016). Rethinking project management: its influence on papers published in the International Journal of Managing Projects in Business. *International Journal of Managing Projects in Business* 9 (4): 716–743.

Wolter, C. and Veloso, F. (2007). The effects of innovation on vertical structure: perspectives on transaction costs and competences. *The Academy of Management Review* 33 (3): 586–605.

12

Understanding Supply Chain Management from a Main Contractor's Perspective

Emmanuel Manu and Andrew Knight

12.1   Introduction

The emergence of supply chain management in construction was predominantly driven by major clients who began to adopt procurement arrangements such as prime contracting, partnering, and framework agreements. These were major clients who had the power leverage to mobilise construction firms for projects that spanned a considerable timeframe. Despite such early efforts towards supply chain management adoption, it increasingly became apparent that numerous subcontractors and suppliers were not fully integrated into these supply chain driven collaborative approaches. This has remained a key shortcoming of such collaborative efforts in the construction sector (Kumaraswamy et al. 2010). This situation has now burdened main contractors with the enormous responsibility of coordinating and managing multilayered tiers of suppliers and subcontractors for production and value addition across multiple projects. Main contractors have taken an interest in supply chain management as an approach for coordinating upstream linkages with demand-side organisations, in addition to the downstream tiers of supply chain firms assembled to deliver projects. However, empirical evidence of supply chain management initiatives and practices that have been implemented by main contractors to manage the lower tiers of the supply chain remains scarce (Pala et al. 2014; Broft et al. 2016).

In this chapter, the multilayered nature of subcontracting in construction will be discussed, together with its ramifications, before reviewing literature on supply chain management principles and practices that can be adopted by main contractors. A case study of supply chain management practices implemented by a large UK main contractor is then used to provide some empirical evidence of supply chain management adoption from a main contractor's perspective. This case study focuses on the 'dyadic' relationship between a Tier 1 main contractor and Tier 2 subcontractors.

Successful Construction Supply Chain Management: Concepts and Case Studies, Second Edition.
Edited by Stephen Pryke.
© 2020 John Wiley & Sons Ltd. Published 2020 by John Wiley & Sons Ltd.

12.2   Multilayered Subcontracting in the Construction Industry

Construction work is specialist in nature and historically has involved input from different specialist firms. Prior to the recognition that the supply chain management concept could be harnessed for the management of these supply chain firms, main contractors had always depended on, and worked with, a pool of subcontractor firms available to them from the market. Main contractors, who are tasked with the ultimate responsibility of carrying out and completing construction projects, have often relied on subcontractors for the provision of specialised works and services. Beyond the need to subcontract specialist aspects of construction works, there has also been a growth in labour-only subcontracting which in the case of the British construction industry, has resulted from the strategic choice by construction companies to emphasise flexibility over productivity as a source of competitive advantage (Winch 1998). This flexibility allows main contractors to cope with constant fluctuations in demand for their services. Evidence suggests that subcontracted work can constitute up to 90% of construction work by value on most projects (Segerstedt et al. 2010). As shown in Figure 12.1, the nature of subcontracting practice in construction is also multilayered (Ronchi 2006; Broft et al. 2016), with Tier 1 contractors depending on a network of Tier 2 and Tier 3 subcontractors and suppliers to achieve production. A similar pattern of multilayered subcontracting also persists amongst in-use suppliers. A typical construction project will usually involve a complex network of interactions between multiple tiers of production suppliers that have to work together to meet the needs of the demand-side organisation(s).

Another characteristic of the construction supply chain is that it is typically a make-to-order supply chain, with every project 'tailor made' to the client's specifications (Ronchi 2006). Information flow across the supply chain commences from the demand organisations (clients and users) in the form of client requirements, through Tier 1 contractors and across the different tiers of the production supply chain. There are consequently material and information flows in the opposite direction, through to the demand organisations and in-use supply chain. Main contractors have the arduous task of managing and coordinating the various subcontractors and suppliers, and managing the associated risks to deliver a successful project. This task of coordinating and managing the complex supply chain network can be challenging and an issue that is often underestimated is the concurrent involvement of specialist contractors on other projects and their inability to manage constrained resources across multiple projects concurrently (Ronchi 2006).

This multilayered nature of subcontracting practice has been attributed to several problems in construction. Inadequate coordination and management of the network of specialist firms that are assembled to deliver projects has been linked to poor planning, lack of information, and faulty materials (Thunberg et al. 2017). Chiang (2009) and Tam et al. (2011) have also suggested that poor quality building products undermine the effectiveness of the multilayered supply chain in Hong Kong. Manu et al. (2013) have highlighted the adverse health and safety performance implications of subcontracting practice and the need to devise strategies that can minimise these adverse health and safety implications. These negative implications of multilayered subcontracting, coupled with the responsibility that main contractors have to mobilise, coordinate, and manage a large number of subcontractors and suppliers on projects has prompted a deepening interest in supply chain management.

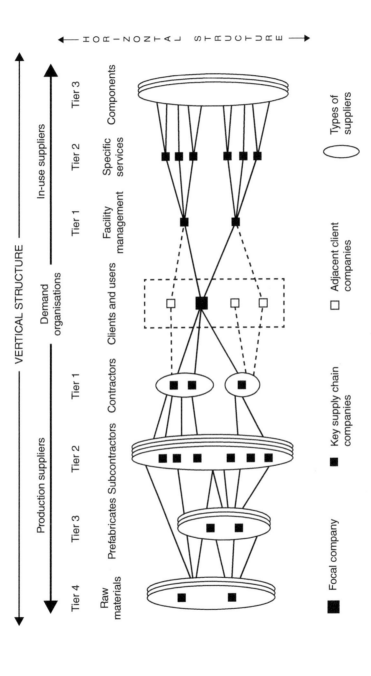

Figure 12.1 Structure of the construction supply chain. Source: Adapted from Ronchi (2006).

The position of main contractors in the construction supply chain has enabled them to adopt and apply supply chain management to the management of subcontractors and suppliers across the downstream tiers, in addition to coordinating upstream linkages with demand-side organisations. Karim et al. (2006) and Smyth (2010) have also argued that the era of shifts towards integrated project delivery and relationship development in the UK construction industry has presented main contractors with an opportunity to increase their role in the market by managing a greater number of stakeholders to facilitate collaborative working. So whilst an unbalanced emphasis has been placed on client-driven supply chain management approaches in construction, it is equally important to examine supply chain management from the main contractor's perspective. This view is supported by Arantes et al. (2015), who has argued that to improve upon construction project delivery through supply chain management adoption, it is important to give specific attention to the relationship interface between main contractors, subcontractors, and suppliers and how these have been influenced by supply chain management adoption.

12.3 Supply Chain Management: Principles and Practices

Supply chain management has been defined widely in the literature, with much more agreement on the definition of 'supply chains' than for definitions of the supply chain management concept (Mentzer et al. 2001). Christopher (2011, p. 13) defined supply chains as 'the network of organisations that are involved through upstream and downstream linkages, in the different processes and activities that contribute value in the form of a product or service delivered to an ultimate consumer'. Christopher then defined supply chain management as 'the management of upstream and downstream relationships with suppliers and customers in order to deliver superior customer value at less cost to the supply chain as a whole' (Christopher 2011, p. 3). The Global Supply Chain Forum has also defined supply chain management as 'an integration of key business processes from end user through original product suppliers with the aim of providing products, services, and information that add value for customers and other stakeholders' (Cooper et al. 1997, p. 2). Lönngren et al. (2010, p. 404) similarly defined supply chain management as 'the task of integrating organisational units along a supply chain and coordinating materials, information and financial flows to fulfil customer demand and improve competitiveness of a supply chain as a whole'.

Supply chain management as a concept extends beyond the logistics thinking as it integrates the management of cooperative relationships with the logistical concerns of material and information flows (Ashby et al. 2012). Mentzer et al. (2001) classified supply chain management as comprising a management philosophy, implementation of a management philosophy, and a set of management processes. The common themes that run through the various definitions imply that supply chain management requires the application of a management philosophy to facilitate the integration of key business processes across supply chain organisations so that they can react as a single entity to market demands, enhance their long-term competitive advantage, and ultimately add value to clients, customers, and end users.

In the UK construction industry, a significant landmark in the evolutionary process of supply chain management adoption was the 'Building Down Barriers' initiative by

Figure 12.2 Seven principles of supply chain management. Source: Holti et al. (2000).

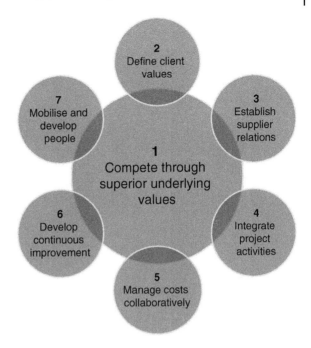

the UK Ministry of Defence (MOD) (Holti et al. 2000). Seven principles were developed based on two pilot projects that were awarded to prime contractors AMEC and Laing in 1997 as an initiative to establish working principles of supply chain integration in construction. These seven principles of supply chain management (see Figure 12.2) were advocated for adoption in the construction industry.

Broft et al. (2016) have thoroughly discussed these seven principles as essential ingredients required for a construction company to function in a supply chain management-driven environment. The first principle 'compete through superior underlying value' requires that the construction supply chain develops a good understanding of the client's perception of value and uses their capabilities to improve quality and reduce unnecessary costs. To achieve this underlying principle, the other six principles will have to be embraced. The second principle 'define client values' requires a rigorous assessment and definition of client value priorities in terms of functional requirements, design character, and whole-life costs. The third principle 'establish supplier relationship' relates to the commitment of forming long-term relationships with a small number of suppliers which can help to improve communication and mutual understanding. The fourth principle 'integrate project activities' relates to collaboration on project activities to resolve design and construction-related issues at the early stages of a project and during the construction phase. The fifth principle 'manage costs collaboratively' advocates for collaborative efforts to develop target costs based on the client's requirements. The sixth principle 'develop continuous improvement' is concerned with the need for long-term efforts to innovate continuously to improve the functionality and value of projects and reduce unnecessary costs, not just across the life of one project but across multiple projects through long-term collaborations. The seventh principle 'mobilise and develop people' relates to the substantial cultural change that is required for successful supply chain management implementation through training in new skills and provision of economic incentives.

These supply chain management principles have driven various practices in the UK construction sector. Fernie and Tennant (2013) have differentiated between client-led and contractor-led supply chain management practices. They likened client-led supply chain management practices to the emergence and use of procurement arrangements like framework agreements to legitimise clusters of organisations led by a principal supply chain partner (contractor) and an associated principal supply chain member (consultant). However, since the number of client organisations that have the financial and operational capacity to procure an ongoing and significant volume of projects can be limited, contractor-led supply chain management practices need evaluating from the perspective of formalised institutions that have been created by main contractors to facilitate integration and collaboration with a network of supply chain firms.

12.4   Supply Chain Management Practices from a Contractor's Perspective

Supply chain management as a concept can be viewed as consisting of processes, management activities, and a supply chain structure (Lambert and Cooper 2000). From a process viewpoint, supply chain management has been applied to business processes like logistics, distribution, and off-site and on-site production and assembly. Beyond the application of supply chain management to logistics and distribution processes, main contractors have used supply chain management to improve management activities associated with the coordination and integration of the supply chain and promoting the levels of cooperation that is required to achieve successful on-site production (Thunberg and Fredriksson 2018). The Egan Report (Egan 1998), which is one of the earliest publications to advocate supply chain management adoption in the UK construction industry, specifically recommended seven features that should be promoted:

1. Acquisition of new suppliers through value-based sourcing.
2. Organisation and management of the supply chain to maximise innovation.
3. Learning and efficiency.
4. Supplier development and measurement of suppliers' performance.
5. Management of workload to match capacity.
6. Incentivisation of suppliers to improve performance.
7. Capture of suppliers' innovations in components and systems.

These supply chain management features underpin to some extent the seven principles of supply chain management that were later advocated by Holti et al. (2000). Other studies have continued to discuss different aspects of supply chain management practice being adopted by construction organisations, all of which fit within the supply chain management principles proposed by Holti et al. (2000) and the above features recommended for supply chain management adoption. For instance, it is considered important that main contractors devise practices for organising and managing the number of firms that constitute their supply chain (Gosling et al. 2010). Category management has emerged as a consistent practice that main contractors adopt for this purpose (Fernie and Tennant 2013). Category management protocols have been implemented by main contractors through the use of explicit labels such as preferred supplier, approved supplier, and pre-approved supplier for categorisation and management of their supplier base (Fernie and Tennant 2013). Ronchi (2006) also differentiated between main contractor practices for managing the supplier base by using the size of the supply base, the existence of an internal classification amongst the suppliers, and the connectivity that is achieved in the network of relationships.

Another important aspect of a contractor-led supply chain management practice relates to how they manage supply chain performance (Cheng et al. 2010; Behera et al. 2015). Fernie and Tennant (2013) revealed the widespread use of formalised prequalification questionnaires (PQQs), key performance indicators (KPIs), and benchmarking by main contractors to categorise and accredit suppliers. The measurement of supplier performance is also strongly linked to continuous performance improvement initiatives with the supply chain, which can only be achieved through supply chain collaboration (Eriksson 2010). Main contractor collaborations with the supply chain has also been linked to planning activities, with supply chain planning emerging as a contractor-led supply chain management practice of involving supply chain actors in the planning process through supply of relevant information for on-site production (Thunberg and Fredriksson 2018). Thunberg and Fredriksson (2018) argued that supply chain planning is an important aspect of supply chain management practice that can improve efficiency of the construction process and facilitate supply chain management adoption. In order to enhance supply chain collaboration, there has also been a significant emphasis on the main contractor's adoption and use of information and communication technologies (ICT), which can improve coordination and information flow amongst supply chain actors (Ronchi 2006; Lönngren et al. 2010; Pala et al. 2014). Pala et al. (2014) revealed that the adoption of relationship management approaches by main contractors, as part of their supply chain management drive, was directly or indirectly facilitated by ICT. There is also a need for main contractors to adopt practices that incentivise and motivate the supply chain to collaborate as part of any supply chain management drive. It has been widely acknowledged that true collaboration in the construction supply chain can only be achieved if benefits are mutually shared. A typical approach is the need to promote fair and prompt payment practices in the supply chain (Manu et al. 2015).

From the above discussion, supply chain management from the main contractor's perspective seems to revolve around practices that support integration and collaboration with supply chain firms through information sharing, performance measurement, continuous improvements, and innovations, and sharing of rewards, with the ultimate view of achieving a competitive edge in the market and delivering better quality projects for clients and end users whilst eradicating unnecessary costs.

12.5   Case Study of a Large UK Main Contractor

A case study approach was used to explore the supply chain management practices of a main contractor in the UK which for the sake of anonymity, is presented here as ABC Construction Company. ABC Construction Company is a major player in the UK construction industry with over £1 billion annual turnover and operates a network of seven regional offices. They employ a considerable number of subcontractors and suppliers across many projects annually and have adopted supply chain management as a strategy for managing the network of firms that they work with to deliver projects. The case study approach was adopted for the study because it allowed for in-depth exploration of supply chain management practices within the main contractor's organisational context, and facilitated evidence gathering from multiple sources (Proverbs and Gameson 2008; Yin 2013). Data on their supply chain management practices was gathered through a supply chain management workshop in ABC Construction Company's premises, direct observations during project meetings, a review of supply chain policy documents and semi-structured interviews with both main contractor and subcontractor personnel. In total, eleven participants were interviewed to explore the supply chain management practices of ABC Construction Company (see Table 12.1). The data was analysed using thematic analysis, which is a core qualitative data analysis method for identifying, analysing, and reporting patterns (themes) within data (Braun and Clarke 2006). The thematic analysis process involved:

- Transcription and initial familiarisation with the data.
- Coding interesting features of the data.
- Collating codes into potential themes.
- Reviewing themes and presenting results.

A main theme that emerged from the analysis was on the 'Main Contractor's Supply Chain Management Features'. This main theme had four categories: 'supply chain management goals', 'supply chain management team', 'supply chain classification', and 'supply chain management practices'. The category on 'supply chain management practices' further consisted of six subcategories that were reflective of ABC Construction Company's supply chain management practices – which will be discussed in the sections that follow.

12.5.1   Supply Chain Management Goals

Having adopted supply chain management as a management philosophy within their business, ABC Construction had to establish clear goals that had to be promoted through supply chain management adoption. The following supply chain management goals emerged from the analysis of the data – to:

- Rationalise the number of subcontractors used across projects.
- Audit health and safety, design, employment policy, financial stability, and other aspects of performance before a supply chain firm works on any project.

Table 12.1 Semi-structured interview participants.

ABC Construction Company	Supply chain subcontractors
1. Supply chain manager	1. Project manager – Mechanical and Electrical (M&E) Services
2. Construction manager	2. Contracts manager – Roofing
3. Senior quantity surveyor	3. Director – Scaffolding
4. Project quantity surveyor	4. Quantity surveyor – Scaffolding
	5. Contracts director – Carpentry and Joinery
	6. Director – Tiling and Mosaic
	7. Operations manager – Panelling

Source: Original

- Have better quality supply chain firms working across projects to deliver innovative and sustainable solutions to clients.
- Collaborate with supply chain firms to promote improved health and safety, build quality, better integration, and materials management, improved sustainability and business ethics, reduced waste, less claims; and fewer defects across projects.
- Allow feedback on project performance to influence future work with the supply chain.
- Share information with subcontractors and to reduce paperwork whilst raising standards.
- Develop greater understanding and trust and improved communication with the supply chain and deliver best results to clients.
- Facilitate regular reviews with key supply chain partners, monitor performance and changes in the business, and set improvement areas.
- Increase spending with strategic and high performing subcontractors where appropriate.

These supply chain management goals all revolved around the development of closer collaboration with supply chain subcontractors that were competent. This could only be achieved by setting goals to select competent supply chain subcontractors based on performance, then collaborate with these firms through increased spending, better communication, information sharing, and integration and continuous improvements across projects, facilitated by performance feedbacks. These goals are very reflective of the supply chain management principles recommended by Holti et al. (2000). In the case of ABC Construction, no explicit goal was set on managing cost collaboratively. However, collaboration was linked to the wider issues of health and safety performance, build quality, better materials management, better integration, improved business ethics, reduced waste, fewer claims, and fewer defects. This range of collaboration issues as set in the supply chain management goals are reflective of ABC Construction's areas of interest as a main contractor. To achieve these goals, ABC Construction had to establish a supply chain management team.

12.5.2 Supply Chain Management Team

ABC Construction appointed Regional Supply Chain Managers across the seven regional branches in the UK. These managers had responsibility for managing a regional team of supply chain management personnel that would act as company contact persons with subcontractors and suppliers that were to become part of their supply chain. As shown in Figure 12.3, the Regional Supply Chain Managers reported to a national (UK) Supply Chain Management Coordinator who had oversight responsibility for setting policy and promoting the supply chain management philosophy within the business. The Supply Chain Coordinator reported to an Executive Director for the construction aspect of the business. A Database Manager was also appointed with responsibility for managing data and information on supply chain firms. The Database Manager had to work closely with the Supply Chain Management Coordinator, an indication of the importance that ABC Construction attached to the management of supply chain data and information as part of their supply chain management practice.

ABC Construction's supply chain management team acted as supply chain management champions to promote and embed the supply chain management practices within

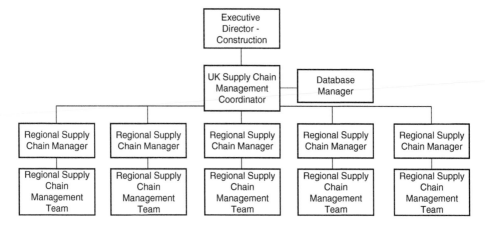

Figure 12.3 Structure of ABC Construction Company's supply chain management team.
Source: Original.

the organisation. As remarked by the Contracts Manager for the Carpentry and Joinery Subcontractor:

> …they [ABC Construction] have a face…[to]..their supply chain, they've got a [supply chain manager] who involves himself with the subcontractors, which I think is a positive thing to have and they..[allocate to]… you.. a person, basically, your contact if anything goes wrong.… not a lot of companies have that. So, it's got structure and I think it certainly seems to work better than most of the other companies out there.

Supply chain subcontractors recognised ABC Construction's commitment to promoting supply chain management within their business through an established and well-structured supply chain management department. Participants from the supply chain firms that were interviewed felt that ABC Construction's supply chain management initiative worked because of the availability of dedicated teams that they could speak to on supply chain management related issues.

12.5.3 Supply Chain Management Classification

ABC Construction also adopted a four-level internal classification system for their supply chain. This classification structure (see Figure 12.4) was institutionalised as the platform for rationalising the number of subcontractors used within the business.

This classification structure provided a platform for implementing the supply chain management practices. As part of the goal to rationalise the number of supply chain firms that ABC Construction work with, this four-level classification was intended to define the number of firms, specify performance requirements, define the nature of relationships that should be promoted, the level of collaboration to pursue, the degree of emphasis to place on continuous performance improvements, and the formality of long-term relationships in the supply chain classes. This classification structure therefore influenced the supply chain management practices discussed in the next section.

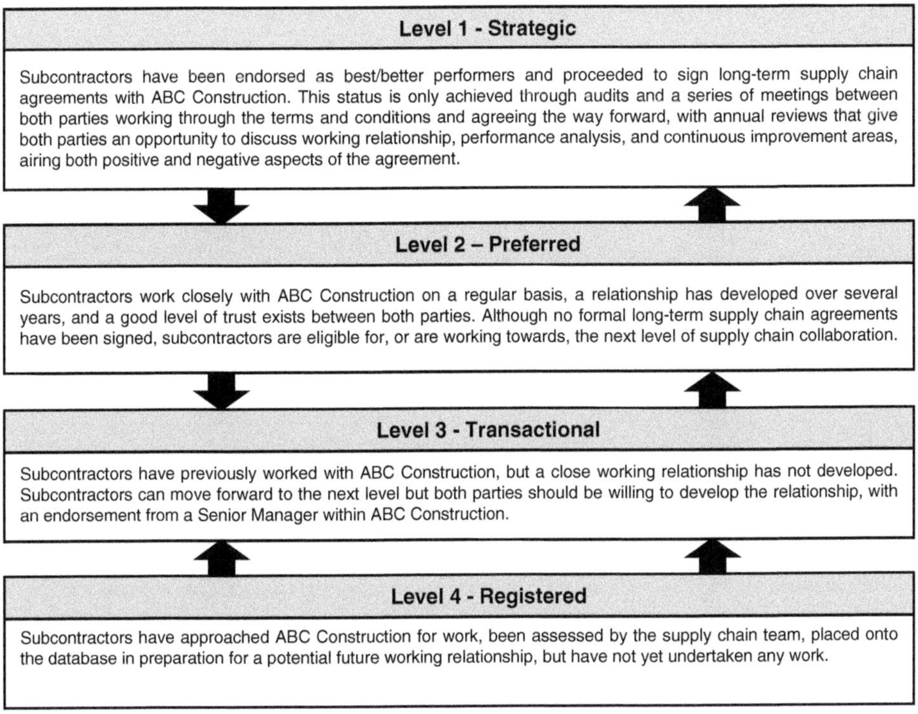

Level 1 - Strategic
Subcontractors have been endorsed as best/better performers and proceeded to sign long-term supply chain agreements with ABC Construction. This status is only achieved through audits and a series of meetings between both parties working through the terms and conditions and agreeing the way forward, with annual reviews that give both parties an opportunity to discuss working relationship, performance analysis, and continuous improvement areas, airing both positive and negative aspects of the agreement.

Level 2 – Preferred
Subcontractors work closely with ABC Construction on a regular basis, a relationship has developed over several years, and a good level of trust exists between both parties. Although no formal long-term supply chain agreements have been signed, subcontractors are eligible for, or are working towards, the next level of supply chain collaboration.

Level 3 - Transactional
Subcontractors have previously worked with ABC Construction, but a close working relationship has not developed. Subcontractors can move forward to the next level but both parties should be willing to develop the relationship, with an endorsement from a Senior Manager within ABC Construction.

Level 4 - Registered
Subcontractors have approached ABC Construction for work, been assessed by the supply chain team, placed onto the database in preparation for a potential future working relationship, but have not yet undertaken any work.

Figure 12.4 Status of supply chain firms on contractor ABC Construction Company's database.
Source: Original.

12.5.4 Supply Chain Management Practices

The 'supply chain management practices' category had six subcategories that encompassed ABC Construction's key supply chain management practices (see Table 12.2). These supply chain management practices were to:

- Audit supply chain firms.
- Use collaborative ICT systems.
- Measure performance of supply chain firms.
- Engage in continuous performance improvement activities.
- Develop long-term collaborative relationships.
- Motivate and incentivise the supply chain.

These supply chain management practices will be discussed in the sections that follow.

12.5.4.1 Audit Supply Chain Firms

Before any subcontractor could undertake work for ABC Construction, they had to be registered onto the supply chain database (see Figure 12.4), which required a rigorous supply chain audit. This audit involved standard company checks such as assessment of company registration information, health and safety documentation, insurance documentation, organisational structure, financial stability checks, key contact persons, and other details that could prove past performance. A standardised

Table 12.2 Supply chain management practices promoted by ABC Construction Company.

Practices	Manifestations
Audit supply chain firms	Supply chain interviews and audits on health and safety, design, employment policy, financial stability, collection of necessary references, commercial checks, and office visits where necessary
Use collaborative ICT systems	Bespoke IT system with a database that held vital details of subcontractors such as trading information, supply chain status, project preferences, performance scores, including key contact persons for each subcontractor. IT system also supported e-tendering
Measure performance of supply chain firms	Performance scoring on health and safety, standard of work, compliance with programme, contractual cooperation, financial cooperation, supervision of work and design input where applicable. Health and safety scorings revealed and discussed with all subcontractors on the project whilst other scores are only revealed and discussed with Level 1 subcontractors during annual review meetings. Performance scores are continuously logged on IT system
Engage in continuous performance improvement activities	Annual meetings with Level 1 (see Figure 12.4) subcontractors to review performance with allocated supply chain contact person and set improvement areas. All subcontractors on database allocated a supply chain point of contact
Develop long-term collaborative relationships	Formal long-term supply chain agreements signed with only Level 1 subcontractors. Tendering assistance is provided by Level 1 supply chain through design input, collaborative planning, and costing
Motivate and incentivise the supply chain	Provided tendering priority based on subcontractor's supply chain status, with on average 50% of annual national spend going to Level 1 supply chain subcontractors and this was as high as 80% on average in some regional branches. Gave out an annual supply chain award to the best performing subcontractor. Strived to promote a 30 days payment arrangement across the supply chain

Source: Original

supply chain questionnaire was used to obtain aspects of the data. In some instances where it became necessary to acquire further information about a subcontractor's office set-up and facilities, ABC Construction's supply chain personnel conducted company visits. Throughout these audits, emphasis was placed on the business initiatives that ABC Construction was promoting such as on building information modelling (BIM), sustainability, and zero harm health and safety targets. Audited and registered supply chain firms were also allocated a supply chain contact person from ABC Construction. This became the starting point of relationship development as part of ABC Construction's supply chain management approach. The allocated supply chain contact persons also acted as supply chain management champions through liaison with ABC Construction's project management team and the supply chain subcontractors.

These audits were strategically used to develop ABC Construction's organisational supply chain rather than for the sole purpose of performing procurement checks for a project. The emphasis that ABC Construction's supply chain team placed on these supply chain audits reflects the importance of the need for careful evaluation and selection of potential supply chain firms (Seth et al. 2018). This was also an important step towards the 'establish supplier relationship' principle of forming long-term relationships with a small number of suppliers which can help to improve communication and mutual understanding (Holti et al. 2000).

12.5.4.2 Use Collaborative ICT Systems

ABC Construction's supply chain management approach was supported by a bespoke supply chain ICT intranet system that facilitated the sharing of data and information about the supply chain across their business. This ICT system hosted the database with records of a subcontractor's previous performance, previous spend levels, and all the relevant documentation obtained during the audit, including key subcontractor contact persons. The database was composed of approximately 5000 supply chain firms nationally. It was a functional intranet system for raising and reviewing important information and performance records about a subcontractor during supply chain procurement at the front end of the project. Project-based staff could review subcontractor performance, conduct localism mapping, and compile a pool of firms to send out tender enquiries. The project team could also review active workload capacity as the ICT intranet system provided an update of work that had been sublet to any supply chain firm within ABC Construction. This was a very useful tool for generating intelligence about supply chain subcontractors to make effective procurement decisions.

ABC Construction also had an ICT project extranet system that facilitated data and information exchanges with the supply chain. This project extranet had e-tendering capabilities. It supported early supply chain involvement in design, specification development, value engineering, planning, and cost compilation. At tendering stage, ABC Construction shared project information with the subcontractors that contributed to their work winning activities. The ICT system was instrumental in facilitating supply chain collaboration during tender preparation and then later, project delivery. These ICT systems were very instrumental for collaboration between ABC Construction and the Tier 2 supply chain subcontractors. Although BIM was mentioned, this was driven on a project-by-project basis and was not relied upon as the main ICT system for supporting supply chain collaboration at the organisational level. However, some evidence from Papadonikolaki et al. (2016) suggests that BIM can play a greater role in fostering supply chain partnering.

12.5.4.3 Measure Performance of Supply Chain Firms

The performance of subcontractors had to be consistently rated by ABC Construction's project team on every project and logged onto the supply chain ICT intranet system. As a requirement, site management staff had to score subcontractors four times a year, depending on their duration on the project. Performance was scored on health and safety, standard of work, compliance with programme, contractual cooperation, financial cooperation, supervision of work, and where applicable, design input. There was also a section for the site management team to provide additional performance-related comments about a subcontractor. Apart from the health and safety performance scores that could be disclosed to subcontractors, only Level 1 subcontractors could review all their performance scores with ABC Construction during annual review meetings. Otherwise, it was for internal company use to review subcontractor performance, determine supply chain classification, and support selection of supply chain firms at the project level. The subcontractor performance scores contributed to the usefulness of ABC Construction's ICT system by providing up-to-date information on performance based on which supply chain procurement decisions were made. The performance information was also used to maintain fluidity in the supply chain, as subcontractors higher up the supply chain levels were always made aware that they could be demoted to lower levels if their performance faltered. This practice was instrumental in preventing supply chain subcontractors from being complacent.

12.5.4.4 Engage in Continuous Performance Improvement Activities
The competitive performance of the supply chain value stream is dependent on continuous learning and development in the supply chain (Hayes 2007). ABC Construction organised supply chain workshops (supplier days) on BIM, health and safety, and sustainability. They worked collaboratively with their supply chain to set standards on reducing the environmental impact of projects, and provided apprenticeships and training for local communities to achieve social value benefits. It was evident that sustainability benefits, particularly social value contributions, could not improve if ABC Construction did not work together with their supply chain to spread understanding of these issues and provide the necessary training. Subcontractors were also provided with training on how to create BIM components, use various BIM tools, and input data into BIM models. These supply chain workshops were to ensure that the entire supply chain as a unit progressed with initiatives that ABC Construction was promoting across their business to improve project delivery. These interactions created knowledge exchange channels that contributed to operational improvements but only the Level 1 supply chain had the additional annual engagements to review performance scores and discuss improvement areas. Due to growing concerns about modern slavery in the construction supply chain, ABC Construction continues to work in partnership with their supply chain to identify risk areas and implement preventive measures.

12.5.4.5 Develop Long-Term Collaborative Relationships
Relationships in the supply chain can be very diverse and have been classified into three distinct typologies by Meng (2012) comprising traditional adversarial, short-term collaborative, and long-term collaborative relationships. Traditional adversarial relationships have a transactional one-off, price-based focus and are driven by a win-lose non-collaborative mentality. This relationship can easily deteriorate due to poor communication in the supply chain, a culture of mistrust and disputes, ultimately yielding substandard performance. Long-term strategic collaborations are used to promote mutual supply chain benefits over the longer term. The nature of relationships that were developed with supply chain subcontractors depended on their status on the supply chain database (see Figure 12.4). ABC Construction developed strategic and formalised collaborative relationships with their Level 1 subcontractors. They formalised the supply chain partnership by signing supply chain agreements to work closely together. There were around 5000 supply chain firms on ABC Construction's supply chain database. At the time of the study only around 250 firms belonged to the Level 1 classification as strategic supply chain partners. These firms were given the opportunity to price for all forthcoming work, although there was no guarantee of winning, and could meet ABC Construction's senior management on an annual basis to review the state of the supply chain collaboration. ABC Construction also counted on the Level 1 firms to provide early supply chain input through tendering assistance such as cost information, design input, and specification development.

There were no formalised supply chain agreements with Level 2, Level 3, and Level 4 subcontractors, although short-term collaborations existed with Level 2 subcontractors, with the potential for progression to the next level – Level 1 subcontractors. To manage any risk of supply chain failure due to overdependence, and to achieve a degree of flexibility, ABC Construction maintained a pool of supply chain subcontractors for each work package across the various levels. This approach has been recommended by Gosling et al. (2010), who discussed the need to maintain a degree of sourcing flexibility by having a pool of suppliers in each category.

12.5.4.6 Motivate and Incentivise the Supply Chain

ABC Construction recognised the need to secure buy-in and commitment from their supply chain by promoting several practices. They created an annual best subcontractor award to recognise and reward subcontractors for exceptional performance. The underlying expectation for subcontractors that had won the best subcontractor award was that recognition for their exceptional performance through the award would contribute to future work winning opportunities and repeat business. Repeat business was the single most important incentivisation for subcontractors. It is not surprising that the opportunity for repeat business was used to secure the most commitment from the Level 1 subcontractors by giving them exclusive access to the pipeline of forthcoming work.

ABC Construction also promoted fair and prompt payment practices in their supply chain. Poor payment practices such as unfair valuations of work for payment, delayed payments, and delayed release of retention deductions, had all been recognised as negative practices that prevail in the construction supply chain. These unethical practices are so culturally embedded in the construction sector that some subcontractors receive payments 90–100+ days after work has been completed and payments have been agreed. This unethical practice in the industry led ABC Construction to promote a 30 days payment policy for all their subcontractors.

12.6   Conclusion

This chapter has focused on supply chain management from a main contractor's perspective. A case study of a large UK main contractor, ABC Construction Company, has been used to evaluate their supply chain management practices. These practices, which comprised an audit of supply chain firms, performance measurement across projects, the use of collaborative ICT systems, engagement in continuous performance improvements, motivation and incentivisation of the supply chain, and prioritisation and promotion of long-term supply chain relationships with best performers, are reflective of supply chain management from a main contractor's organisational context. ABC Construction identified and built collaborative long-term supply chain relationships with the few trusted and committed supply chain subcontractors that provided them with a competitive advantage. These supply chain subcontractors were relied on for work winning support during tendering through early supply chain involvement in pricing, specification development, design contributions, and planning. All the supply chain management practices were adopted to promote integration and collaboration and enhance mutual competitive advantage by reacting as a single entity to maximise value creation for clients. A visible supply chain management team also acted as supply chain management champions in the organisation. What worked well with ABC Construction's supply chain management practice was the level of priority that they gave to supply chain subcontractors with Level 1 status:

> since we've been a [Level 1], pretty much guaranteed at least £1m worth of turnover…We know what they've got coming in their pipeline, which is another reason for being a [Level 1], obviously you get exposure to that, we can then build in to our business plan for any one year
>
> Contracts Manager, Carpentry and Joinery

Despite these efforts, some aspects of ABC Construction's supply chain management practice could have worked much better. It was not always possible to award contracts to supply chain subcontractors who had sometimes made significant tendering support at pre-contract stage. The main contractor had to sometimes audit new firms from the marketplace to obtain cheaper prices. The Construction Manager for ABC Construction, who appreciated the challenge with this approach during the projects remarked:

> I think the biggest thing that can be done to manage the supply chain is actually to have fewer [firms] and work with them closely and actually negotiate jobs with them. The problem you have is that, in this marketplace it's difficult to do, but in the future, that's certainly the way that I'd like to see it happening…

Another weakness is that ABC Construction's supply chain management practices are characteristically dyadic as they focus mainly on Tier 2 subcontractors. It did not extend to the other complex network of firms at the lower tiers of the supply chain (Tier 3, 4, and 5 subcontractors). This is largely reflective of the low level of supply chain management maturity in construction and the inability of main contractors to maximise their role as supply chain managers, despite the opportunity. For main contractors like ABC Construction to have a greater control of the extended tiers of supply firms further down the supply chain tiers, they will have to take a leap from dyadic to triadic sourcing arrangements. Triadic sourcing arrangements are tripartite arrangements that can be established between one buyer and two suppliers (Dubois and Fredriksson 2008). The buyer, as in this instance a Tier 1 contractor, establishes and manages two buyer–supplier relationships (one with the Tier 2 subcontractor and the other with the Tier 3 subcontractor), in addition to the direct relationship between the two suppliers (Tier 2 and Tier 3 subcontractor). Considerable effort will be needed to fully embed supply chain management practices in such triadic arrangements as part of the evolutionary journey towards supply chain management adoption by Tier 1 contractors.

Another area of improvement is the unexploited opportunity to incentivise the supply chain through prompt payments by facilitating seamless transactions using collaborative ICT systems. The emergence of blockchain technology and smart contracts, which are still at the early stages of evolution, can provide a significant opportunity for main contractors like ABC Construction to extend the capabilities of their collaborative ICT systems so that firms across the supply chain network can receive payments immediately a precoded condition has been met. While such an ICT system could alter the current power dominance of main contractors and perhaps initiate another business model, it would represent a significant step towards mutually beneficial supply chain collaboration.

For the industry to prosper, a clearer drive towards ethical practice is required. This has recently been highlighted by the much-publicised collapse of Carillion plc, a British multinational facilities management and construction services company. Members of the Business, Energy and Industrial Strategy and Work and Pensions Committees (House of Commons 2018) criticised the overall corporate culture. The collapse of Carillion plc clearly had a profound impact on its associated supply chains.

> Carillion relied on its suppliers to provide materials, services, and support across its contracts, but treated them with contempt. Late payments, the routine quibbling of invoices, and extended delays across reporting periods were company policy. Carillion was a signatory of the Government's Prompt Payment Code, but its standard payment terms were an extraordinary 120 days. Suppliers could be paid in 45 days but had to take a cut for the privilege
>
> (House of Commons 2018 p. 87).

It is obvious that merely signing up to codes of best practice and paying lip-service to business ethics will have no effect if sections of the construction industry, and specifically prominent and/or large organisations, continue to focus on their own short-term benefit. Those who lose the most in these supply chains are frequently the lower tier subcontractors. More ethical and inclusive practices need to be demanded by clients to ensure that cultures change, otherwise the imbalances in power along the supply chain will continue to lead to exploitation and nonlegitimate risk transferals to those least able to manage it.

References

Arantes, A., Ferreira, L.M.D., and Costa, A.A. (2015). Is the construction industry aware of supply chain management? The portuguese contractors' perspective. *Supply Chain Management: An International Journal* 20 (4): 404–414.

Ashby, A., Leat, M., and Hudson-Smith, M. (2012). Making connections: a review of supply chain management and sustainability literature. *Supply Chain Management: An International Journal* 17 (5): 497–516.

Behera, P., Mohanty, R., and Prakash, A. (2015). Understanding construction supply chain management. *Production Planning & Control* 26 (16): 1332–1350.

Braun, V. and Clarke, V. (2006). Using thematic analysis in psychology. *Qualitative Research in Psychology* 3 (2): 77–101.

Broft, R., Badi, S.M., and Pryke, S. (2016). Towards supply chain maturity in construction. *Built Environment Project and Asset Management* 6 (2): 187–204.

Cheng, J.C., Law, K.H., Bjornsson, H. et al. (2010). Modeling and monitoring of construction supply chains. *Advanced Engineering Informatics* 24 (4): 435–455.

Chiang, Y.-H. (2009). Subcontracting and its ramifications: a survey of the building industry in Hong Kong. *International Journal of Project Management* 27 (1): 80–88.

Christopher, M.L. (2011). *Logistics and Supply Chain Management*, 4e. Edinburgh: Pearson Education Ltd.

Cooper, M.C., Lambert, D.M., and Pagh, J.D. (1997). Supply chain management: more than a new name for logistics. *International Journal of Logistics Management, The* 8 (1): 1–14.

Dubois, A. and Fredriksson, P. (2008). Cooperating and competing in supply networks: making sense of a triadic sourcing strategy. *Journal of Purchasing and Supply Management* 14 (3): 170–179.

Egan, J. (1998). *Rethinking Construction: The Report of the Construction Task Force*. London: Department of the Environment, Transport and the Regions (DETR).

Eriksson, P.E. (2010). Improving construction supply chain collaboration and performance: a lean construction pilot project. *Supply Chain Management* 15 (5): 394–403.

Fernie, S. and Tennant, S. (2013). The non-adoption of supply chain management. *Construction Management and Economics* 31 (10): 1038–1058.

Gosling, J., Purvis, L., and Naim, M.M. (2010). Supply chain flexibility as a determinant of supplier selection. *International Journal of Production Economics* 128 (1): 11–21.

Hayes, J. (2007). *The Theory and Practice of Change Management*. Palgrave Macmillan Basingstoke.

Holti, R., Nicolini, D., and Smalley, M. (2000). *Building Down Barriers: Prime Contractor Handbook of Supply Chain Management*. Ministry of Defence.

House of Commons (2018) Energy and Industrial Strategy and Work and Pensions Committees – Carillion. Report, together with formal minutes relating to the report: https://publications.parliament.uk/pa/cm201719/cmselect/cmworpen/769/769.pdf (accessed August 2019).

Karim, K., Marosszeky, M., and Davis, S. (2006). Managing subcontractor supply chain for quality in construction. *Engineering, Construction and Architectural Management* 13 (1): 27–42.

Kumaraswamy, M.M., Anvuur, A.M., and Smyth, H.J. (2010). Pursuing "relational integration" and "overall value" through "rivans". *Facilities* 28 (13/14): 673–686.

Lambert, D.M. and Cooper, M.C. (2000). Issues in supply chain management. *Industrial Marketing Management* 29 (1): 65–83.

Lönngren, H.M., Rosenkranz, C., and Kolbe, H. (2010). Aggregated construction supply chains: success factors in implementation of strategic partnerships. *Supply Chain Management: An International Journal* 15 (5): 404–411.

Manu, P., Ankrah, N., Proverbs, D., and Suresh, S. (2013). Mitigating the health and safety influence of subcontracting in construction: the approach of main contractors. *International Journal of Project Management* 31 (7): 1017–1026.

Manu, E., Ankrah, N., Chinyio, E., and Proverbs, D. (2015). Trust influencing factors in main contractor and subcontractor relationships during projects. *International Journal of Project Management* 33 (7): 1495–1508.

Meng, X. (2012). The effect of relationship management on project performance in construction. *International Journal of Project Management* 30 (2): 188–198.

Mentzer, J.T., DeWitt, W., Keebler, J.S. et al. (2001). Defining supply chain management. *Journal of Business Logistics* 22 (2): 1–25.

Pala, M., Edum-Fotwe, F., Ruikar, K. et al. (2014). Contractor practices for managing extended supply chain tiers. *Supply Chain Management: An International Journal* 19 (1): 31–45.

Papadonikolaki, E., Vrijhoef, R., and Wamelink, H. (2016). The interdependences of bim and supply chain partnering: empirical explorations. *Architectural Engineering and Design Management* 12 (6): 476–494.

Proverbs, D. and Gameson, R. (2008). Case study research. In: *Advanced Research Methods in the Built Environment* (eds. A. Knight and L. Ruddock), 99–110. Chichester, UK: Wiley.

Ronchi, S. (2006). Managing subcontractors and suppliers in the construction industry. *Supply Chain Forum: An International Journal* 7: 24–33.

Segerstedt, A., Olofsson, T., Hartmann, A., and Caerteling, J. (2010). Subcontractor procurement in construction: the interplay of price and trust. *Supply Chain Management: An International Journal* 15 (5): 354–362.

Seth, D., Nemani, V.K., Pokharel, S., and Al Sayed, A.Y. (2018). Impact of competitive conditions on supplier evaluation: a construction supply chain case study. *Production Planning & Control* 29 (3): 217–235.

Smyth, H. (2010). Construction industry performance improvement programmes: the UK case of demonstration projects in the 'continuous improvement' programme. *Construction Management and Economics* 28 (3): 255–270.

Tam, V.W.Y., Shen, L., and Kong, J.S.Y. (2011). Impacts of multi-layer chain subcontracting on project management performance. *International Journal of Project Management* 29 (1): 108–116.

Thunberg, M. and Fredriksson, A. (2018). Bringing planning back into the picture–how can supply chain planning aid in dealing with supply chain-related problems in construction? *Construction Management and Economics*: 1–18.

Thunberg, M., Rudberg, M., and Karrbom Gustavsson, T. (2017). Categorising on-site problems: a supply chain management perspective on construction projects. *Construction Innovation* 17 (1): 90–111.

Winch, G. (1998). The growth of self-employment in British construction. *Construction Management and Economics* 16 (5): 531–542.

Yin, R.K. (2013). *Case Study Research: Design and Methods*, 5e, vol. 6. Sage Publications, Inc.

13

Lean Supply Chain Management in Construction: Implementation at the 'Lower Tiers' of the Construction Supply Chain

Rafaella Dana Broft

Supply chain management as a concept has received more and more attention within the construction industry. Why? It is believed that successful supply chain management can be related to outstanding performance and 'business' excellence – a way of leading an organisation with its suppliers, or in other words, a chain of organisations, to high productivity and fast delivery. Supply chain principles from manufacturing have therefore been reconceptualised and applied to the specific context of construction. However, examples and initiatives within our industry, particularly at the lower tiers of the construction supply chain, continue to remain limited, and mainly project-specific. Supply chain management emerged from repetitive manufacturing, and both academics and practitioners sometimes question its relevance in construction, an industry delivering almost exclusively 'unique' projects. Some argue that principles should be investigated further and reconceptualised more carefully. Others think that the industry's characteristics should be changed to resemble manufacturing. Some even consider it appropriate to move away from supply chain management and look for other concepts, ideal for a project-based environment?

This chapter deals with this debate. It focuses on the challenges at the lower tiers of the construction supply chain and how, using the philosophy of lean production, the lower tiers, with the main contractor in its central role, can create an environment in which supply chain principles might possibly flourish. Lean and supply chain management are two concepts that belong together and as a combination can be seen as an interesting alternative for the management of construction production.

13.1 Supply Chain Management in a Project-Based Environment

13.1.1 The Supply Chain Management Concept

Supply chain management is considered a way of thinking about management and processes, where the delivery of customer value is seen as a direct result of how supply chains are coordinated, or how the associated relationships are managed

Successful Construction Supply Chain Management: Concepts and Case Studies, Second Edition.
Edited by Stephen Pryke.
© 2020 John Wiley & Sons Ltd. Published 2020 by John Wiley & Sons Ltd.

(Pryke 2009; Fulford and Standing 2014). Supply chain management core concepts primarily originate from the Japanese automotive sector, known as the Toyota Way or the Toyota Production System (TPS). Toyota has used unique processes to manage and operate its own supply chain effectively, where value creation is truly believed to be a joint effort – the long-term philosophy is to create value for customers, suppliers, and society. Innovation and continuous improvement have been shown to be essential in these processes.

Various supply chain concepts in generic theory and manufacturing practice have emerged in parallel over the last decades – all highly related, leading to a high level of ambiguity between the definitions of the different concepts, and reflecting the cross-functional nature of supply chain management (Ellram and Cooper 2014). Most of them have originated from logistics and materials management. Supply chain management as a term first appeared in the early 1980s and was used to describe the connection between logistics and other internal functions or external organisations – its development was initially along the lines of physical distribution and transport, reflected in this description: 'supply chain management covers the flow of goods from supplier through manufacturing and distribution chains to the end user' (Christopher 1992, p. 13). However, this flow of materials and the associated flow of information, has already been managed differently to the approaches that had been known before. Nowadays, logistics management is considered a subset of supply chain management, specifically focused on planning, implementing, and controlling work to move and position inventory throughout the supply chain (Schniederjans et al. 2010). Definitions of supply chain management have gradually evolved towards broader approaches to the supply chain, including additional aspects such as supplier involvement in product development and marketing, and a collaborative customer focus and value creation. As mentioned before, a great variety of related concepts have emerged – supply chain collaboration, relationship management and partnering, supply chain integration and so on – all emphasising a different aspect or perspective of supply chain management. There has not been an agreed-upon definition of supply chain management (Chicksand et al. 2012) and 'it is precisely *this* broad perspective and coverage of supply chain management that makes the concept so difficult to study' (Ellram 1991, p. 21) – and possibly deriving from this, it also seems difficult to implement, at least in some industries…

13.1.2 The Project Focus in Construction

The interest in adopting supply chain management techniques has been growing in the construction industry (Segerstedt and Olofsson 2010) – mainly because of the lack of performance. The search for new and more integrated approaches to the supply chain have taken on a renewed importance for many organisations within the wider construction industry (Holti et al. 2000). But, despite the successful examples of supply chain management initiatives at the higher tiers of the construction supply chain, relationships at the lower tiers seem to remain traditional and the supply chain maturity of construction firms continues to be low (Broft et al. 2016). There is a paucity of properly documented examples of successfully implemented supply chain management initiatives, particularly examples at the lower tiers.

In some industries, such as in aerospace and car manufacturing, supply chain management has become a central strategy, dealing with total business excellence

(Lambert et al. 1998). Intense and often global competition, high technological standards, and rapidly changing market demands have pressed manufacturing industries to manage processes throughout the supply chain in an effective and efficient way (Cagliano et al. 2006). The high levels of alignment and repetition that characterise these supply chains have led to highly productive and fast operating strategic coalitions of firms (Kim 2006). The construction industry, in contrast, involves high levels of fragmentation and low levels of repetition. Many of the applications of supply chain management in construction have been limited to the logistics associated with construction materials and long-term arrangements with suppliers (Vrijhoef 2011). Supply chain management in construction is often seen solely as a project-specific approach.

13.1.3   The Lower Tiers of the Construction Supply Chain

The lower tiers of the construction supply chain comprise many organisations – subcontractors, material suppliers, and service suppliers – all directly or indirectly linked with a main contractor. Almost all construction projects are divided into parts that are subcontracted to individual enterprises, resulting in a fragmented supply chain. A supply chain can be described as a network of interconnected – through upstream and downstream linkages – organisations that are involved in the different processes and activities that produce value in the form of products and services to the ultimate consumer (Harland 1996; Dainty et al. 2001; Christopher 2005). The main contractor, the principal construction organisation that manages a construction project, executes only a small part of the product with its own personnel and its own production facilities (Dubois and Gadde 2000). This means that the quality of a main contractor's relationships, and specifically the ones with its suppliers and subcontractors, affects the main contractors' ability to perform on projects (Kale and Arditi 2001) – a company's processes and activities are interrelated with the processes and activities of its surrounding companies (Broft et al. 2016) – which inevitably has direct consequences on project outcomes.

Unfortunately, the relationships required for the successful delivery of the constructed product among main contractors and suppliers seem weak and difficult to manage (King and Pitt 2009). The collaboration remains rather problematic and traditional, characterised with highly opportunistic and adversarial approaches to relationships (Greenwood 2001) that are riddled with mistrust and scepticism (Akintan and Morledge 2013). This relationship might actually be considered the most important collaboration because it can be argued that most value is being added at the construction site where all the lower tier parties come together and most of the product is being assembled or executed … it is where people are actually building a product!

This difficult collaboration is largely a result of the fragmented nature of the industry and its dependence on subcontracting and competitive pricing (Morledge et al. 2009). Here, rapid technological development in both products and services has driven main contractors to look for external suppliers rather than to develop in-house capabilities. However, unlike other industries, this does not 'automatically' lead to productive or efficient supply chains. The share of subcontracting in construction has risen, and the variety of different subcontracting companies has also grown (Hughes et al. 1997), encouraged by the low barriers to entry; the construction industry is comprised of many small and medium-sized construction-related organisations.

Supply chains have become more and more disjointed (Eriksson 2015) – relationships are often opportunistic, with main contractors competing to win work through competitive pricing while reducing the quality of the end product in order to improve profit margins (King and Pitt 2009). Relationships with suppliers can be regarded as transactional. The relational component, therefore, is missing. The ability to build collaborative relationships is hindered by the prevalent adversarial relationships brought in by opportunism, lack of trust, and inequitable allocation of risk. While suppliers are often regarded as individualistic and only motivated by profit, main contractors are viewed as opportunistic when it comes to winning bids, usually transferring risk to the lower tiers of the supply chain (Cox and Ireland 2002). Some hold the view that existing initiatives for improved collaborations are adopted by main contractors in order to increase their profitability at the expense of other members of the supply chain (Tommelein and Ballard 1998; Dainty et al. 2001). The consequences are poor production processes, limited ability or willingness to innovate due to lack of investment, late project delivery, and budget overrun (Morledge et al. 2009). The largely sequential approach typically supports a lack of integration between all the necessary processes, leading to inefficiencies, inferior value, and poor margins (Holti et al. 2000).

Fragmentation, however, must not be seen as wholly problematic. The involvement of many different specialised firms in projects does not necessarily cause low levels of efficiency. On the contrary, it has been claimed that this could just as easily increase the efficiency of resource allocation and speed of information exchange between parties (Pryke 2002). This might suggest an important task within the supply chain: how to redevelop a construction supply chain and create a supply chain that derives benefit from fragmentation?

13.1.4 A Main Contractor's Position and Role in the Construction Supply Chain

Contractually, main contractors are responsible for the construction of projects, but as mentioned before, they rely heavily on suppliers to execute the works. Subcontracting has been adopted as a procurement strategy following from the uncertainty faced by main contractors in obtaining continuous work – it seems to be a way to reduce overhead and operating costs, improve efficiency, and achieve a more economic delivery of projects (Arditi and Chotibhongs 2005). Then there is also a need to accommodate the different, increasingly specialised and complex requirements of each project (Tam et al. 2011). Main contractors require capabilities and knowledge which do not reside in their own core competencies and which need to be purchased from suppliers. In other words, main contractors increasingly depend on their suppliers, both for realising projects and for achieving the required performance in these projects (Bemelmans et al. 2012). The converse of this is that, owing to the disconnected processes and the large number of suppliers, main contractors are needed to coordinate operations to provide focus and integration of the varied parts (Akintan and Morledge 2013). This is based on the premise that main contractors have an influence on the organisation of the project and on the performance and quality of the work of its suppliers (Latham 1994).

The supply chain has become more and more complex and fragmented, and therefore the need for supply chain management is inevitable. The main contractor's central position offers great potential in having a leading role in initiating the integration of the supply chain – a focal company that coordinates and ties together all flows through the supply chain as if it were an extended enterprise – an organisation that plays an important role in the management of supply chains (Broft and Pryke 2016). However, implementation of supply chain management by main contractors is relatively slow (Green et al. 2005). In addition, within a main contractor's organisation, the management function is typically disconnected from the production function on site as if it were two separate organisations: 'one for the management function and one for getting the work done. The two organisations do not coordinate their work, and they are characterised by different goals and viewpoints' (Applebaum 1982, p. 229). Moreover, projects are not always organised as part of a programme.

In practice, main contractors have a low supply chain maturity and inability to play the essential role of supply chain managers (Broft et al. 2016). For this reason, the main contractor's position and role within the construction supply chain is more and more questioned (see Practitioner Interviews 13.1).

Practitioner Interviews 13.1 The Role and Position of a Main Contractor?

A Discussion between Four Main Contractors and Four Suppliers

Potential for improvement in the role system seemingly lies within the main contractor's role because main contractors are the most significant risk carriers. Almost all suppliers acknowledged this coordinating role of main contractors (Supplier 05; Supplier 06; Supplier 08) and emphasised that the ideal role for a main contractor would be to provide clarity regarding the client's requirements and to perform a timely selection of suppliers. They articulated that main contractors currently seem '*to lack all specialist knowledge, including knowledge regarding logistics, needed to fulfil a "linking" role*' (Supplier 05) and that price calculations were mainly performed without transparency and just aiming to achieve the lowest costs possible – '*the contractor seems to pursue an inappropriate role that ignores openness, but rather involves a "divide and rule" tactic*' (Supplier 05). In general, the choice for suppliers is made too late (Main Contractor 01) and can be '*random*' (Main Contractor 03), a simple result of the type of construction (Main contractor 02) or a result of a search for common business drivers (Main Contractor 04). '*The relationship that follows seems a trial and error process with problems with discipline and holding onto agreements*' (Main Contractor 03) – by suppliers perceived as '*missing professionalism*' (Supplier 05).

However, the increasing percentage of project turnover which is outsourced and therefore spent on buying goods and services provides opportunities for contractor–supplier collaboration, and emphasises the importance and significance of managing suppliers (Bemelmans et al. 2012). Main contractors are willing to develop closer relationships, but implementing supply chain management seems a long-term, complex process and requires a certain level of understanding and therefore learning throughout the supply chain (Broft et al. 2016). Furthermore, some characteristics of the industry are not believed to be suitable for implementing supply chain management (Cox and Ireland 2006). This has been attributed to the inherent nature of the industry, to traditional practices, and to traditional attitudes (Elfving and Ballard in press). Why?

13.2 The Characteristics of Construction

13.2.1 Construction from a Production Perspective

Construction is about the delivery of construction products, such as houses, commercial buildings, and infrastructure. It is a type of production, just a slightly different one – 'Production is the action of making or manufacturing from components or raw materials, or the process of being so manufactured' (Oxford Dictionaries 2018). For many reasons construction products are considered complex products that need to be managed through projects. Repetition, standardisation, variability, and so on, are terms that play an important role in the way production is designed, managed, and controlled as production theory and practice have sprung from thinking about repetitive manufacturing. The low levels of repetition in construction increase the unpredictability of the flow of work and contribute to the way construction production, construction management, and supply chains are organised. In construction, a typical supply chain shows the following production-related characteristics (Broft 2017; Broft and Koskela 2018):

- Logistics converging to a common and fixed point in the supply chain: the construction site where the 'construction factory' is located – the object is assembled from incoming materials, coming from different supply units, and through different services (Luhtala et al. 1994).
- Temporary and non-repetitive, or in other words, one-off construction projects that are produced through repeated reconfiguration of project organisations (Vrijhoef and Koskela 2000), delivered to single end customers – construction is prototype production.
- Multiple and concurrent projects.
- A number of studies have linked construction with the characteristics of the engineer-to-order production strategy (Segerstedt and Olofsson 2010; Gosling et al. 2012). Engineer-to-order projects are described as having high levels of customisation and are typically managed on a project basis.
- Construction is mainly based on two types of processes: small batch processes and job processes (Krajewski et al. 2015). Sometimes a project is considered a separate production system (Figure 13.1) as it seems to be a more fundamental form of production system than the factory.
- The uniqueness of schemes (Souza de Souza 2015).
- Construction can show part and work station congestion, where one location can be worked on by several work stations at the same time and the work is carried out in suboptimal conditions, with lessened productivity (Koskela 1999).

The degree of repetition seems to differentiate between products and the system that is used to produce these products. It defines the standard typology of production systems, which could be dominated by either (one-off) designing or (repetitive) making. Construction, however, involves projects organised around the delivery of a building. Or in other words, it is project-based with the project being the primary business mechanism for coordinating and integrating all the main business functions of the firm (Lundin and Söderholm 1995; Hobday 2000). A project has three features: it is unique, novel, and transient – 'in business there are repeat objectives, which require us to do repetitive things, and there are new objectives which require us to do unique, novel, and transient things' (Turner 2009).

> *There is no truth in the construction industry. All projects are unique and therefore, an exception and each project comprises a new belief system and set of rules. This inhibits the recognition people are able to find in solutions that are offered to the industry.*
>
> (Main contractor 01)

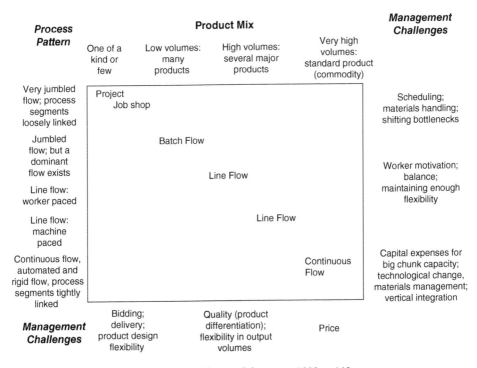

Figure 13.1 Types of production systems. Source: Schmenner 1993, p. 140.

A project involves a temporary organisation and process set-up to achieve a specified goal under the constraints of time, budget, and other resources (Shenhar and Dvir 2007). The parties involved in construction projects have been interpreted as 'organisational units joining and operating together as a single production organisation when it is advantageous' (Harland et al. 2001) or a coalitions of firms, described as 'a number of independent firms coming together for the purpose of undertaking a single building project and that coalition of firms having to work as if it were a single firm, for the purposes of the project' (Winch 1989). The focus in temporary organisations, however, is believed to be on transition rather than production – when transition becomes necessary within a permanent organisation, temporary organisations are often created to deal with it (Lundin and Söderholm 1995). A project has no autonomous capability – we rely entirely on mobilising the resources supplied by clients and the organisations in the construction industry for their existence (Winch 2010). For this reason, a project could also be described as a temporary organisation to which resources are assigned to do work to deliver beneficial change (Turner 2009), as a coalition of resource bases (Winch 2010), or as a system for organising labour and materials (Pryke 2017).

13.2.2 Construction: True Peculiarities?!

The characteristics of construction, as described above, are often seen as 'peculiarities' (see Practitioner Interviews 13.2) – peculiarities that are considered to differentiate between project-based industries and repetitive manufacturing, and prevent the attainment of flows as efficient as in manufacturing (Koskela 1992). The most differentiating characteristics are: 'one-of-a-kind nature of projects, site production, and temporary multiorganisation'. Despite the fact that other types of production also possess one or several of these characteristics, it is the specific combination of characteristics that might define construction as 'peculiar' (Ballard and Howell 1998; Segerstedt and Olofsson 2010). Construction outputs possess two characteristics which together uniquely define them: (i) they belong to the category 'fixed position manufacturing', and (ii) they are rooted in place.

There are different ways to categorise types of manufacturing (Ballard 2005). One of these methods differentiates types of manufacturing in terms of the primary determinant of flow through the process. Factories that produce one or a few products have flows designed specifically for those products, whereas process-based flow is found in factories organised to perform certain types of operations on many different products. The third type identified by this method is fixed position manufacturing, where the objects are wholes assembled from parts. In the assembly process, the product being manufactured eventually becomes too large to be moved through work stations, so the work stations or crews have to move through the product, adding pieces to the emerging whole as they move (Ballard 2005). This happens in construction. Some degree of site production, at minimum final assembly, is a necessary aspect of construction. This rootedness-in-place brings with it uncertainty and differentiation (Ballard and Howell 1998). Every location is different and, therefore, its conditions can vary widely from place to place.

Practitioner Interviews 13.2 Construction Peculiarities?

A Discussion with Supply Chain Members On-site

Different opinions exist regarding the peculiar characteristics of the construction industry. Supply chain members on-site seem to agree that, despite their admitted unfamiliarity with other industries, the weather conditions can be seen as the biggest peculiarity (Supplier B; Supplier D). However, only when one actually works outside (Supplier E)! *'There are a lot more parties involved, but not many of them are truly working simultaneously'* (Supplier B). *'Sometimes one has more projects at the same time, but that seems easily manageable'* (Supplier C). Supplier D states that *'products might differ from each other, but it does not seem to be too different in other industries'*, whereas Supplier G says *'that always the same thing is being made'*. The product might be different, but the perception regarding its uniqueness relates to the amount of experience a supply chain member has in making it (Supplier X). The location does change, but work does not differ – *'it offers the chance to see a lot more of the country'* (Supplier C). Supplier F specifically adds the aspect of freedom to decide on the working method used, to which Supplier G seems to agree: *'one knows what to make but is flexible on how to make it'*. Supplier B mentions the great amount one can learn when working in construction.

The organisation of production and the supply chains is strongly adapted to these peculiarities, these basic characteristics (Koskela 2000; Broft 2017) that come with this project thinking and seem to increase complexity. Supply chain management, however, was born in repetitive manufacturing (Elfving and Ballard in press) where, as mentioned before, project-exceeding characteristics, as opposed to the characteristics in construction, exist. Proponents of supply chain management in construction argue that a basic shift from a one-off approach to a project-independent approach in the construction supply chain is necessary in order to stabilise the project and production environment, and achieve operational and competitive improvements across projects and organisational boundaries (Vrijhoef and Koskela 2005). (Others, through these peculiarities, and the relationship to manufacturing as described above, challenge the acceptance of construction as a separate industry.)

The most important role of a production manager in construction includes the management of this complexity because the effects of complexity are variable schedules or variable production (Kenley 2005). Complexity management can be improved if the nature of complexity is identified, necessary complexity accepted, and unnecessary complexity decreased (Pennanen and Koskela 2005). A portion of complexity can be regarded as waste. The first issue to consider is whether any peculiarities could be eliminated or at least reduced. The rule is really not to accept any peculiarity – and the related complexity – unless necessary and appropriate (Vrijhoef and Koskela 2005). With the 'elimination' of peculiarities, its related complexity can be reduced. The remaining complexity has to be embraced through appropriate managerial concepts and tools – and this brings us to lean production. Lean production contributes to tackling both relative and objective unnecessary complexity (Saurin et al. 2013) resulting in a more repetitive environment. In other words, lean highlights repetitive characteristics and might make implementation of supply chain management thinking in construction more feasible.

13.3   Lean Supply Chain Management in Construction

13.3.1   An Introduction to Lean

The TPS has not only been known for the way it collaborates with suppliers, but also for how its production process is organised and managed – 'lean'. 'Lean' emerged at a time of great interest in Japanese production and management methods generally, and particularly innovative management ideas at Toyota. The term was first used in Krafcik's (1988) (a researcher from the Massachusetts Institute of Technology (MIT) working on the International Motor Vehicle Programme) article to describe the TPS. The word was selected to capture the essence of the far less resource-hungry TPS compared with typical western, so called 'buffered', production systems (Figure 13.2). It was developed during the 1950s and 1960s as a result of intense postwar competition and is considered a decontextualisation of a new dominant paradigm that is displacing mass production (introduced by Ford and also referred to as the Ford Production System or 'Fordism') in the search for methods to compress time and increase flow (Samuel et al. 2015). The system consists of a complex cocktail of ideas including continuous improvement, flattened organisational structures, teamwork, elimination of variation, efficient use of resources, and cooperative supply chain management (Green 1999).

The primary goal of lean production is the elimination of waste – the logic of lean production describes value-adding processes unencumbered by waste, where lean is a means or systematic approach of waste identification in operations so that it can be eliminated for greater efficiency – to ensure that value flows swiftly and smoothly to the customer (Schniederjans et al. 2010; Samuel et al. 2015):

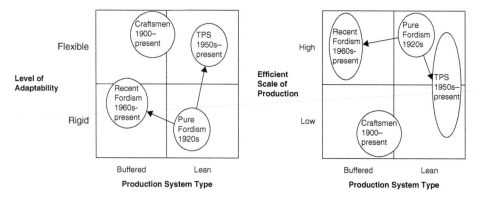

Figure 13.2 Toyota Production System (TPS) versus Ford Production System. Source: Krafcik (1988), p. 44.

- It implies a total view of the process – from raw material source to end consumer, and perhaps beyond, through recycling of materials (Lamming 1996).
- It does not limit the focus to traditional assumptions on 'necessary' or 'unnecessary' activities: wasteful practices must be defined anew in the search for lean systems (Lamming 1996).
- It could be argued that by eliminating unnecessary elements and interactions, lean production contributes to eliminating unnecessary complexity (Saurin et al. 2013).

13.3.2   The Role of Lean in Combination with Supply Chain Management

Lean construction represents a way of implementing important elements of lean production in construction. It includes many tools that a main contractor can use to improve predictability, eliminate variability in production through standardisation, and reduce lead time. Essentially, it could be seen as an adaptation of the lean philosophy to a project environment – as the application of lean thinking to construction. But what happens if we would like to move away from project thinking and focus on supply chains and supply chain management?

The role of the supply system (or supply chain, network, stream, etc.) – comprising the purchasing activities of the assemblers and the supply activities of the component manufacturers (Lamming 1996) – received attention from the outset of the discussion on lean production. It may be defined as an integrated sociotechnical system whose main objective is to eliminate waste by concurrently minimising supplier, client, and internal variability (Shah and Ward 2007). 'This restructuring process has strong inspiration as a philosophy of organising production: lean manufacturing' (Lima and Zawislak 2003).

Lean challenges us to fundamentally rethink value from the customer. This includes the identification of the 'entire' value stream. The value stream is the set of all the specific actions required to bring a specific product or service through the critical management tasks of any business. In supply chain management, all supply chain actors need to be able to make a full contribution to ensure that the client's needs are fulfilled, and that value creation is maximised (Broft et al. 2016). This implies a collaborative customer focus and a higher quality of delivery for each subprocess. Lean thinking must go beyond the organisation to look at the whole: the entire set of activities entailed in creating and producing a specific product. Flow is created in order to accomplish tasks continuously from raw material to finished good rather than departmentalised, in batches (Womack and Jones 2003).

From this flow perspective, the object of supply chain management is 'to integrate and manage the sourcing, flow and control of materials using a total system perspective across multiple functions and multiple tiers of suppliers' (Mentzer et al. 2001). A supply chain, encompassing all the subprocesses (Broft and Koskela 2018), is conceptualised as a production flow (rather than a series of transactions or contracts) – it covers the flow of goods from the different suppliers through manufacturing and distribution chains to the end user (Christopher 2005). While conventional construction management focuses on the project, supply chain management emphasises the product and the tasks or subprocesses organised around this product as a network. The concept acknowledges the interdependency between subprocesses and includes integration (Broft and Koskela 2018).

In lean supply chain management, the entire flow from raw materials to a customer is considered as an integrated whole (Lamming 1996). Interfaces between the different organisations of the supply chain are thus seen as artificial – as a result of the economic arrangement of assets governed by many other factors rather than as natural transformation stages in the development (or addition of value) (Lamming 1996). The combination of lean applied to the management of supply chains can generate outstanding business performance. It involves the application of all the lean principles within a supply chain management context (Schniederjans et al. 2010). Some authors refer to this combination as supply chain best practices (Blanchard 2010), as a lean supply chain (Wincel 2004), or a way of harnessing value in the supply chain (Banfield 1999).

Lean supply chain management could be defined as 'the use of a highly transparent, trusting and long-term relationship between the buyer and supplier to create a physically efficient supply chain through the reduction of waste in processes or responsiveness in product delivery' (Schniederjans et al. 2010). Value creation, that has become a function of the network of iterative and transient relationships between actors, is central.

13.3.3 Lean and Supply Chain Management in Construction

Supply chain management plays an important role in achieving project-independent construction. From a main contractor–supplier perspective this must be achieved through altering strategies (Vrijhoef and Koskela 2005) in production management. Supply chain management offers an alternative way for this production management, involving the control and optimisation of subcontracted activities, where suppliers are invited to focus on the efficiency of their subprocess (through perhaps, standardisation and prefabrication) and to eliminate unnecessary costs (with the help of the production manager) (Broft and Koskela 2018). One of the supply chain principles from manufacturing that could be reconceptualised and applied to the specific context of construction (Vrijhoef 2011) includes the introduction of the role of the supply chain integrator in the supply chain. Fundamental to the theory of supply chain management is the notion of exercising control of an identified sequence of activities from a vantage point. This vantage point is usually occupied by the firm or organisation conducting the last significant transformation of the product before it reaches the customer (through the downstream supply chain) (Lamming 1996). One of the critical phenomena lacking in the construction industry is the recognition of a generally accepted focal company initiating the integration of the supply chain. The members of a supply chain include all organisations with whom the focal company interacts directly or indirectly through its suppliers or customers, from point-of-origin to point-of-consumption (Lambert et al. 1998). A main contractor, with its central position within the supply chain, might be a suitable candidate for this role (Broft 2017).

Supply chain management considers the relationships between suppliers because it involves elements such as the creation of a more permanent production process through long-term relationships with suppliers (Broft and Koskela 2018). As opposed to transaction cost economics that treats each transaction separately (make-or-buy), supply chain management includes the systems' benefits of organising clusters of related transactions as supply chains are introduced – related transactions are grouped and managed as chains (Williamson 2008). This offers alternative ways of minimising (transaction) costs (Pasquire et al. 2015).

The current approach in construction might still be sequential, but supply chain management in construction should be seen as the management of a network of interconnected organisations that are involved in the different processes and activities that produce products and services to the customer (Dainty et al. 2001; Christopher 2005). Owing to the still disconnected processes and the large number of suppliers, main contractors are needed to coordinate operations to provide focus and integration of the varied parts (Akintan and Morledge 2013): supply chain management is 'the task of integrating organisational units along a supply chain and coordinating materials, information and financial flows in order to fulfil customer demand with the aim of improving competitiveness of a supply chain as a whole' (Stadtler 2000, p. 11). This integration between subprocesses focuses on the overall efficiency of the entire supply chain, through the use of important flow-related principles – the reduction of the lead time of the product through the elimination of waste within the overall production process and the reduction of variability (Berliner and Brimson 1988; Koskela 2000). It relates to the synchronisation of an organisation's processes with those of its suppliers and customers to match the flow of materials, services, and information with customer demand (Krajewski et al. 2015). Again, from a lean perspective, the interfaces between organisations are considered artificial. Management objectives have moved away from the attention focused on the finite domain of a single organisation to deliver competitive advantage. Attention is now focused on ensuring competitive advantage for the integrated supply chain (Green et al. 2005).

The project-focus within the 'project-based production system' as seen in construction, supporting fragmentation and the perceived uniqueness of each project, seems to overrule this product-focus and the advantages of flow, necessary to improve customer value. A manufacturing system, which involves the flow of material through a plant, is an objective-oriented network of processes through which entities (the parts to be manufactured) flow (Hopp and Spearman 1996). Besides the flow of materials, construction involves two other flows: location flow and assembly flow, which are related to the characteristics of construction as described earlier. Production in construction is of assembly-type, where different material flows are connected to the end product on-site. Due to the size of the product of construction, an intermediate workflow arises where all installation locations proceed through the installation work station (Koskela 2000).

The attitude of the construction industry tends to be oriented towards conformance with contractual specifications rather than gaining additional financial benefits or competitive strength from quality improvement (Vrijhoef 2011). As a result, construction seems oriented more towards production and getting the work done on time and within budget (Lai and Cheng 2003). Lean supply chain management in construction means focusing on value and rethinking value. It is a constant reflection on the questions: Who is the client? What are his or her expectations from the entire supply chain? – a collaborative focus rather than a focus on internal value delivery.

13.4 Conclusion

Supply chain management is often presented as suitable for efficient management of construction production. Its successful implementation in the industry, however, remains limited to the improvement of logistics and inventory, whereas in some industries supply chain management has become a central strategy, dealing with total business excellence.

Implementation of supply chain management at the lower tiers of the construction supply chain can be continually improved when we focus on the similarities that construction has with repetitive manufacturing. The chapter shows that it should not be that different when focusing on the product to be delivered. Lean plays an important role in this implementation. With the help of lean, main contractors might be able to move their focus from projects to supply chains and create characteristics similar to repetitive manufacturing.

At some point, the actual value added by the role of the main contractor might be questioned. Is a main contractor part of production delivery, a production manager, or just an organisation translating through demand to the suppliers that deliver the value? Supply chain management plays an important role in achieving project-independent construction and it offers an alternative way for production management. In this chapter, I suggest that the combination of lean applied to the management of supply chains can generate outstanding business performance. This can be achieved through the exploitation of the potential of the central role of the main contractor and strong links with the lower tiers of the supply chain.

References

Akintan, O.A. and Morledge, R. (2013). Improving the collaboration between main contractors and subcontractors within Traditional Construction Procurement. *Journal of Construction Engineering* Art. ID: 281236.

Applebaum, H.A. (1982). Construction management: traditional versus bureaucratic methods. *Anthropological Quarterly* 55 (4): 224–234.

Arditi, D. and Chotibhongs, R. (2005). Issues in subcontracting practice. *Journal of Construction Engineering and Management* 131 (8): 866–876.

Ballard, G. (2005) "Construction: one type of project production system." In: Proceedings IGLC-13, July, Sydney, Australia.

Ballard, G. and Howell, G.A. (1998) "What kind of production is construction?" In: *Proceedings 6th Annual Lean Construction Conference* (IGLC-6) 13–15 August 1998, Guarujá, Brazil.

Banfield, E. (1999). *Harnessing Value in the Supply Chain*. New York, NY: Wiley.

Bemelmans, J., Voordijk, H., and Vos, B. (2012). Supplier-contractor collaboration in the construction industry: a taxonomic approach to the literature of the 2000-2009 decade. *Engineering, Construction and Architectural Management* 19 (4): 342–368.

Berliner, C. and Brimson, J.A. (1988). *Cost Management for Today's Advanced Manufacturing*. Boston: Harvard Business School Press.

Blanchard (2010). *Supply Chain Management: Best Practices*. Oxford: Wiley.

Broft, R.D. (2017). "Exploring the application of lean principles to a construction supply chain." In: *Proceedings of the 25th Annual Conference of the International Group for Lean Construction (IGLC)*, 9–12 July, Heraklion, Crete, Greece.

Broft, R.D. and Koskela, L. (2018). Supply chain management in construction from a production theory perspective. In: *IGLC 2018 – Proceedings of the 26th Annual Conference of the International Group for Lean Construction: Evolving Lean Construction Towards Mature Production Management Across Cultures and Frontiers*, vol. 1 (ed. V.A. Gonzalez), 271–281.

Broft, R.D. and Pryke, S. (2016). Who should be leading in the process of successful supply chain management implementation in construction? In: *Proceedings of the CIB World Building Congress 2016*, Tampere, Finland, 575–588.

Broft, R.D., Badi, S. and Pryke, S. (2016). Towards SC Maturity in Construction. *Built Environment Project and Asset Management*, 6 (2): https://doi.org/10.1108/BEPAM-09-2014-0050.

Cagliano, R., Caniato, F., and Spina, G. (2006). The linkage between supply chain integration and manufacturing improvement programmes. *International Journal of Operations and Production Management* 26 (3): 282–299.

Chicksand, D., Watson, G., Walker, H. et al. (2012). Theoretical perspectives in purchasing and supply chain management: An analysis of the literature. *Supply Chain Management: An International Journal* 17: 454–472.

Christopher, M. (1992). *Logistics and Supply Chain Management: Strategies for Reducing Cost and Improving Service*. London: Pitman/Prentice Hall.

Christopher, M. (2005). *Logistics and Supply Chain Management: Creating Value-Adding Networks*. Harlow: Pearson Education.

Cox, A. and Ireland, P. (2002). Managing construction supply chains: the common-sense approach. *Engineering, Construction and Architectural Management* 9 (5/6): 409–418.

Cox, A. and Ireland, P. (2006). Relationship management theories and tools in project procurement. In: *Management of Complex Projects: A relationship Approach* (eds. S.D. Pryke and H.J. Smyth). Oxford: Blackwell.

Dainty, A.R.J., Briscoe, G.H., and Millett, S.J. (2001). New perspectives on construction supply chain integration. *Supply Chain Management: An international Journal* 6 (4): 163–173.

Dubois, A. and Gadde, L.E. (2000). Supply strategy and network effects – purchasing behaviour in the construction industry. *European Journal of Purchasing and Supply Management* 6 (3–4): 207–215.

Elfving, J. and Ballard, G. (in press). Supplier Development in the Construction. *Journal of Purchasing and Supply Management*.

Ellram, L.M. (1991). Supply chain management: the industrial organization perspective. *International Journal of Physical Distribution and Logistics Management* 21: 13–22.

Ellram, L.M. and Cooper, M.C. (2014). Supply chain management: it's all about the journey, not the destination. *Journal of Supply Chain Management* 50 (1): 8–20.

Eriksson, P.E. (2015). Partnering in engineering projects: four dimensions of supply chain integration. *Journal of Purchasing and Supply Management* 21: 38–50.

Fulford, R. and Standing, C. (2014). Construction industry productivity and the potential for collaborative practice. *International Journal of Project Management* 32 (2): 315–326.

Gosling, J., Towill, D.R., and Naim, M.M. (2012). Learning how to eat an elephant: implementing supply chain management principles. In: *Proceedings of the 28th Annual ARCOM Conference*, 3–5 September, Edinburgh, UK.

Green, S.D. (1999). The missing arguments of lean construction. *Construction Management and Economics* 17: 133–137.

Green, S.D., Fernie, S., and Weller, S. (2005). Making sense of supply chain management: a comparative study of aerospace and construction. *Construction Management and Economics* 23 (6): 579–593.

Greenwood, D. (2001). Subcontract procurement: Are relationships changing? *Construction Management and Economics* 19 (1): 5–7.

Harland, C.M. (1996). Supply chain management: relationships, chains and networks. *British Journal of Management* 7 (S1): S63–S80.

Harland, C.M., Lamming, R.C., Zheng, J., and Johnsen, T.E. (2001). A taxonomy of supply networks. *Journal of Supply Chain Management* 37 (4): 21–27.

Hobday, M. (2000). The project-based organisation: an ideal form for managing complex products and systems? *Research Policy* 29: 871–893.

Holti, R., Nicolini, D., and Smalley, M. (2000). *The Handbook of Supply Chain Management: the Essentials*. London: Construction Industry Research and Information Association and The Tavistock Institute.

Hopp, W. and Spearman, M. (1996). *Factory Physics: Foundation of Manufacturing Management*. Boston: Irwin/McGraw-Hill.

Hughes, W., Gray, C. and Murdoch, J. (1997). Specialist Trade Contracting – A Review. *CIRIA Special Publication 138*. CIRIA, London.

Kale, S. and Arditi, D. (2001). General contractors' relationships with subcontractors: a strategic asset. *Construction Management and Economics* 19 (5): 541–549.

Kenley, R. (2005). Dispelling the complexity myth: founding lean construction on location-based planning. In: *Proceedings 13th International Group for Lean Construction Conference*, July 2005, Sydney, Australia.

Kim, S.W. (2006). The effect of supply chain integration on the alignment between corporate competitive capability and supply chain operational capability. *International Journal of Operations and Production Management* 26 (10): 1084–1107.

King, A.P. and Pitt, M.C. (2009). Supply chain management: a main contractor's perspective. In: *Construction Supply Chain Management: Concepts and Case Studies* (ed. S. Pryke), 182–198. Oxford: Wiley Blackwell.

Koskela, L. (1992). Application of the New Production Philosophy to Construction. Technical Report 72. CIFE, Stanford University, Stanford, CA.

Koskela, L. (1999). Management of production in construction: a theoretical view. In: *Proceedings 7th Annual Conference of the International Group for Lean Construction*, 26–28 July, Berkeley, California, USA.

Koskela, L. (2000). *An Exploration Towards a Production Theory and its Application to Construction*. Finland: VTT Publications.

Krafcik, J.F. (1988). Triumph of the lean production system. *Sloan Management Review* 30 (1): 41–52.

Krajewski, L.J., Malhotra, M.K., and Ritzman, L.P. (2015). *Operations Management: Processes and Supply Chains*. Pearson.

Lai, K.H. and Cheng, T.C.E. (2003). Initiatives and outcomes of quality management implementation across industries. *Omega* 31: 141–154.

Lambert, D.M., Cooper, M.C., and Pagh, J.D. (1998). Supply chain management: implementation issues and research opportunities. *International Journal of Logistics Management* 9 (2): 1–19.

Lamming, R. (1996). Squaring lean supply with supply chain management. *International Journal of Operations and Production Management* 16 (2): 183–196.

Latham, M. (1994). Constructing the Team: Joint Review of Procurement and Contractual Arrangements in the UK Construction Industry: Final Report. HMSO, London.

Lima, M.L.S.C. and Zawislak, P.A. (2003). A produção enxuta como fator diferencial na capacidade de fornecimento de PMEs. *Revista Produção* 13 (2): 57–69.

Luhtala, M., Kilpinen, E., and Anttila, P. (1994). LOGI Managing Make-to-Order Supply Chains. Report. Helsinki University of Technology, Finland.

Lundin, R.A. and Söderholm, A. (1995). A theory of the temporary organisation. *Scandinavian Journal of Management* 11 (4): 437–455.

Mentzer, J.T., DeWitt, W., Keebler, J.S. et al. (2001). Defining supply chain management. *Journal of Business Logistics* 22 (2): 1–25.

Morledge, R., Knight, A., and Grada, M. (2009). The concept and development of supply chain management in the UK construction industry. In: *Construction Supply Chain Management: Concepts and Case Studies* (ed. S. Pryke), 23–41. Oxford: Wiley-Blackwell.

Oxford Dictionaries (2018). https://en.oxforddictionaries.com/definition/production (accessed August 2019).

Pasquire, C., Sarhan, S., and King, A. (2015). A critical review of the safeguarding problem in construction procurement: unpicking the coherent current model. In: *23rd Annual Conference of the International Group for Lean Construction*, 308–318.

Pennanen, A. and Koskela, L. (2005). Necessary and unnecessary complexity in construction. In First International Conference on Complexity, Science and the Built Environment, 11–14 September 2005 in Conjunction with Centre for Complexity and Research, University of Liverpool, UK.

Pryke, S.D. (2002). Construction coalitions and the evolving supply chain management paradox: progress through fragmentation. In: *Proceedings COBRA*, 5 September 2002, Nottingham, UK.

Pryke, S. (2009). *Construction Supply Chain Management: Concepts and Case Studies*. Oxford: Wiley Blackwell.

Pryke, S. (2017). *Managing Networks in Project-Based Organisations*. Oxford: Wiley.

Samuel, D., Found, P., and Williams, S.J. (2015). How did the publication of the book The Machine That Changed The World change management thinking? Exploring 25 years of lean literature. *International Journal of Operations and Production Management* 35 (10): 1386–1407.

Saurin, T.A., Rooke, J., and Koskela, L. (2013). A complex systems theory perspective of lean production. *International Journal of Production Research* 51 (19): 5824–5838.

Schmenner, R.W. (1993). *Production/Operations Management*, 5e. New Jersey: Prentice Hall.

Schniederjans, M.J., Schniederjans, D.G., and Schniederjans, A.M. (2010). *Topics in Lean Supply Chain Management*. Singapore: World Scientific Publishing Company.

Segerstedt, A. and Olofsson, T. (2010). Supply chains in the construction industry. *Supply Chain Management: An International Journal* 15 (5): 347–353.

Shah, R. and Ward, P. (2007). Defining and developing measures of lean production. *Journal of Operations Management* 25: 785–805.

Shenhar, A.J. and Dvir, D. (2007). *Reinventing Project Management: A Diamond Approach to Successful Growth and Innovation*. Boston: Harvard Business School Press.

Souza de Souza, D.V. (2015). A Conceptual Framework And Best Practices For Designing And Improving Construction Supply Chains. PhD thesis. University of Salford.

Stadtler, H. (2000). Supply chain management – an overview. In: *Supply Chain Management and Advanced Planning – Concepts, Models, Software and Case Studies* (eds. H. Stadtler and C. Kilger), 3–28. Berlin: Springer.

Tam, V.W., Shen, L.Y., and Kong, J.S. (2011). Impacts of multi-layer chain subcontracting on project management performance. *International Journal of Project Management* 29 (1): 108116.

Tommelein, D. and Ballard, G. (1998). Coordinating specialists. *Journal of Construction Engineering and Management (ASEC)*: 1–11.

Turner, J.R. (2009). *The Handbook of Project-Based Management: Leading Strategic Change in Organizations*. USA: McGraw-Hill.

Vrijhoef, R. (2011). *Supply Chain Integration in the building industry*. Delft University Press.

Vrijhoef, R. and Koskela, L. (2000). The four roles of supply chain management in construction. *European Journal of Purchasing and Supply Management* 6 (3–4): 169–178.

Vrijhoef, R. and Koskela, L. (2005). A critical review of construction as a project-based industry: identifying paths towards a project-independent approach to construction In: Kazi, A.S. (ed.). *Proceedings CIB Combining Forces*, 9: Learning from Experience: New Challenges, Theories and Practices in Construction, 13 June, Helsinki, VTT, Espoo, 13–24.

Williamson, O.E. (2008). Outsourcing: transaction cost economics and supply chain management. *Journal of Supply Chain Management* 44 (2): 5–16.

Wincel, J.P. (2004). *Lean Supply Chain Management*. New York: Productivity Press.

Winch, G. (1989). The construction firm and the construction project: a transaction cost approach. *Construction Management and Economics* 7 (4): 331–345.

Winch, G. (2010). *Managing Construction Projects: An information Processing Approach*. Oxford: Wiley Blackwell.

Womack, J. and Jones, D. (2003). *Lean Thinking*. NY: Simon and Schuster.

14

Knowledge Transfer in Supply Chains
Hedley Smyth and Meri Duryan

14.1 Introduction

Knowledge sharing is a well-trodden concept in theory and practice across the construction industry. As part of the broader theory about knowledge management, research has consistently shown that knowledge sharing practices are poorly implemented. There are good reasons for this. Firstly, the financial management of construction companies leads senior managers to keep investment in new capabilities and competencies to a minimum in order to manage the vagaries of economic cycles, and keep project transaction costs to a minimum. Secondly and flowing from this, supply-side programme management, where knowledge management systems reside, is underdeveloped. This inhibits the knowledge sharing between projects and across the supply chain. Thirdly, procurement processes use minimum benchmarks, which incentivise low bids, typically without a cost contingency to share lessons within a project and across a client programme. This chapter addresses these barriers in detail. The result is transactional processes that map into cultural norms that solicit *defensiveness* and *learned helplessness.*

The chapter goes on to consider how procurement from leading clients can be dynamic and incentivise improvements and break down silos. Dynamic procurement practices can drive forward internal improvements and supplier competition, and stimulate engagement with HR practices to support the improvement, whereby staff selection, promotion, and annual reviews embed knowledge sharing as an important part of progression and pay. The chapter also identifies interorganisational communities of practice as a means of facilitating the necessary culture to enable knowledge sharing and constructing meaning through collaborative mechanisms. The case study on which this chapter was based involved an interorganisational community located in the client organisation; it was set up towards the end of the research.

Empirical evidence is provided from research into one client undertaking large programmes of work and how the barriers in the supply chain are in evidence. Evidence comes from both the client and from part of its supply chain. Supply chain members include consultants, main contractors, and subcontractors. It considers further the remedies that were brought forward for consideration and how these were selected for implementation.

Successful Construction Supply Chain Management: Concepts and Case Studies, Second Edition.
Edited by Stephen Pryke.

14.1.1   The Supply Chain Issue

Clients have the choice to make or buy the things they need as organisations. The advantage of outsourcing (buying) is twofold:

1. *Risk* – Subcontracting potentially passes the risk of provision to another party, providing they deliver.
2. *Specialist expertise* – The benefit of specialised resources, know-how, and capability of the supplier.

The same applies to the Tier 1 main contractor as the customer for its subcontractors.
The 'buy' decision puts the client in the commanding position in terms of market power, reinforced by a common feature that those higher in the food chain tend to be larger organisations. However, adversarial features are well-rehearsed in the literature (e.g. Bresnen and Marshall 2000; Humphreys et al. 2003; Pryke and Smyth 2006) and in industry (in the UK see for example, Latham 1994; Egan 1998; IPA 2017).
The benefits of in-house provision are at least twofold:

1. *Cooperation* – A potentially cooperative relationship, although in-house subsidiaries can be treated as if they are any other subcontractor.
2. *Coordination* – Configuring and delivering supply with the parent's provision in efficient and effective ways.

The in-house service provision or 'make' decision can lead to complacency and higher costs, which have to be weighed up against the contractual transaction costs (Macneil 1980; Williamson 1975). In construction, the perceived risk carries potential costs that are perceptually added to the anticipated transaction costs to provide an additional outsourcing incentive.

Managing the supply chain is the endeavour to achieve the best of both worlds. The aim is to solicit better value for money by having more control over quality and value through intervention into the supply chain and its own management. This puts good subcontractors on a similar basis to in-house suppliers without the complacency of the captured market that in-house provision can induce. Indeed, Campbell (1995) writing from the management perspective terms this collaborative relationship 'subcontracting'. He states that the interdependencies grow and the outsourced supplier becomes 'domesticated'. This flows through effective relationships (Christopher et al. 2002).

This has echoes of the Tavistock Report (1966) on the industry, which drew attention to the benefits of (and indeed difficulties associated with) interdependencies in the construction supply chain. It also reflects the work of Söderlund (2012), who distinguished between cooperation as an effective relational characteristic and coordination as effective resource and task alignment for project activities. Söderlund, therefore, contrasted the *relational* approach to the *transactional* one as many others have done (e.g. Pryke and Smyth 2006). Effective relationships have the potential to deliver added value (Christopher et al. 2002). Projects, whether sourced in-house or outsourced, can be bundled into programmes for ease of rationalising their management. In both cases the strength of relationships is important. Improvements are limited to the extent of relationship management to increase effectiveness and efficiency (Smyth 2015). For a more *transformational* approach to improvements in management, strategies through innovation, technical and technological advances are needed. This poses serious challenges to contractors and subcontractors, who are driven by the bidding process and other financial management drivers to keep investment to a minimum. This is part of the broken business model among contractors (Smyth 2018). There has to be a shift for contractors, especially for main contractors to survive. Clients are motivated to drive change (e.g. Ive 1995), yet contractors tend to throw innovation and improvements 'over the wall' to the next party along the supply chain (Smyth 2006). The supply chain needs to be proactive and responsive, for as Fernie and Tennant comment:

> Largely taken for granted within the UK construction sector has been a view that supply chain management theory is robust, relevant and reliable
> (Fernie and Tennant 2013, p. 1038).

Those that do respond will survive and grow. In other words, contractors and subcontractors will have to be more transformative to meet market demand, especially for the largest and most complex projects, embedding the developments in the firm for future programmes and projects.

One challenging yet low cost way to begin to transform performance and improve delivery is through learning and knowledge transfer. It is the interplay between the client driving changes and the response of the supply chain that is the focus of this chapter. Knowledge management in the supply chain is the specific area for examining transformational change to meet challenging and complex demands. To what extent are practitioners engaged with such an interplay and seeking transformations through learning and applying knowledge in and across supply networks? To what extent are supply chain actors, particularly lower tier actors, *given the opportunity* to engage in this way?

14.1.2   Learning and Knowledge Transfer

Learning is central to developing service provision and delivering valuable outputs. Learning needs to be managed systematically (e.g. Senge 1990). This is ambitious and is multidimensional, having organisational, interpersonal, and individual dimensions or layers of activity and behaviour. At the organisational level, one of the main building blocks is knowledge transfer, which requires deliberate management efforts and incentives and a robust knowledge management system. While knowledge management has been given considerable research attention in project environments, research findings and leaders across project sectors acknowledge considerable shortfalls in practice (e.g. Koskinen 2000; Anumba et al. 2005; Kivrak et al. 2008; Kelly et al. 2013). The Tier 1 main contractor has received most attention in construction (e.g. Kivrak et al. 2008; Smyth 2010; Kelly et al. 2013). What has received less attention is the role of the client on the one hand and the activities of the lower tiers of the supply chain on the other hand. The dual focus and the interplay between these actors provide the focus for this chapter.

Learning in the supply chain and knowledge transfer on large projects, within a client programme and across projects undertaken by subcontractors, is passive and therefore weak to the extent known, yet is also underexplored (Bresnen 2009). Interorganisational knowledge transfer not only requires collaboration, which is a general alignment, it requires cooperation, which is deeper, to facilitate knowledge transfer (Anvuur and Kumaraswamy 2008; Bresnen 2009). This is because projects have no memory (Dubois and Gadde 2002; Love et al. 2005). Therefore, project-based organisations have to be prepared to share with the trust and confidence that they are better off doing so than being defensive (Smyth and Edkins 2007). Defensiveness stifles learning and sharing as it keeps the organisation in a passive and reactive mode of operation.

To what extent are clients driving change through their supply chains to induce proactivity towards cooperation around learning and knowledge transfer? To what extent are supply chains responding positively and actively sharing knowledge across projects and between projects on large client programmes? These are two questions addressed in the chapter through consideration of a large project-based client organisation undertaking a multibillion pound programme of work, which is outsourced to a supply chain of national and international contractors and subcontractors.

In order to examine to what extent practitioners are engaged with seeking transformations through learning and applying knowledge in and across supply networks, the chapter first looks in more detail at the relevant literature prior to setting out the methodology and methods used in the research. The findings are presented followed by conclusions and recommendations flowing from the analysis of the findings.

14.2   What Is Known – A Summary Review of the Literature

14.2.1   The Supply Chain Ecosystem

The term 'supply chain' implies a linear set of arrangements. This may be an accurate depiction for certain projects where serial subcontracting for a hierarchy of tasks induces a chain of layered suppliers. However, most project-based organisations have a network of suppliers that provide a range of provision in interdependent and interlocking ways. This can be termed a network (e.g. Dubois and Gadde 2002; Pryke 2009) or project ecosystem (Xu 2020; cf. Grabher 2004), which in value terms is expressed as a service ecosystem (Akaka et al. 2013; Chandler and Vargo 2011; Vargo and Lusch 2016). Knowledge has the potential to enhance the service experience and potentially enhance the value of the project delivered and realised in use (Smyth 2015).

Whether knowledge leads to additionality in practice depends upon an effective learning environment and management capabilities at both the project and project-based organisation levels as well as between the two (Davies and Brady 2000; Brady and Davies 2004; Love et al. 2005). The extent of integration of the supply chain is the responsibility of the client, which needs to appoint an effective systems integrator as the main contractor, and for that main contractor to be an efficient and effective integrator for the project (e.g. Hobday et al. 2005; Davies et al. 2007). However, it is commonplace for there to be a lack of integration within the supply chain because the client and main contractor respectively only take responsibility for the next tier down (London et al. 1998) and not always in ways that encourage cooperation and transformation of the service and content (cf. Bresnen 2009).

Effective client or supply organisations embed learning capabilities and provide integrated services. In so doing, they are not operating alone; they build interdependencies through their networks or ecosystem. Yet, the balance of market power is variable between the organisational actors (cf. Campbell 1995). At one level the organisational boundaries are more artefacts of accounting and financial management and may be barriers to effective and cost-effective integration. The internal accountability flowing from transactional financial management can act to inhibit learning and knowledge management; indeed it can even impede service provision. Gummesson (2002) has alluded to this, which raises how learning is applied; he puts it this way:

> Nobody has seen a corporation! ... we mistake the phenomenon for its tangible representation. ... Our perceptions about organisations take over, and we become slaves instead of making them our servants. Inferior quality, disinterest in the customer, erroneous decisions and inertia are blamed on organisation and system: 'I'm sorry, I can't do anything about it'. This has been called 'learnt helplessness'.
>
> (Gummesson 2002, p. 259)

14.2.2 Supply Chain Learning and Knowledge Management

Knowledge practitioners agree that the great challenge in developing effective and systematic application of learning through knowledge management and engagement with that system lies in the organisational and cultural dimensions (e.g. Davenport et al. 1997; Davenport and Prusak 1998). Cultural values shape patterns of interactions, influencing the willingness and behaviour for tacit knowledge sharing and explicit knowledge transfer (Gray and Densten 2005). The values help shape the organisational mental model (Blackman and Henderson 2005). A culture where knowledge sharing is the norm encourages people to collaborate and rewards good practices through praise, pay, and promotion. Communication systems and IT platforms only support the culture for knowledge sharing and application (Bloom 2000; Love et al. 2005).

Dawson (2000) cites the need for firms to capture knowledge for organisational benefit. Actionable knowledge is created on and within projects (Sexton and Lu 2009), where knowledge is produced and held collectively (Love et al. 2005). Formal knowledge procedures and systems enable leverage of explicit and tacit knowledge beyond the organisational boundaries of any one project. This is important in construction where effective practices are nested in projects around managing change and problem solving (Love et al. 2005; Senaratne and Sexton 2008; Kenley 2012). Yet, awareness of the need to support and engage with knowledge sharing is variable (Kivrak et al. 2008). The absence of an enabling culture leads to a passivity or one where barriers are raised, essentially feeding the concept of behavioural *learned helplessness* that becomes structured into systematised ways that raise further barriers (Abramson et al. 1978). Learned helplessness which starts as individual and team responses to passive management becomes embedded as a negative habit in the project organisation and construction firm. Winter (2013) argues that organisational habits are embedded as routines, which then become rigidities that require active management to change or overcome (Leonard-Barton 1992; Gilbert 2005).

While many construction companies engage in some formal and informal 'lessons learned' practices, partial and frequently untimely capture and a failure to validate, disseminate, and reapply knowledge, is commonplace. The systems are partial, management focus and leadership is typically passive, and bottom-up engagement is minimal (e.g. Sage et al. 2010; Smyth 2010). Successive waves of partial implementation, coupled with blame in corporate culture, creates risk aversion among people and raises more barriers for knowledge exchange through defensiveness and *learned helplessness*. Partiality is unsurprising. If it was easy, all organisations would be investing in it and implementing knowledge management effectively. The social and physical dislocation of projects renders the challenge greater in construction (Pryke and Smyth 2006), on top of which and contributing to the challenge is the inherent 'stickiness' of knowledge in its cultural and operational context (Szulanski 2000; Love et al. 2005).

In large infrastructure programmes, more relevant knowledge is generated outside the client and main contractor organisational boundaries than inside (Anumba et al. 2005). Thus, collaboration with supply chain members and engagement with external institutions is one of the key sources of shared expert knowledge (Smyth and Longbottom 2005). Organisations working together in supply chains and the broader network are likely to spread and share best practices and the results of research and development. One mechanism within and across organisational boundaries is to apply communities of practice, which promote interaction and create meaning through negotiation (Lave and Wenger 1991; Wenger 1998) in complex supply chains where knowledge sharing is essential for effective operations and outcomes (Foss 2007; Sanaei et al. 2013).

A survey of large construction organisations demonstrated that communities of practice are the most widely used technique for knowledge sharing (Carrillo et al. 2002). Gray (2001) argues that communities of practice provide a key means to encourage knowledge flow and address stickiness. They provide a forum for sharing tacit knowledge and making it explicit in a way that IT platforms and social media cannot adequately achieve due to the low levels of engagement and impartation. It has been argued that knowledge application is difficult in all organisations and especially so in construction (e.g. Smyth and Longbottom 2005; Pryke and Smyth 2006). There are benefits from manager meeting the challenge and overcoming the barriers to knowledge application. Low investment levels and project on costs are important barriers. Firstly, the financial management of construction companies leads senior managers to keep investment in new capabilities and competencies to a minimum in order to manage the vagaries of economic cycles, and keep project transaction costs to a minimum. Secondly and flowing from this, supply-side programme management where knowledge management systems reside is underdeveloped. This inhibits the knowledge sharing between projects and across the supply chain. Thirdly, procurement processes incentivise low bids, typically without a cost contingency to share lessons within a project and across a client programme. Arguably, construction business models have become outmoded in relation to market demands and firm survival (Smyth 2018).

These factors are enduring if the client side, whether it is the main contractor or project sponsor, does not demand rates of improvement above static thresholds. The prequalification and bidding process are the places where serial improvements can be evaluated against theory and measured in practice.

14.2.3 Prequalification and Bidding Processes

Procurement proceeds on the basis of exerting market power, and emphasises value and price or value for money. In practice, price typically overrules the value proposition. However, supply chain management practices endeavour to improve the value proposition through more cooperative relationships (e.g. Campbell 1995).

Yet procurement is a complex process and commences with prequalification. Prequalification practices have become more sophisticated over recent times, qualifying contractors and supply chain members against a range of factors such as health and safety practice, sustainable sourcing, financial stability and other industry key performance indicators (Doloi 2009). However, the criteria for qualification are minimum ones, which can probably be best described as *static thresholds*. They lack a dynamic assessment of improvement, for example based upon rates of improvement. Knowledge management potentially introduces a dynamic element in terms of inputs application and performance indicators providing some measurable outputs.

Bidding criteria tend to be more project specific, with very little evidence or analysis required of past performance or strategies in relation to future performance improvements. Indeed, the criteria to be used by clients are not always clear overall or in the weighting for decision making (cf. Edkins and Smyth 2016). Again, there is a lack of a dynamic element to selection criteria according to the literature, which might suggest that the extent to which practitioners engage with transformational practices concerning knowledge management is of little importance and that change is slow. Yet this cannot be presumed.

To what extent are practitioners engaged with transforming performance – interacting to induce an interplay around knowledge and learning? How active and committed are management in this endeavour? This will be explored through using a £5bn infrastructure programme as a detailed case. The original research employed a combination of engaged and action research.

14.3   Methodology and Methods

An interpretative methodology was used, which was appropriate for a topic embracing explicit and tacit aspects of knowledge sharing and application. The research is more specifically conducted through engaged methods. A two-year action research study in a major client organisation was supported through Innovate UK, a government-funded body. The client and its supply network operate in a multiorganisational environment of new provision and other project work. The new provision is divided into programmes of work containing multiple projects, many of which are megaprojects (those in excess of the equivalent of 1 billion US dollars). There are overlapping and interlocking operational systems reaching into the supply chains and up to institutional bodies that also directly link to the supply as part of the broader ecosystem. The research reaching into the supply chain was largely conducted through engaged research that tried to influence practice yet was not organisationally embedded.

The method was to employ semistructured interviews, which were analysed according to the main emergent themes as well as events of significance (Smyth and Morris 2007). Cognitive mapping was employed as an additional means of analysis – a visual technique to show perceptions, patterns, and causal relations between the issues (Ackermann and Eden 1994). Cognitive mappings are graphs comprising nodes and interconnections between the perspectives of individual actors. Decision Explorer software (Brightman 1998) is used to aid analysis of the maps, the prime tools being head, centrality, domain, hierarchical cluster, potency, and cotail analyses.

The data was solicited from six supply chain members, comprising two consultants, two contractors and two subcontractors, with 23 semistructured interviews being conducted in the supply chain. Defensiveness in the supply chain led to uneven access across the organisations, which is a finding in its own right because defensiveness is a symptom of a static picture and suggests barriers existing to transformational practices in the supply chain, in this case around knowledge management.

14.4 Findings

One main contractor summed up the overall tenor of the findings by saying, 'There's a lot of ad hoc stuff around lessons learned'. There was a prevailing reliance upon individuals to take the initiative. The companies were not coherently managing knowledge, and nor was the client, in a consistent and systematic way. The knowledge that was generated within one project was buried in unread reports to the extent that these were required to be written at the end of projects, or, knowledge was lost because people moved to other projects, left the employer, or retired.

It was repeatedly reported by respondents that there were low levels of engagement with the client and supplier IT platforms for knowledge management, although not all supply chain members had platforms. The programme firms did not 'own' or embed the explicit lessons or knowledge at the project and firm levels of organisation. A learning culture is an important issue for effective knowledge transfer and application.

While a degree of egalitarian conduct existed around internal tasks and activities, hierarchical accountability was invoked and a propensity to blame others was prevalent within the client and supply chain. This was underpinned by an extensive management reliance on transactional risk and cost control at the expense of transformational practices. There was also a low level of investment to develop capabilities in programme management on the supply side.

Knowledge management did not occur in real time. Project reports, where required, were completed at handover. The timely transfer on the project and to other parallel projects in the programme became lost opportunities, and knowledge capture was time-lagged resulting in lessons being lost in the delay and through personnel reallocation. Resultant documents were inaccessible and hard to interrogate on the IT platform and the platforms ware unsupported by parallel human systems.

To the extent that knowledge management was conducted, it was primarily decentralised to the project level of operations. There was an absence of programme management on the supply side among supply chain members. Programme management is the level where capabilities would, or arguably should, be located to enable knowledge management across projects in general and is necessary when undertaking serial projects for a client with its own programme. The absence of programme management and, at the project level, competitive pricing and a lack of contingency budgets, inhibited knowledge capture transfer during projects. The finance department and the commercial directors applied transactional management to project and functional budgets, failing to understand the transformational transition needed for complex projects. One exception was that the client held six-monthly conferences to serve the purpose of 'socialising ideas'. This encouraged a degree of strategic knowledge sharing at intrafirm level, yet lacked operational follow through.

It was found that there was common agreement about the shortcomings of knowledge management practices across the organisations. There was also hope for improvement, which was evident in the rhetoric that was ahead of the practices. The linkages and interdependencies between the issues are depicted in a cognitive map for the supply chain (Figure 14.1). The map indicates the shared thinking across organisational boundaries and the desire to see improvement. Based on the results of the map analysis, the nodes 'improve knowledge management' and 'improve collaboration between the client and supply chain' are the heads of the map. They are the goals expressed in terms of

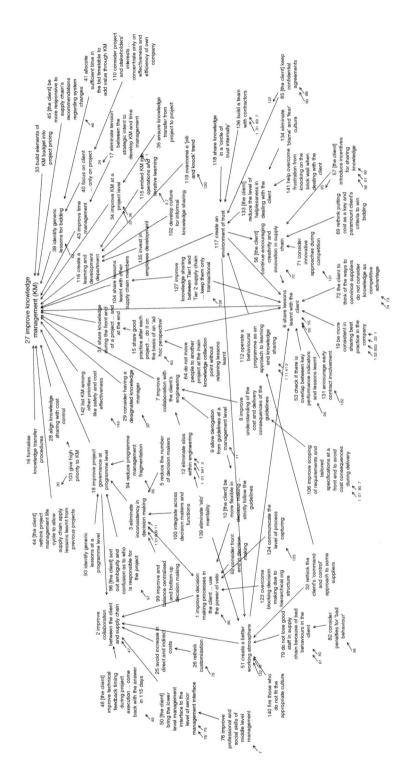

Figure 14.1 Cognitive map of interviews with client supply chain (dotted lines represent the links to hidden nodes). Source: Original.

final ends or effects. This means that supply chain members agree that there is an urgent need to improve their collaboration with the client and the way they manage knowledge. Domain and centrality analyses of the map were used to identify the key strategic directions leading to the goals (see Table 14.1). The first ten commonly raised strategic issues are shown in rank order in Table 14.2 with numbered crossreferencing to Figure 14.1. Therefore, personal or individual motivation is not in question. It is a matter of management.

The improvement of decision-making processes at the client firm was among the strategic objectives aimed at improvement of collaboration and knowledge exchange

Table 14.1 Key strategic options in descending order.

Rank	Key strategic objectives	Reference on the map (Figure 14.1)
1.	Share lessons learnt with the client	6
2.	Create a better working atmosphere	51
3.	Create an environment of trust	117
4.	Improve project governance at programme level	18
5.	[The client] continue encouraging innovation in supply chain	56
6.	Improve collaboration with the client's engineering	7
7.	Improve knowledge management on a project level	34
8.	Identify generic lessons for bidding	39
9.	Be more consistent in sharing best practice in the company	19
10.	Improve decision making processes of the client rather than use the power of veto	1

Source: Original.

Table 14.2 The most influential options in descending order.

Rank	Key potential options	Reference on the map (Figure 14.1)
1.	Eliminate inconsistency in decision making	3
2.	Allow derogation from guidelines at a management level	9
3.	Eliminate 'silo' mentality	139
4.	Rethink the client's 'command and control' approach to some suppliers	32
5.	[The client] keeps confidential agreements	85
6.	Be strategically more proactive especially at the front end	78
7.	[The client] introduces incentives for sharing knowledge	57
8.	Make knowledge from site operations across projects more explicit	37
9.	Share good practice after each project rather than do it on the basis of an 'ad hoc perspective'	15
10.	Improve knowledge sharing between Tier 1 and Tier 2 supply chain rather than keep them only transactional	127

Source: Original.

between the client and the supply chain (Table 14.1). The respondents mentioned that the management of the client firm should eliminate inconsistency in decision making and be more flexible in dealing with the supply chain rather than using the power of veto. From the perspective of the respondents, in order to encourage the supply chain to share knowledge, the firms need to exhibit speedy, efficient, and thorough execution of promises.

The cognitive map, shown in Figure 14.1, displays the concepts that supported the achievement of more than one goal. Potency analysis is applied on the basis that the more goals a concept supports, the more potent this concept is. It allows prioritising options that have consequences for the bigger number of key issues. The most influential issues that have consequences for the strategic objectives in practice are listed in Table 14.2.

The concepts that support the achievement of more than one goal are mainly in the 'collaboration' domain (see Table 14.2). Network relationships and collaboration are key means for knowledge creation and sharing. From the perspectives of respondents, more scope and influence in decision making, better collaboration across and along supply chains, supported by a better knowledge transfer, can reduce tensions between organisations and actors. In the long run, knowledge transfer breaks down silos and associated 'silo' mentalities internally and defensiveness interorganisationally. The client firm needs to take the lead by cooperatively managing relationships to encourage supply chain members to share learning that solves problems, hence making things better, cheaper, and faster. On the ground, 'man marking' was employed under the rhetoric of collaboration, which in practice is perceived as increasing monitoring, accountability, and costs rather than enabling and facilitating. Indeed, 'man marking' was found to encourage defensiveness.

At the front end, the client and entire supply chain use static threshold criteria for qualifying suppliers. Additional dynamic criteria to include knowledge management to assess improvement rates were absent. Main contractors reported 'improvements' were uncritically cut and pasted across prequalification and bid documents without acting upon the stated improvement agenda. HR policies in the supply chain lacked knowledge sharing and application criteria in staff selection, induction, training, annual reviews, and promotion. These factors created a gap between the rhetoric about what the firms claimed to be doing and what happens on the ground on top of the barriers incurred from client behaviour.

The client tended to commence delivery prior to completing the scoping, defining, and specifying of each project. The main contractors described this as a major disincentive to engage with effective knowledge management due to the subsequent and constant change. Activities were delayed with the consequence that timely knowledge sharing and application were eradicated as a consequence.

During delivery, the consultants tended to be better than other firms, yet this is largely the result of employing 'knowledge workers' and being 'reflective practitioners' rather than strategic and tactical commitment from the firms' senior management. Main contractors displayed some effective knowledge management practices. These were largely focused upon cost savings and efficiency gains for the firm to improve profitability. They were not directly about adding value to serve or save costs for the client. While indirect benefits may accrue at times, this is fortuitous rather than through designed service improvement. The focus was, therefore, self-interested and inward facing.

The respondents identified a series of barriers to effective knowledge sharing and application:

1. Insufficient time is allowed for early contractor involvement and for bid managers to apply lessons learned.
2. Untimely and confused client decision making during execution because of poor programme and front-end management of the projects.
3. Client confusion between collaboration and intervention to manage projects which reduces the room for flexible responses among suppliers.

There were as many internal barriers as external. Finance departments were serving agendas of survival and profit declaration by keeping costs and investment to a minimum. This is part of the cash flow management practice regarding return on capital employed to declare dividends at the expense of the long-term interests of the firm, their clients, and in this case making a portion of the profits from taxpayers. Senior management of the client organisation lived with the conflict of interest associated with the need to support the willingness of operational personnel and project teams to be transformational, and needing to comply with the governance imposed by the head of financial management and the commercial directors.

There was a lack of a common professional language or a common perspective on knowledge management across functions by the supply chain actors and across the organisations. The client exacerbated this issue with their own project management methodology and associated terms. The lack of a common professional language increased the complexity of creating a collaborative environment internally and among the key stakeholders.

At the individual level, there appeared to be scant appreciation of the difference between generic solutions and tailored solutions to context and the role of knowledge to initially identify the generic and later apply knowledge to tailor to context. There was reliance upon the assumed notion of 'project uniqueness' as if this is justifiably self-explanatory – taken for granted thinking in the culture. The consultants as specialist providers and the specialist subcontractors as solution suppliers were more aware than the main contractors, although it is the latter that is responsible for knowledgably configuring integrated solutions and service innovation.

Knowledge management is more difficult for contractors due to site dislocation, exacerbated by a temporary workforce that infrequently receives robust induction. In most industries, the knowledge management process is a six-stage process:

Identify → Capture → Process → Store → Disseminate → Apply

For projects, it depends upon double processing to sift on site and tease out the generic knowledge and lessons at programme management level, followed by the problem of identifying applications on new projects, which then usually involves tailoring and customising the generic lessons to produce contextually applicable solutions. This is a nine-stage process that is more costly with low levels of repeatability:

Identify → Capture → Process on site → Process in the firm → Store →

Disseminate → Identify → Adapt → Apply

This is a considerable barrier to developing knowledge sharing and applying lessons learned. This is why investment in programme management is necessary.

Overall, programme management and the strategic project front-end were driven on the client side by external and organisational factors in ways that constrained knowledge management in the supply chain. The client had developed some strategies for sharing, which needed to be (but was not in practice) cascaded down the supply chain for implementation. Knowledge application at the programme management level in the supply chain and network was highly constrained, resulting in internal transactional policies, especially low investment, and process barriers. The derived cultural norms resulted in defensiveness and a sense of learned helplessness, which can be described as a nuanced position between adverse and collaborative behaviour.

The suppliers had an extremely defensive culture in the challenging context of infrastructure provision, client actions, and market drivers. The management of a public client firm which is constrained by the regulatory, government, and public policy environment and by the pressure to deliver on time and within budget is not always able to respond to the problem situation in a prompt and efficient manner. They tend to blame external circumstances and parties for their inability to act. This is the client side of *learned help-lessness* that shapes the culture of the organisation and strongly influences the responses and engagement levels of the supply chain. Learned helplessness is internalised and embedded as passivity and endemic behavioural responses through departmental and functional divisions. Responsibility for addressing problems and challenges is allocated to others, including blame when matters go wrong. From the supply chain perspective, this renders liaison and collaboration with the client more challenging. Eventually the supply firm develops a standard response about the difficulty of taking timely and decisive action, especially where crossfunctional and interorganisational working is needed. Initiatives get stifled in the process. Therefore, the supply chain itself becomes defensive and delays decisions and action until clarity of direction is known.

This is exacerbated in the supply chain by a lack of leadership commitment. Finance departments do not invest in supporting knowledge management implemented during programme management. HR and procurement departments do not embed practices. Bid management, supply chain, commercial, and project management functions do not apply knowledge management along the project lifecycles. Altogether, these features contribute to *learned helplessness* in the supply chain. The lack of support in the supply chain firms for knowledge management, coupled with their defensiveness in response to client behaviour induces a cocktail of behaviours contributing to this type of helplessness in the supply chain.

14.5 Conclusions

There are a number of patterns emerging from the research, summarised as follows:

1. There were commonly held perceptions about the importance of knowledge sharing in collaborative relationships of trust and robust governance, which is a positive basis upon which management can build knowledge management.
2. Lessons in practice were assimilated and transferred on an ad hoc basis, relying upon individuals and small groups taking responsible action.
3. There was a lack of investment in management capabilities, and programme management.

4. There was a management perception that IT platforms provided solutions yet had very low levels of engagement at the operational level.
5. Firms applying more systematic knowledge sharing were self-interested rather than seeking to enhance the service and value for money for the client and other stakeholders.
6. The culture was transactional and very defensive with a focus on risk around time and cost control.
7. Departmental functions, especially finance, HR, and procurement failed to perceive the significant potential role they could play in facilitating knowledge sharing.
8. Senior management failed to show commitment to and leadership on knowledge sharing and application.

While there was evidence of individual and small group/team willingness to be transformational bottom-up, there were considerable barriers to facilitating knowledge sharing and application. This emanated from the top and in particular the client. Investment and leadership was absent, especially at the level of programme management on the client side and within the supply chain. The pattern of findings aligns with the condition of 'learned helplessness'. This abnormal condition (Abramson et al. 1978) becomes pathological in the organisational setting (Gummesson 2002, p. 259) where individuals and teams end up saying in effect, 'I'm sorry, I can't do anything about it'.

A more proactive and analytical approach to procurement practices might address the metrics for prequalification as panel members and the basis for continued patronage; this would set a positive basis for transitioning practices for projects. Such an approach addresses how rates of improvement can supplement simple static threshold criteria. A dynamic approach using rates of improvement needs to be evidence-based. This could be designed to drive forward internal improvements and supplier competition. HR practices need to address the issues of staff selection, promotion, and annual reviews, embedding knowledge sharing as an important part of progression and pay. Such an approach begins to change the culture by instilling learning and knowledge transfer into daily practices. In order to maintain competitive advantage in a knowledge economy, construction organisations are characterised by a significant number of multidisciplinary specialists. They produce a great deal of tacit knowledge and need to build a knowledge sharing culture internally and across their supply chain. Interorganisational communities of practices can contribute to that by becoming the spaces of risk-free tacit knowledge exchange among expert teams working to span silo boundaries and boost collective learning and innovation. One was instigated in the client organisation towards the end of the research and had scope for development.

References

Abramson, L.Y., Seligman, M.E., and Teasdale, J.D. (1978). Learned helplessness in humans: critique and reformulation. *Journal of Abnormal Psychology* 87 (1): 49–74.

Ackermann, F. and Eden, C. (1994). Issues in computer and non-computer supported GDSSs. *Decision Support Systems* 12: 381–390.

Akaka, M.A., Vargo, S.L., and Lusch, R.F. (2013). The complexity of context: a service ecosystems approach for international marketing. *Journal of International Marketing* 21 (4): 1–20.

Anumba, C., Egbu, C., and Carrillo, P. (2005). *Knowledge Management in Construction*. Oxford: Wiley-Blackwell.

Anvuur, A. and Kumaraswamy, M. (2008). Better collaboration through cooperation. In: *Collaborative Relationships in Construction: Developing Frameworks and Networks* (eds. H.J. Smyth and S.D. Pryke), 107–128. Oxford: Wiley-Blackwell.

Blackman, D.A. and Henderson, S. (2005). Know ways in knowledge management. *The Learning Organisation* 12 (2): 152–168.

Bloom, H. (2000). *The Global Brain*. New York: Wiley.

Brady, T. and Davies, A. (2004). Building project capabilities: from exploratory to exploitative learning. *Organisation Studies* 25 (9): 1601–1621.

Bresnen, M. (2009). Learning to co-operate and co-operating to learn: knowledge, learning and innovation in construction supply chains. In: *Construction Supply Chain Management: Concepts and Case Studies* (ed. S.D. Pryke), 73–112. Oxford: Wiley-Blackwell.

Bresnen, M. and Marshall, N. (2000). Partnering in construction: a critical review of issues, problems and dilemmas. *Construction Management and Economics* 18 (2): 229–237.

Brightman, J. (1998). *An Introduction to Decision Explorer*. Banxia Software, Ltd.

Campbell, N. (1995). *An Interaction Approach to Organisation Buying Behaviour: Relationship Marketing for Competitive Advantage*. Oxford: Butterworth-Heinemann.

Carrillo, P.M., Robinson, H.S., Al-Ghassani, A.M. and Anumba, C.J. (2002). Survey of Knowledge Management in Construction: KnowBiz Project Technical Report. Department of Civil and Building Engineering, Loughborough University, Loughborough.

Chandler, J.D. and Vargo, S.L. (2011). Contextualization and value-in-context: how context frames exchange. *Marketing Theory* 11 (1): 35–49.

Christopher, M., Payne, A., and Ballantyne, D. (2002). *Relationship Marketing: Creating Stakeholder Value*. Oxford: Butterworth-Heinemann.

Davenport, L. and Prusak, H. (1998). *Working Knowledge: How Organisations Manage What They Know?* Boston: Harvard Business Press.

Davenport, T.H., De Long, D.W., and Beers, M.C. (1997). *Building Successful Knowledge Management Projects*. Center for Business Innovation/Ernst & Young LLP.

Davies, A. and Brady, T. (2000). Organisational capabilities and learning in complex product systems: towards repeatable solutions. *Research Policy* 29 (7–8): 931–953.

Davies, A., Brady, T., and Hobday, M. (2007). Organizing for solutions: systems seller vs. systems integrator. *Industrial Marketing Management* 36: 183–193.

Dawson, R. (2000). *Developing Knowledge-Based Client Relationships*. Oxford: Butterworth Heinemann.

Doloi, H. (2009). Analysis of pre-qualification criteria in contractor selection and their impacts on project success. *Construction Management and Economics* 27 (12): 1245–1263.

Dubois, A. and Gadde, L.-E. (2002). The construction industry as a loosely coupled system: implications for productivity and innovation. *Construction Management and Economics* 20: 621–631.

Edkins, A.J. and Smyth, H.J. (2016). Business development and bid management's role in winning a public private partnership infrastructure project. *Engineering and Project Organizing Journal* 6 (1): 33–34.

Egan, S.J. (1998). *Rethinking Construction*. London: HMSO.

Fernie, S. and Tennant, S. (2013). The non-adoption of supply chain management. The non-adoption of supply chain management. *Construction Management and Economics* 31 (10): 1038–1058.

Foss, N. (2007). The emerging knowledge governance approach. *Organisation* 14: 29–52.

Gilbert, C.G. (2005). Unbundling the structure of inertia: resource versus routine rigidity. *Academy of Management Journal* 48: 741–763.

Grabher, G. (2004). Architecture of project-based learning: creating and sedimenting knowledge in project ecologies. *Organisation Studies* 25 (9): 1491–1514.

Gray, P. (2001). A problem-solving perspective on knowledge management practice. *Decision Support Systems* 31 (1): 87–102.

Gray, J.H. and Densten, I.L. (2005). Towards an integrative model of organisational culture and knowledge management. *International Journal of Organisational Behaviour* 9 (2): 594–603.

Gummesson, E. (2002). *Total Relationship Marketing*. Oxford: Butterworth-Heinemann.

Hobday, M., Davies, A., and Prencipe, A. (2005). Systems integration: a core capability of the modern corporation. *Industrial and Corporate Change* 14 (6): 1109–1143.

Humphreys, P., Matthews, J., and Kumaraswamy, J. (2003). Pre-construction project partnering: from adversarial to collaborative relationships. *Supply Chain Management: An International Journal* 8 (2): 166–178.

IPA, 2017. Transforming Infrastructure Performance, Infrastructure and Performance Authority. http://www.gov.uk/government/publications/transforming-infrastructure-performance (accessed August 2019).

Ive, G. (1995). The client and the construction process: the Latham report in context. In: *Responding to Latham: The Views of the Construction Team* (ed. S.L. Gruneberg). Ascot: CIOB.

Kelly, N., Edkins, A.J., Smyth, H.J., and Konstantinou, E. (2013). Reinventing the role of the project manager in mobilising knowledge in construction. *International Journal of Managing Projects in Business* 6 (4): 654–673.

Kenley, R. (2012). Managing change in construction projects: a knowledge-based approach. *Construction Management and Economics* 30 (2): 179–180.

Kivrak, S., Arslan, G., Dikmen, I., and Birgonul, M.T. (2008). Capturing knowledge in construction projects: knowledge platform for contractors. *Journal of Management in Engineering* 24: 87–95.

Koskinen, K.U. (2000). Tacit knowledge as a promoter of project success. *European Journal of Purchasing and Supply Management* 6 (1): 41–47.

Latham, S.M. (1994). *Constructing the Team*. London: HMSO.

Lave, J. and Wenger, E. (1991). *Situated Learning: Legitimate Peripheral Participation*. Cambridge: Cambridge University Press.

Leonard-Barton, D. (1992). Core capabilities and core rigidities: a paradox in managing new product development. *Strategic Management Journal* 1 (S1): 111–125.

London, K., Kenley, R., and Agapiou, A. (1998). Theoretical supply chain network modelling in the building industry. In: *Proceedings of the 13th Annual ARCOM Conference*, 369–379. University of Reading.

Love, P., Fong, P.S.W., and Irani, Z. (eds.) (2005). *Management of Knowledge in Project Environments*. Oxford: Butterworth-Heinemann.

Macneil, I.R. (1980). *The New Social Contract: An Inquiry Into Modern Contractual Relations*. New Haven, CT: Yale University Press.

Pryke, S.D. (2009). *Construction Supply Chain Management: Concepts and Case Studies.* Oxford: Wiley-Blackwell.

Pryke, S.D. and Smyth, H.J. (2006). Scoping a relationship approach to the management of projects. In: *Management of Complex Projects: A Relationship Approach* (eds. S.D. Pryke and H.J. Smyth), 21–46. Oxford: Wiley-Blackwell.

Sage, D.J., Dainty, A.R., and Brookes, N.J. (2010). Who reads the project file? Exploring the power effects of knowledge tools in construction project management. *Construction Management and Economics* 28 (6): 629–639.

Sanaei, M., Javernick-Will, A.N., and Chinowsky, P. (2013). The influence of generation on knowledge sharing connections and methods in construction and engineering organisations headquartered in the US. *Construction Management and Economics* 31 (9): 991–1004.

Senaratne, S. and Sexton, M. (2008). Managing construction project change: a knowledge management perspective. *Construction Management and Economics* 26 (12): 1303–1311.

Senge, P. (1990). *The Fifth Discipline: The Art and Practice of the Learning Organisation.* London: Century Books.

Sexton, M. and Lu, S.L. (2009). The challenges of creating actionable knowledge: an action research perspective. *Construction Management and Economics* 27 (7): 683–694.

Smyth, H.J. (2006). Competition. In: *Commercial Management of Projects: Defining the Discipline* (eds. D. Lowe and R. Leiringer), 22–39. Oxford: Wiley Blackwell.

Smyth, H.J. (2010). Construction industry performance improvement programmes: the UK case of demonstration projects in the "continuous improvement" programme. *Construction Management and Economics* 28 (3): 255–270.

Smyth, H.J. (2015). *Relationship Management and the Management of Projects.* Abingdon: Routledge.

Smyth, H.J. (2018) *Castles in the Air? The Evolution of British Main Contractor.* www.ucl.ac .uk/bartlett/construction/castles-in-the-air (accessed August 2019).

Smyth, H.J. and Edkins, A.J. (2007). Relationship management in the management of PFI/PPP projects in the UK. *International Journal of Project Management* 25 (3): 232–240.

Smyth, H.J. and Longbottom, R. (2005). External provision of knowledge management services: the case of the concrete and cement industries. *European Management Journal* 23 (2): 247–259.

Smyth, H.J. and Morris, P.W.G. (2007). An epistemological evaluation of research into projects and their management: methodological issues. *International Journal of Project Management* 25 (4): 423–436.

Söderlund, J. (2012). Theoretical foundations of project management: suggestions for a pluralistic understanding. In: *The Oxford Handbook of Project Management* (eds. P.W.G. Morris, J.K. Pinto and J. Söderlund), 37–64. Oxford: Oxford University Press.

Szulanski, G. (2000). The process of knowledge transfer: a diachronic analysis of stickiness. *Organisational Behavior and Human Decision Processes* 82: 9–27.

Tavistock Institute (1966). *Interdependence and Uncertainty: A Study of the Building Industry.* London: Tavistock Publications.

Vargo, S.L. and Lusch, R.F. (2016). Institutions and axioms: an extension and update of service-dominant logic. *Journal of the Academy of Marketing Science* 44 (4): 5–23.

Wenger, E. (1998). *Communities of Practice: Learning, Meaning and Identity.* Cambridge: Cambridge University Press.

Williamson, O.E. (1975). *Markets and Hierarchies: Analysis and Antitrust Implications.* New York: Free Press.

Winter, S.G. (2013). Habit, deliberation, and action: strengthening the microfoundations of routines and capabilities. *The Academy of Management Perspectives* 27 (2): 120–137.

Xu, J., in press (2020). Understanding trust in construction supply chain relationships. In: *Construction Supply Chain Management* (ed. S.D. Pryke). Oxford: Wiley-Blackwell.

15

Understanding Trust in Construction Supply Chain Relationships
Jing Xu

15.1   Introduction

Trust is a social phenomenon that enables collaboration among actors and organisations. In a context such as construction, involving increasing complexity and uncertainty, increased specialism and need for collaboration, trust is required not least because of the call for nonadversarial working and integrated supply chains, for example Egan (1998) and its influential UK successors (e.g. Egan 2002; Wolstenholme et al. 2009). Trust helps leverage better service standards for actors and organisations involved. A decade or so ago, Smyth (2008) presented a framework of trust drawing conceptual, philosophical, and methodological elements together to deepen the understanding of trust in and between project businesses. While trust among project-based organisations is essential, it is particularly challenging to develop and sustain it in projects, especially between contractors and suppliers. This is a management challenge. Main contractor and second-tier subcontractor relationships have been identified as being worse than client and main contractor relationships (Alderman and Ivory 2007). This chapter aims to deepen the understanding of trust in construction supply chain relationships, particularly in the project delivery. This helps academia and practitioners understand how they can lever value in supply chain relationships through the constitution of trust.

Trust, in the broadest sense, sustains institutional, social, and organisational life (Luhmann 1979; Giddens 1990; Kramer and Tyler 1996). Increased division of labour means the need for more collaboration between actors and organisations to create integrated solutions (Pryke 2009). In this vein, interorganisational trust is believed to be an appropriate governance mechanism for enhancing communication quality, reducing transaction costs and increasing project efficiency (Zaheer et al. 1998). The view on trust as a governance mechanism or organisational form further leads to studies promoting trust relations among organisations through relational marketing, relational contracting, relational governance, and partnering projects (e.g. Nooteboom et al. 1997; Doloi 2009; Ling et al. 2013). Once established, trust may generate an environment of integrity and openness where actors are willing to share risks, commit resources, and work jointly (Kadefors 2004; Smyth and Thompson 2005), which in turn increases the chances to enhance trust. Pryke (2017) has suggested that trust is

Successful Construction Supply Chain Management: Concepts and Case Studies, Second Edition.
Edited by Stephen Pryke.
© 2020 John Wiley & Sons Ltd. Published 2020 by John Wiley & Sons Ltd.

an important enabling factor in the establishment and maintenance of supply chain networks. The virtuous cycle of trust nurtures relational norms such as reciprocity and equity that stabilise relationships (Macneil 1980).

Despite collaborative mechanisms and tools in construction project management research, their effects on relationships and trust vary (cf. Cicmil and Marshall 2005; Brady and Davies 2014). One barrier to improving supply chain relationships is the institutional logic of goods dominant, called the goods-dominant logic and the project counterpart, which is termed here the project-focused logic, which emphasises trans-actional efficiency and promotes practices maximising short-term profits and assessing performance on a project-by-project basis (Kadefors 2004; Smyth 2015a). Under a goods-dominant logic and project-focused logic, partnering projects and collaborative mechanisms simply move the singular transactions to multiple transactions over time in order to profit from supply chains (Alderman and Ivory 2007; Smyth 2015b). The lack of empirical studies on supply chain relationships (Bygballe et al. 2010), especially the process of relationship development in project delivery, might also contribute to the difficulties of implementing partnering arrangements in practice (Bresnen and Marshall 2000). It has been identified that second-tier subcontractors and suppliers have less understanding of collaborative mechanisms and question the benefits they can get through collaborating with main contractors (Mason 2007).

In summary, to promote collaborative relationships and trust in construction supply chains, construction project management research need to:

- Open the black box of trust development in projects;
- Shift the institutional logic of goods-dominant logic and project-focused logic so to;
- Shed light on the value of trust for those involved, including main contractors and subcontractors, as well as a broader view on value beyond points of transactions towards value-in-use over time (Vargo and Lusch 2004; Saxon 2005).

This chapter draws on the above points from the perspective of structuration theory (Giddens 1984) and service-dominant logic (Vargo and Lusch 2004, 2008, 2016). Based on data from the case study, this process-based research demonstrates how trust, from the main contractor to second-tier subcontractor, develops and how the value of trust unfolds while trust develops.

15.2 Towards an Understanding of Trust in Construction Supply Chains

15.2.1 Towards a Service-Dominant Logic View

The division of labour and subcontracting systems enable previous in-house production to become a type of service and main contractors as system integrators to provide the service of management for clients. Construction projects then become less about pro-duction and more about the establishment of delivery channels of service for the benefit of end-users such as building occupants and road users. In the project management field, the changing place of service has gained increasing interest (Brady et al. 2005; Edvards-son et al. 2005) and distanced the delivery of projects from a goods-dominant logic and project-focused logic (Table 15.1).

Table 15.1 Goods-dominant logic, the project-focused view and service dominant logic in the delivery of construction projects.

	The delivery of construction projects	Goods-dominant logic /project-focused logic perspective	Service-dominant logic perspective
Creation of the value proposition	Before delivery	After production	A balancing mechanism that links actors at different times and positions in the service ecosystem
	Little can be accurately quantified	Quantifiable evidence of transactional value, such as cost, programme, and quality	
	Continuously shaped before and during the delivery	Static promises from the producer to customer	
Role of customer	Main contractor as the intermediate customer of specialist projects participating in the delivery and providing service	Reactive recipient of goods and services	Customers can passively accept the value proposition or participate in the creation of value proposition
	Some clients participating in the delivery	An operand resource to be profited from	An operant resource to co-create value with
Firm–customer interaction	Intensive interactions between the main contractor and supply chains in delivering the project	Active producer and reactive customer in making value propositions and production	Interchangeable role of service provider between the customer and producer in the creation of value proposition
	Main contractor and subcontractor may facilitate the use process post completion	Producer may provide facilitating services as added value	Service beneficiary co-creates value in use; experience through direct or indirect interactions with the service provider as well as other actors in the service ecosystem
Firm–customer relationship and trust	An integral part of project success	Not addressed in a goods-dominant logic	Inherently relational
	Related to repeat businesses and economics of repetition and recombination	Temporary and an operand resource for project efficiency in a project-focused logic	An operant resource determining the meaning of goods and services

Source: Original.

Under a goods-dominant logic, value is added in goods and services by producers before completion and determined at the point of making a transaction or a series of transactions, that is value-in-exchange. Value propositions are promises of value to be delivered by producers and are treated as 'quantifiable evidence' of value (Skålén et al. 2015). From a goods-dominant logic perspective, value propositions are active from the producers' side (Lanning 1998) and customers reactively accept the offering or not. The role of the customer in creating value propositions is not implicitly addressed by a goods-dominant logic. The focus is on the exchange of manufactured goods and services (Skålén et al. 2015). Therefore, the interactions and relationships between customer and producer in the production are largely neglected in a goods-dominant logic.

Nevertheless, the delivery of construction projects requires intensive interactions between contractors and supply chain members. Supply chain organisations are selling and bidding, the formation of value propositions occurring prior to production or delivery. Value propositions can be shaped through interactions and early involvement of main contractors and supply chains at the front end (Cova and Salle 2008). Main contractors, more often than not, participate in the service provision. Interactions between main contractors and subcontractors continue beyond the formation of value propositions and might even become intensified in project execution. In order to manage changes and uncertainties, both main contractors and subcontractors may take the role of service provider during service delivery. In construction, exchange is a process, broken down into stage payments. How this is managed is an integral part of the relationship and trust is a crucial part of the mutual service experience.

On the other hand, a project-focused logic emphasises the temporary nature and uniqueness of projects (Lundin and Söderholm 1995) and in this vein, trust relations are viewed as a vehicle to increase project efficiency and terminated with the project; projects have no memory and trust is mostly built from scratch (Dubois and Gadde 2002b). Yet, this is far from the case, especially in supply chain networks and where repeat business is commonplace and sometimes the norm. The shadow of the past and future, rules and resources at multiple levels of service ecosystems influence perceptions, actions, practices, and power relations in current interactions and hence their performance in knowledge transfer and capability building (Brady and Davies 2004; Manning and Sydow 2011).

A service-dominant logic shifts the focus away from projects, goods, and services towards service and relationships in service ecosystems (Vargo and Lusch 2008, 2016). This inherently relational feature links service-dominant logic with construction project management. The central argument of a service-dominant logic is that in an actor-to-actor context the basis of exchange is service; "service is exchanged for service" (Bastiat et al. 1964, as quoted in Vargo and Lusch 2004, pp. 6–7). In a service-dominant logic, service is more than what was traditionally meant in a goods-dominant logic as an activity or a set of input activities resulting in a singular output aimed at assisting the customer's practice, but the "application of specialised competencies (skills and knowledge) through deeds, processes, and performances for the benefit of another party or the entity itself" (Vargo and Lusch 2004, p. 2). A service-dominant logic views goods and services as operand resources, on which an operation or act is performed to produce an effect and operant resources that are employed to act on operand resources as well as other operant resources (Vargo and Lusch 2004). In this vein, relationships and hence trust are not operand resources to be profited from; they are operant resources and determine how a certain resource can be efficiently used or easily accessed. A service-dominant logic regards relationship value beyond bringing repeated transactions; "a service-centred view is inherently beneficiary oriented and relational" (Vargo and Lusch 2016, p. 4). Relationships are embedded with social capital that can be transformed to other types of capital and help lever value for those involved (Bourdieu 1986; Coleman 1988).

Under a service-dominant logic, main contractors and subcontractors can only create value propositions, namely construction projects and components. Service beneficiaries such as building occupants and road users, while they use the service delivered through the construction project, co-create value with other actors in service ecosystems and determine value-in-use of the service. In some cases, main contractors and subcontractors, together with the client, facilitate the end-users' systems post completion (Grönroos 2008; Lusch et al. 2010). Therefore, value-in-use depends on two processes:

1. Service experience of delivering value propositions.
2. Use experience of realising value propositions as value-in-use (Smyth 2015a).

From the perspective of main contractors and subcontractors, it is the first process, service experience of creating value propositions, that levers value, instead of actualising such value propositions. By participating in service interactions, actors have more opportunities for mutual learning, knowledge sharing, and relationship building, hence gaining resources and service rights that lever value for future service exchanges. Between selling and project completion, the components of value propositions, perceptions of value, and power relations may alter, resulting in value realisation enhanced for some yet reduced for other organisational actors. This points to the importance of complementary objectives, mutual understanding on the basis of shared meanings, resource commitment, and access to mobilising other actors' resources and reflexive learning in experience (Lusch et al. 2010; Kowalkowski et al. 2012). All these are founded upon trust-based interactions at different levels of service ecosystems that align actors' value expectations as reciprocal promises *"to and from suppliers and customers seeking an equitable exchange"* (Ballantyne and Varey 2006, p. 344).

Value co-creation implies reciprocity and equity for those involved (Ballantyne and Varey 2006; Aarikka-Stenroos and Jaakkola 2012). However, the majority of service-dominant logic research focuses on the practices of and value for customers (e.g., van der Valk and Wynstra 2012). Service value, particularly for suppliers, is somewhat neglected, with the exceptions of Smyth et al. (2016) and Walter et al. (2001). This imbalance of research interests is consistent with the client-centred focus of construction project management. While many studies have focused on the in-use phase, the experience of service provision at the meso and micro levels attracts little interest (Karpen et al. 2017). In short, despite the argument that service is a process, the service-dominant logic community lacks empirical research on how this service aspect enacts and levers value for those involved – pointing to the necessity for a process-based study on service provision and value-in-use for both customer and supplier.

15.2.2   Towards a Process-Based View

Service-dominant logic provides a lens for viewing resources and value beyond both the transactional sense and project duration. This section moves the ontological foundation for this research towards a process-based view. Most trust research in construction has centred on:

- Identifying antecedents and outputs of trust in projects (e.g. Wong et al. 2008).
- Trust in governance structures at the project and corporate level, relational contracting, and the design of collaborative mechanisms and tools (e.g. Rahman and Kumaraswamy 2012).

- The atmosphere and culture derived from specific trusting behaviour conducted by actors in interactions (e.g. Smyth and Thompson 2005).

Despite the enlightenment on trust in construction project management, most studies have taken static snapshots of trust. Like Zeno's arrow, an individual snapshot gives a glance of trust in a given point of time and space but sheds little light on the understanding of the essence of trust. Also, trust research in construction project management neglects the relational and social context where actors and organisations are situated. Trust is foundational to relationships, meaning that trust exists, develops, and functions through relationships among people and organisations (Smyth et al. 2010). Individuals would have no occasion nor need to trust apart from engaging in relationships (Lewis and Weigert 1985). Trust cannot be fully understood on either a psychological or institutional level alone. Zooming out to a broader picture shows that supply chain relationships are nested in multiple levels of service ecosystems that equip project actors with rules of interpretations and legitimation as well as resources of power (Manning 2008; Sydow 2017). Time also matters. Despite the fact that projects have an *ex-ante* defined duration, actors and firms have earlier experience and future expectations that influence current perceptions and behavioural orientations towards others.

To capture the dynamics of trust requires taking a 'becoming' ontology (Chia 2002), bringing interorganisational relationships to the centre of analysis and theorising on the basis of contexts over the course of time. This move towards a 'becoming' ontology requires a more open, dynamic, and reflexive management approach and a broader view of organisational theory (Sydow 2017). Structuration theory (Giddens 1984) is one of the theories able to reconcile the interplay of structures and agency in process studies. Specifically, the concept of duality of structure provides a dynamic lens for viewing the recurrent interactions between structures and interaction processes. Structure, according to Giddens' structuration theory, consists of rules and resources. Rules of signification and legitimation constitute the interpretative and normative aspects of structure, and resources are constituted from authoritative and allocative resources from social systems. Rules and resources in the institutional, social, and organisational environment then constrain and enable practices in interactions. Knowledgeable actors reflexively choose among multiple rules and resources in interactions with others and in this way reproduce or transform rules and resources. This research does not include a full review of structuration theory here (see Bresnen et al. 2004). The point is that structuration theory provides a theoretic perspective of viewing how service interactions help constitute trust and how trust influences service process and outcomes.

In construction projects, interpretative and normative rules and resources are divergent and are imposed by rules and regulations at the level of organisational fields, industrial norms at the network level, and organisational policies at the organisational level (Manning 2008). Such rules and resources form structural conditions of trust. On the other hand, the management of construction projects is highly decentralised (see for example, Pryke et al. 2018), giving managers power to make decisions at the local level. Actors are also able to generate norms, form routines, and resources at the project level. As such, construction actors have various interpretative schemes and facilities of gaining resources, depending on rules and demands from different stakeholders. Actors play an active role in choosing forms of communication, whether and how to use power, and ways of sanction. Their decisions depend on interests of the parent organisation but also power relationships and resources in the project as well as the institutional environment at the project location (Bresnen et al. 2004). Through the lens of structuration theory, generating trust enables the following (Sydow 1998):

- The raising of the perception of trustworthiness (interpretative rule).
- The using of trust relations to allocate resources (facility of resources).
- The legitimising of relational norms that constrain opportunism and encourage trusting and trustworthy behaviour (normative rule).

The phenomenon of trust, in return, influences communication, power relations, and sanctioned behaviour in interactions (Sydow 1998), hence service process and project performance.

To recap, a process-based view is used, viewed through the service-dominant logic, and demonstrates whether and how trust, from the main contractor to second-tier subcontractor, develops. It also addresses the value of that trust (see Figure 15.1). Based on extant research (Mayer et al. 1995; Rousseau et al. 1998; Smyth et al. 2010; Sydow 1998), this study uses the working definition of trust as:

An actor's current intention to rely on the actions of, or to be vulnerable to, another actor, based on the expectation that the other actor can reduce risks and co-create value in a relationship.

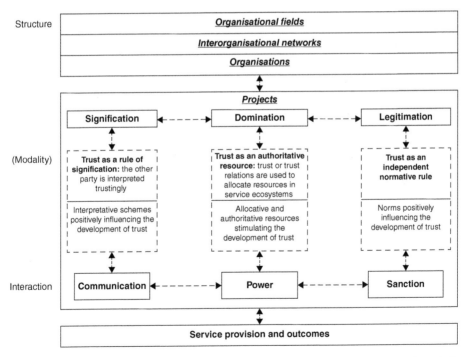

Figure 15.1 The constitution and value of trust: a structurationist view. Source: Adapted from Sydow (1998, p. 40) and Manning (2008, p. 33).

A relationship can be interpersonal. It can also be interorganisational based upon the sum of the key interactions and individuals.

15.3   Methodology and Methods

The study on which this chapter is based on is an interpretive and process-based approach and focused on the informants' view of the topic and individuals' perceptions of events and processes (Smith 2004). The case study offered a 'zooming-in' opportunity for investigating processes in a local situation (Yin 2009). One interorganisational relationship was selected to demonstrate how trust develops or decays in supply chain relationships and how trust influences service value in the project. Case Gamma involved an internal supply chain relationship between a main contractor, referred to herein as Build Gamma, and a foundation and piling subcontractor referred to herein as Found Gamma. Case Gamma was selected because of the dual relationship feature between Build Gamma and Found Gamma. On the one hand, Build Gamma and Found Gamma are functional units of the same parent company, Gamma UK, and structural conditions at the organisation level may influence service interactions and trust at the project level. On the other hand, the two parties have separate profit goals, business plans, and different organisational structures. For instance, Found Gamma needed to bid as an external subcontractor for Build Gamma's projects. In this sense, the relationship between Build Gamma and Found Gamma is similar to an interorganisation relationship that may be influenced by the interorganisational networks. Taking the Build Gamma: Found Gamma relationship as the unit of analysis offers an opportunity for investigating the interplays of trust and interaction process at the project level but also structural conditions at different levels of ecosystems. Table 15.2 summarises key features of the case.

To investigate the sequence and flow of events in order to understand processes in the course of time, the research was enlightened by the longitudinal study method. Data collection involved three-round visits and interviews at preconstruction and procurement, then execution stage, and finally completion stage of the piling project so as to

Table 15.2 Case Gamma overview.

Case study information	
Project context	Piling project for a multiple-use high-rise building in a city centre
Piling duration	Overall including procurement, design, and resource mobilisation: 16 months Execution: 6 months
Relationship nature	Functional units of Gamma UK
Main procurement and contract	Two-stage procurement JCT 2011[a], management fee (preconstruction agreement), 30-day payment
Piling procurement and subcontract	Limited bid invitation JCT 2011[a], lump sum, 42-day payment

a)   Joint Contracts Tribunal 2011.
Source: Original.

capture both historical and contemporary processes (Pettigrew 1990). By doing so, the author gathered data about past experiences that dated back three years. Conducting repeated, multiple-wave interviews mitigates the possibility of bias due to incomplete, misinterpreted, and mistakenly reported memories. Twenty-one interviews were conducted, with informants across different organisation levels and functional units from both the main contractor and subcontractor. Bias due to the same functional roles and organisations were mitigated. All interviews were semistructured, guided by a protocol and open-ended questions. Structuration theory provided general guidelines and mechanisms to explain local findings, but these mechanisms and concepts are subjective to revision (Dubois and Gadde 2002a).

All interviews were recorded and transcribed by the researcher. Transcription was stored and managed through MAXQDA 12.

15.4 Case Study

This section is to illustrate the context, events, and processes of case Gamma and analyse the case study on the basis of theoretical mechanisms. Specifically, the case study findings are summarised in Boxes 15.1–15.4, each box followed by theoretical interpretations of the influences on, or of, trust.

Box 15.1 Project Participants, Service Ecosystems, and the Shadow of the Past

Project Participants

The parent company is Gamma UK, a major construction company in the UK. Found Gamma is the piling and ground engineering unit and Build Gamma is the building construction unit, each having their own managing director and leadership team. The whole company is supported by a number of enabling functions such as health and safety and communications. The client is a private property company. Several consultants were in partnership with the client, providing quantity surveying, design and engineering services.

The Shadow of the Past

Before Project Gamma, Build Gamma and Found Gamma just delivered 'Lemon project'. Both parties perceived the experience of Lemon project as negative, in which communication at the project level became *'just to execute obligations'*. After project completion, both parties denied their own responsibilities to some extent and wanted to *'cut ties'* with each other. When the piling project was completed, the two companies were in extensive disputes. Build Gamma charged Found Gamma 25% of contract value for delay, and Found Gamma asked for compensation for additional works.

Service ecosystems

At the organisation level, Gamma UK had a collaborative policy, referred to herein as 'One Gamma', to encourage service integration between internal units and increase the competence

---

(Continued)

Box 15.1 (Continued)

and reputation of Gamma UK at the board level. After Lemon project, the executive management team drove the initiative of relationship development 'top-down' to the senior management and middle management. Also, the two companies shared common enabling systems, procedures, and standards for joint activities. At the middle management level, the biannual business-to-business (B2B) meeting between project directors of Build Gamma and Found Gamma was the main event for relationship building after the Lemon project. Gamma UK's organisational structure, on the other hand, allowed functional units to competitively procure piling and ground engineering services, despite owning the specialist unit Found Gamma. This autonomy was to pursue lowest cost and satisfy clients' requirements.

At the network level, the client used two-stage procurement to encourage early contractor and subcontractor involvement and an integrative team. The main stakeholder of Project Gamma was a local government body responsible for transport, herein referred to as City Ltd. The site was close to underground stations and Build Gamma and City Ltd had frequent communications regarding the risk of disturbing underground operations.

15.4.1 Context

15.4.1.1 Assessing the Shadow of the Past

Prior to the front end, actors assessed the other party's competence and intentions on the basis of the two companies' past experiences. Structures at the organisational level influenced trust in two ways. Firstly, the executive management team of Gamma UK drove communication between the senior management and middle management of two units to improve relationships. Secondly, routines between internal units such as the biannual B2B director meeting provided the opportunity for joint learning.

Despite the initiative and joint activities, the Lemon project negatively affected Build Gamma's perception of Found Gamma's trustworthiness as the middle management failed to establish shared learning and invoke actions to exploit the lessons of the past for future benefits. The challenge of forming shared understanding was attributed to the lack of first-hand experience and thus a belief in the other party's competence and intentions since directors got involved in the project when disputes had already occurred. As a consequence, directors at both 'Build' and 'Found' avoided, rather than dealt with, problems. Both parties denied their own mistakes. Financial disputes caused by the Lemon project exacerbated the relationship at the firm level. According to the project director of Found Gamma:

> It is a very contractual and transactional relationship. It shouldn't be, but it is…And because we are the subcontractor, we are always at the bottom of the food chain…So at the very beginning of the job [Project Gamma], there was some resistance from… [Build Gamma] to use us because of… [the Lemon project]. And it took some higher-level people to say 'No, … [Found Gamma] is our in-house company. We can't dismiss them because of… [the Lemon project]. We have to fix it and move on.

15.4.1.2 Organisational Structure and Policy: Forming a Sense of Unfairness

The narrative just described pointed to the second factor influencing trust between internal companies – the organisational structure of Gamma UK. Specifically, the structure enabled the contractual relationship to dominate the in-house relationship between Found Gamma and other functional units as main contractors. The strategies of internal units' businesses, although seeming to mitigate complacency and maintain the competence of the in-house service, was to capture short-term profits, even if it was at the expense of long-term benefits across projects. The organisational structure in place at the time promoted discreteness and facilitated the project-focused view and goods-dominant logic practices. Under a project-focused logic and a goods-dominant logic, internal actors tended to ignore the benefits from trusting and well-structured internal relationships. No strategic relationship was established between Found Gamma and other functional units.

Further, the organisational structure and institutionalised, transaction-based actions constrained the effect of 'One Gamma' policy, especially on Found Gamma. 'One Gamma', on the broadest level, was conducive to integrated service and resource efficiency. As 'One Gamma' was disseminated within the organisation and continuously communicated between individuals, it could influence individuals' interpretations such as the meaning of internal relationships, ways of utilising internal resources, and the intention of collaborating with internal companies. Despite the top-down approach, interpretations were affected by individuals' experiences. Under the structure in place at the time, specifically after the Lemon project, 'One Gamma' was perceived by Found Gamma as employing a rhetoric of using their resources to facilitate other units as main contractors. Looking at this as a whole, the management of Gamma UK might have tried out of self-interest to secure their profit by not adequately paying Found Gamma. Transactional relations dominated over the in-house relations and generated the perception of unfairness, which dramatically hindered the development of relationships and trust.

Box 15.2 Procurement and Preconstruction

The main contract was open to two-stage tendering in August 2015. To satisfy the client's requirement for an early start with piling, in October 2015, Build Gamma invited three piling contractors to competitively tender. Inviting Found Gamma was a normative practice between internal units of Gamma UK and Found Gamma offered the lowest price. Following that, the two companies engaged to jointly develop the main bid. Based on experience with piling for surrounding buildings and established relationships with City Ltd, Found Gamma helped Build Gamma to win the first stage and, in this way, they added value to the main bid, and helped mitigate risks in main contract terms.

In January 2016, Build Gamma was awarded a preconstruction agreement (PCA), which was originally for six months. During the PCA, the client paid a monthly fee to Build Gamma who were responsible for site management and the management of piling. After Build Gamma had been awarded the PCA, Found Gamma introduced an operations manager and a project manager into the project, and continued to help develop the terms and conditions of the main contract and had early involvement in design. A joint risk workshop was initiated where engineers and managers of both parties identified risks together. Based on ground information from past experiences, Found Gamma reduced risks and main bid price. Found Gamma were involved in the meetings between Build Gamma and the client.

(Continued)

Box 15.2 (Continued)

Meanwhile, the main contractor, Build Gamma and the piling subcontractor, Found Gamma started to negotiate the piling contract. Learning from the Lemon project, Found Gamma refused any changes in their programme. Build Gamma supported Found Gamma's programme and assisted Found Gamma's site preparation and welfare. In April, Found Gamma and Build Gamma signed the contract with a lump sum price of £4 million.

15.4.2 Procurement and Preconstruction Stage

15.4.2.1 Early Involvement: Forming a Sense of Security and Familiarity

The experience of the Lemon project increased the perceived risks of interdependence, and both parties preferred to interact with a transactional approach so as to reduce relational elements. At the front end, project organisations were governed by the price mechanism. From the perspective of Build Gamma, the early inquiry was to use Found Gamma's resources to optimise main bids and reduce the risk of disturbing underground operations. Found Gamma was perceived as an operand resource to be profited from, 'an asset to win jobs… to help for the technical systems… to pass on the risk' (Commercial manager, Build Gamma). From the perspective of Found Gamma, early involvement in the main bid development meant more chances to direct the main contract content to the benefits the firm could deliver and hence increase their own influence in the project. In other words, early involvement was driven mainly by self-interest.

Nevertheless, competence trust increased as Found Gamma helped improve the value proposition. Through Found Gamma's technical solutions, advice, and joint activities such as risk workshops, Build Gamma learnt about Found Gamma's specialist capabilities, although this was not an intended consequence. As commented by the project director of Build Gamma:

> …they did give us advice on the logistics and programme, which we used in our first-stage tender submission. So, we put that information in our first stage tender submission… [Found Gamma] are very educated. So, they understand risks more… [In contract negotiation] … [Found Gamma] would ask searching questions. At the end of the day, it is good because it protects everybody.

Found Gamma's competence, past experiences in the local area, and relationships with City Ltd gave Build Gamma a sense of security. Moreover, the early introduction of an operations manager and project manager at the front end nurtured a sense of familiarity at the individual level and mitigated inconsistency between the front end and execution stage.

15.4.2.2 Two-stage Procurement: Creating a Sense of Equity

Throughout the piling procurement stage, both companies maintained transactional relationships to reduce perceived risks of interdependence. Actors from both parties repeatedly stressed that Found Gamma was awarded the contract because of the lowest price, rather than the internal relations. Contract negotiation was more about building safeguards, especially from Found Gamma's side, indicating a sense of insecurity in collaboration with Build Gamma.

The sense of insecurity was alleviated by perceived equity in the exchange. Two-stage procurement constrained the use of power by Build Gamma since the main contract was not awarded by the time of piling procurement. Build Gamma could not control terms and conditions for self-interest. The piling contract was jointly determined, and the value proposition was reciprocal to both parties. Co-determination added a dose of equity between Found Gamma and Build Gamma. Perceived equity was further strengthened by Build Gamma's support for the piling programme. Equity reduced Found Gamma's safeguard and encouraged collaboration in the execution, which laid the foundation for trust development.

15.4.2.3 The Value of Trust

Competence trust motivated Build Gamma to collaborate with Found Gamma. This is evident in joint risk workshops, bringing Found Gamma into the meetings with the client and their support for Found Gamma's programme. From the perspective of Build Gamma, collaborating with Found Gamma increased the client's trust and the effectiveness of communication with the client as Found Gamma was able to explain risks and technical solutions better. From the perspective of Found Gamma, they obtained quality information about the project and the client, built the relationship with the client, and reduced risks at an early stage by direct communication with the designer and the client. Effective communication improved the value proposition that potentially brought about a good project. An expanded resource base with broader networks of relationships and information benefited future business and thus leveraged improved service value.

Box 15.3 Project Execution

Piling execution started in June 2016. Found Gamma and Build Gamma maintained consistent teams. The two parties established project routines and maintained regular communication. Supervisors had daily reports and diaries, and project managers and engineers had weekly progress meetings on site. Found Gamma's operations manager met Build Gamma's project manager fortnightly. At the middle-management level, project directors of both parties were also scheduled to meet regularly. The two parties agreed weekly programmes, resource plans, site records and risk assessment, and method statements. Found Gamma also used rolling accounts to evaluate and predict the final account after each change and informed Build Gamma of their prediction. In this way, both parties were able to jointly monitor project progress and deal with issues immediately rather than leaving them to the final account. As internal companies, Found Gamma and Build Gamma did joint inspection on health and safety and environmental issues. Found Gamma achieved programmes, crafted effective technical solutions, and shared their knowledge to help main bid submission and initiate relationships between Build Gamma and City Ltd. Relationships with City Ltd were especially important for Build Gamma; City Ltd was the key stakeholder for their next project. The way of collaborating was repeated and gradually routinised.

In September 2016, as the client and Build Gamma could not reach an agreement on price, the main contract was back to tender. Build Gamma agreed to extend the PCA until the piling was completed. To save costs, Build Gamma reduced resources and maintained key actors at the project level. Despite the resource reduction, the two parties showed higher solidarity so as to increase joint power relative to the client. They jointly solved problems to ensure both parties could get benefits. Found Gamma also had more flexibilities in their operations and piling issues

---

(Continued)

Box 15.3 (Continued)

were decided by both parties. In circumstances where Found Gamma delayed outstanding information, Build Gamma chose to discuss the issue by phone or face-to-face meetings, rather than contractually by formalising letters or emails. Similarly, Found Gamma did extra-mile works without charge.

15.4.3 Execution Stage

15.4.3.1 Structuring the Project: Maintaining Security and Familiarity

Relating, Controlling, and Monitoring Consistent teams sustained relationships between key actors and shared understanding established in the procurement, contract negotiation, and design stages. Structuring project routines initiated formal relationships and ensured regular direct communication between two parties, which enabled continuous learning through monitoring or through joint activities. Further, exchanging legitimised boundary objects, such as weekly programmes and rolling financial reporting, reduced misunderstanding and ambiguity in communications. Such formal mechanisms enabled the ability to control through structural influence. Through monitoring project tasks and performance, actors from Build Gamma increased their competence trust in Found Gamma as the latter continued to comply with the programme.

Routinising Apart from actualising the value proposition, Found Gamma's service enabled Build Gamma to obtain knowledge and establish relationships with City Ltd. The effectiveness of service process and outcomes increased competence trust, which in turn sustained the collaboration. This way of collaborating was reproduced and gradually routinised, embedding competence trust in the service interactions and relationships between Build Gamma and Found Gamma. In other words, competence trust and collaboration formed a self-reinforcing cycle where competence trust served as the medium and outcome of collaboration. Build Gamma became more willing to rely on Found Gamma and specifically on their solutions and advice. Interdependence increased between the two parties.

Relating, controlling, and routinising sustained familiarity at the individual level, and security in operations laid the foundations for generating trustworthiness and further using trust as a facility for coordinating resources.

15.4.3.2 Joint Activities: Forming the Interpretations of Trustworthiness

Collaborating and Shared Learning Shared systems, standards, and common knowledge facilitated joint activities between Build Gamma and Found Gamma. Joint activities enabled the co-presence of both parties, and actors were able to identify, understand, and solve project issues in shared experiences, hence reducing misunderstanding.

> We do rely on them to do a good job, and we trust them to deal with what they are supposed to do in terms of quality, health and safety, everything else…because you know they are… [Gamma UK], they follow the same standard as we follow, we rely on them to make sure that if there are procedures to follow we trust they will do it. We rely on them to do it.
>
> (Construction manager, Build Gamma).

The Role of Internal Relations The role of internal relations was twofold here. On the one hand, the internal relation reduced perceived risks in collaboration and thus supported the virtuous cycle of competence trust and collaboration. On the other hand, the belief in internal relations substituted intention trust. Found Gamma was believed not to be opportunistic because of the internal relations and associated obligations, rather than experiential and reflexive learning about Found Gamma's actions.

15.4.3.3 Using Trust Relations in Resource Coordination: Bounded Solidarity and Economic Reciprocity

Build Gamma's failure in the main bid and reduction in resources changed power relations among the main contractor, subcontractor, and the client and promoted bounded solidarity between Build Gamma and Found Gamma. Bounded solidarity was driven by the recognition of their own powerlessness relative to the client and the economic reciprocity from collaborating with each other. Internal relations facilitated the formation of solidarity.

> [Internal relationship] A bit better with collaborating…kind of against the client…working together to make sure that both companies are achieving as much as they can and not make any mistakes that are going to affect another company.
>
> (Construction manager, Build Gamma).

Bounded solidarity, on the basis of competence trust and internal relations, meant that Build Gamma and Found Gamma formed shared intentions of protecting collective benefits from the client in Project Gamma. Social orientations emerged at this point as actors became concerned about their own benefits but also those of the other party. Within shared intentions, bounded solidarity constrained contractual and opportunistic behaviour towards the other party and thus encouraged the use of trust relations in resource allocation and integration, rather than hierarchical authority or market price.

To use trust relations required mutual service and reciprocal value propositions, which nurtured economic reciprocity in the relationship between Build Gamma and Found Gamma. Economic reciprocity started with small actions with short-term returns, specifically economic returns, completing the programme on time, and getting fair payment for instance. The repeated reciprocation of small actions promoted actions with less specified payback. The balance of the exchange was expected in a longer term, though still within the duration of the project. Contractual elements were mitigated, as mentioned by the project engineer of Found Gamma:

> We are trying to be as helpful as possible. We never said 'no, we cannot do it'…Also in terms of … [Build Gamma] as a company on this site, they have been very helpful and supportive. And they really try to understand what it is, what we really do, what we need…we do our best to be helpful, and they do their best to be helpful. So, we will continue to do that.

The narrative above illustrates the phenomenon of collaborating beyond merely integrating service. Collaborating for co-creating value involved social orientation, mutual service, and reciprocal value propositions. Bounded solidarity and economic reciprocity supported the reproduction of value co-creating and hence competence trust as a rule of signification and trust relations as facilities of obtaining resources; the more trust-based service reproduced, the more trust became embedded.

15.4.3.4 The Value of Trust

Competence trust encouraged more enquiries and information sharing but also gave Build Gamma confidence in reducing resources and delegating some authorities to Found Gamma. Although partly driven by the external environment, reducing resources depended upon the positive path created by the virtuous cycle of trust and collaboration and the belief in the nature of internal relations. From the perspective of Found Gamma, Build Gamma's openness and delegation of authorities signalled Found Gamma's trust and increased their confidence in communication. They became more proactive in resource sharing. From the perspective of Build Gamma, trust created a learning atmosphere where actors asked questions, shared knowledge, and jointly solved problems. In other words, the self-reinforcing cycle of trust and collaboration promoted closer collaboration where actors were able to use trust as social capital in pursuit of value. The use of trust relations induced mutual service and reciprocal value propositions that made service experiences of both parties more flexible and effective. Moreover, the shared intention of protecting collective benefits from the client made the relationships more cohesive.

Box 15.4 Project Completion

At the project level, Found Gamma and Build Gamma continued to deliver an integrated service, Found Gamma taking the responsibility of delivering service contents and Build Gamma providing information from the client about design changes and requirements of City Ltd. Build Gamma gave more flexibilities to Found Gamma in piling operations, allowing Found Gamma to manage piling on their own. Both parties maintained core staff. As actors became more familiar with each other, they were willing to share resources and learn in the project. Found Gamma engineers introduced Build Gamma to different types of pile and the requirement for their implementation; and Build Gamma shared their knowledge and experience as a main contractor. This knowledge helped Found Gamma's future business that the project required Found Gamma to deliver integrated solutions including piling and temporary works. The recurrent collaborative behaviour formed relational norms that actors complied with to maintain stability of relationships and ensure continuous benefits.

At the firm level, Found Gamma and Build Gamma secured future business. Communication increased and was future-oriented. Directors were more involved with each other. They inquired and advised each other about the future project. Build Gamma invited Found Gamma to their internal director forum so as to increase mutual understanding and identify potential business opportunities.

The piling was completed in December 2016, two weeks prior to the programme. The experience of Project Gamma largely improved the relationship between Found Gamma and Build Gamma. By the end of the project, actors from both parties had a shared understanding of 'One Gamma UK' as comprising openness, honesty, flexibility, listening, understanding, sharing, and non-blaming. Found Gamma gained a 15–25% increase in value of the work and improved their status internally in Gamma UK. They also gained reputation as engineers and were invited to give a presentation by the client's consultant.

15.4.4 Completion Stage

15.4.4.1 Stabilising the Relationship: Trust as a Rule of Legitimation

At the project level, bounded solidarity, economic reciprocity, and equity of the service process and outcomes motivated actors to maintain relationship stability and the trusting and trustworthy way of interactions. As mentioned by the quantity surveyor of Found Gamma:

> ...I will say in any project there's a chemistry of people managing it. So, the chemistry I would say is good enough to manage and establish trust, that follows that you actually say what you would do. You are going to say as you do. Trust - you build on that. You just become a far better working relationship.

Actors nurtured norms of conduct in their day-to-day interactions, which in return constrained opportunism and encouraged trust and trustworthiness. In this manner, actor-generated norms were legitimised in the recurrent pattern of trust-based behaviour, generating trust as a rule of legitimation. Further, as actors referred to the norms, they were more likely to use trust as a mechanism for resource coordination than apply coercive power and opportunism. In this vein, trust as a rule of legitimation also strengthened trust as a rule of signification and trust as a resource of domination.

15.4.4.2 The Shadow of the Future: Social Reciprocity

Secured future projects meant that resources gained in Project Gamma might be reused and/or recombined in the future, which helped maintain actor-generated norms at the project level. The engagement at the firm level, such as service exchanges for future businesses and the project director forum, indicated an element of social reciprocity in the service provision. Social reciprocity induced actions with no specified return within the duration of the current project. The balance of exchange was expected in future businesses. The shadow of the future potentially extended social orientations and value co-creating beyond the project level.

15.4.4.3 The Value of Trust

As trust was recursively constituted as social capital in service interactions, the initiatives of sharing and learning increased, especially on Found Gamma's side. The scope of communication extended beyond problems and tasks within Project Gamma. At the project level, Found Gamma managers and engineers introduced their partners technical knowledge and learnt about main contractors' businesses. At the firm level, directors of both companies discussed their future projects and identified business opportunities. The aim was to increase mutual understanding of each other's operations and organisation. On Build Gamma's side, they maintained responsiveness to problems and fairness in the service process and outcomes. Actors knew more about each other, technically, organisationally, and relationally, which made service experiences more informative, flexible, and effective. As the project manager of Found Gamma mentioned:

> I think from both sides we were both quite open and honest with each other as to what the requirements were, or what was important to the project. So, we were able to very quickly come to the best solution between all of us, (a) for the project,

and (b) for each party, which avoids the conflict…It [this good relationship] made it an easier place to work. Everyone knew what everyone did. It made it an enjoyable project to build.

Furthermore, trust as a rule of legitimation stabilised collaboration and increased relationship cohesion. The cohesion of relationship is evident as actors of both parties recognised the benefits of trust and trust relations with the other party and had a shared understanding of 'One Gamma UK'.

Better service experiences enabled actors to lever service value and achieve higher performance. As actors became more informative, they gained resources that could be used in the future projects, such as knowledge and relationships with City Ltd for Build Gamma and main contractors' operations and businesses for Found Gamma. The value proposition became more viable and acceptable to the client, which improved the piling programme, cost, and quality. From the perspective of Build Gamma, being able to deliver integrated solutions also increased their status and reputation in the broader market. Successful delivery of Project Gamma demonstrated Found Gamma's capabilities and increased their status and reputation within Gamma UK. Relationship value increased as both companies benefited more than delivering piling as an end product.

15.5 Discussion

15.5.1 The Constitution of Trust

This chapter demonstrates five types of trust-generating interaction processes – learning, relating, controlling, collaborating, and routinising (Figure 15.2). From the beginning of the front end to completion, these processes recursively constituted trust by influencing actors' interpretative schemes, encouraging the use of trust to allocate resources and legitimising relational norms sustaining collaboration.

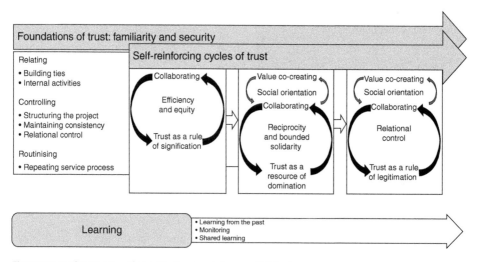

Figure 15.2 The process of constituting trust. Source: Original.

Learning is the underlying process of constituting trust. To generate the perception of trustworthiness, use trust relations to gain resources, and value and legitimate relational norms, actors need to learn about their partners as well as the environment. The perception of trustworthiness could be a consequence of intentional assessment of past experiences but also an unintended consequence of discursive learning such as monitoring programme and shared learning about problems and tasks.

To generate trust requires a sense of familiarity and security. Familiarity and security help breed trust by furnishing a sense of assurance in repeated interactions and encouraging the leap of faith in the face of uncertainties (Luhmann 1988; Gulati 1995). We identified that relating, routinising, and controlling at the structure level could raise familiarity and security in construction supply chain relationships. *Relating* provides relational ties and informal socialisation mechanisms that help sustain communication between two parties. Informal socialisation mechanisms help reconcile discrepancies in meanings, goals, and value (Grant 1996; Carlile 2004; Lawson et al. 2009; Ballantyne et al. 2011). The case study found controlling occurring mainly at the structure level, through establishing and adapting project structures for instance. Formal roles and positions, and the adoption of standard procedures and project routines regularised expectations and increased predictability in operations. In this vein, controlling can form confidence and positive expectations on the other party's behaviour because of structural influences (Möllering 2005; Bachmann and Inkpen 2011). Formal mechanisms create common knowledge, frame of reference, and collectively accepted norms of conduct (Olson et al. 2002; Bechky 2006; Maurer 2010; Enberg 2012), which 'can hardly be (mis-)used by them (individual actors) for opportunistic strategies' and 'can foster the efficient production of a high level of trust in trans-organizational relations' (Bachmann 2001, pp. 358–359). Maintaining consistent core members of project teams, communication, and service quality throughout the project lifecycle sustained shared meanings. *Routinising* effective and efficient service processes further strengthens the reliability of procedures and processes. On the basis of security and familiarity, collaboration creates shared experiences within which trust has embedded rules and resources. Project efficiency was the first driver for collaboration between organisations in interorganisational projects. Furthermore, to sustain collaboration required perceived equity to make 'fair dealing' (Ring and Van de Ven 1994, p. 93), in which organisations seek benefits proportional to their investments, with the condition of maintaining social relationships. A sense of reciprocity and bounded solidarity emerged as actors and organisations continuously exchanged service in a trusting and trustworthy way. Reciprocity and bounded solidarity tie organisations together by forming identities of each other and recognising the limits of both parties relative to a third party (Portes 1998), forming the desire to uphold the collaboration and use trust relations to allocate and integrate resources (Pervan et al. 2009; Swärd 2016). Relational norms emerged as collectives of actors continued to collaborate in a trusting and trustworthy way, which formed relational control and sustained the collaboration by refraining from opportunism and encouraging trustworthiness. Relational norms control the behaviour of those involved by generating a sense of responsibility, which induces care, empathy, and appreciation in interactions. In this vein, trust is not only about risk mitigation but also risk sharing. The various effects of controlling, including controlling at the structure and interaction levels and relational control, indicate a dynamic relation between trust and control, which can be both complementary and substitutive (Woolthuis et al. 2005). Trust, in return, sustains existing collaborating and promotes closer collaboration specifically under uncertainties. In other words, trust and collaboration form self-reinforcing cycles.

15.5.2 The Value of Trust

Trust in the other party's competence first increases the intensity of service communication as actors become more open to each other. From the perspective of the main contractor, they are more willing to share information and acquire advice from the subcontractor. From the perspective of the subcontractor, the main contractors' openness is perceived as one of the first signals of trust, which gives them confidence in advising and sharing project information. As perceived trustworthiness increases, actors exchange information beyond the current project, so the communication becomes thicker. Competence trust also helps form a new relationship between previously unacquainted individuals, such as the relationship between the client and Found Gamma in this case, which increases the efficiency and effectiveness of communication in the project networks. By doing so, the phenomenon of trust forms a learning and sharing atmosphere in a network of relationships that makes service experiences more *informative* for both main contractor and subcontractor.

Trust also enables the delegation of authorities and tolerance of uncertainties. Trust mitigates perceived risks for those who take the first step of using trust as social capital and fulfilling their obligations in the service exchange (Coleman 1988). Trust creates the conditions for expecting serial equity that reduces the need for instantaneous and equal compensation. Moreover, trust in the other party's specialist capabilities helps establish clear and specialised roles between the two parties. Hence, each party can concentrate on their own specialities. In this vein, both parties have a certain level of *flexibility* to programme works, make decisions, and control their own operations. Cycles of trust and collaboration form relational norms, shared intentions, and meanings that supply relational thinking and expectations and guide collectively accepted practices, which helps maintain a stable and cohesive experience.

Where actors and organisations have better service experiences, this may further increase the *effectiveness* of service provision and outcome as they are able to understand the changing context and each other's expectations quickly, improve solutions and the value proposition to each other's requirements and preferences, and exchange operant resources beneficial to future businesses. Figure 15.3 illustrates the value of trust.

15.5.3 Conditions of Trust: Influences of Ecosystems and Time

At the organisation level, shared systems and internal collaborative policies from the parent organisation furnish common knowledge, facilitate joint activities, and therefore form a conducive environment for trust development between internal companies. The positive effects of the internal relations are constrained by injecting market elements into the structure of the internal relationship because it:

- Discourages the structuring of strategic relationship and hence trust between internal companies.
- Helps routinise the transactional view on the internal relationship and its value.
- Creates the paradox between the meaning of the internal relationship and practices between internal companies.

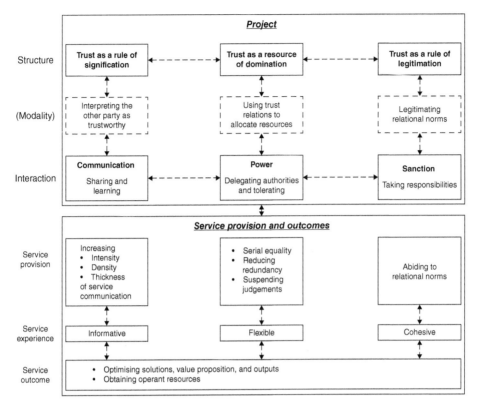

Figure 15.3 The value of trust. Source: Original.

Moreover, such structure dramatically hinders trust development because:

- It leads to the use of internal companies as operand resources and/or the perception of unfairness, and
- Internal companies rely on structural arrangements to form collaboration and replace intention trust with the belief in the internal relationship.

At the network level, the procurement system, including both the main contract procurement and subcontract procurement, influences trust constitution. Compared with single-stage procurement for the main contract, two-stage procurement encourages the supply chain relationship as the main contractor and subcontractor can have earlier involvement in the procurement and jointly develop the value proposition before main contract submission, which mitigates potential conflicts of contractual contents between the main contract and subcontract. Two-stage procurement also balances the power relation between the main contractor and subcontractor, creating a sense of equity in interactions.

Past experiences and future opportunities between two companies affect interpretations of the other party's trustworthiness through the process of learning. The shadow of the past generated an initial condition for interactions. The path from prior history to trust and present projects was not direct. Rather, the influence of past experiences depends on whether supply chain partners can cognitively and behaviourally learn from the past lessons and experiences (Poppo et al. 2008; Elfenbein and Zenger 2014; Buvik and Rolfsen 2015). The shadow of the future led to future-orientated learning, relational investment, and social reciprocity since relationships and knowledge obtained in the present might be transformed and reused in the future (Ebers and Maurer 2016).

15.6 Conclusions and Recommendations

This chapter has endeavoured to deepen the understanding of trust in the project lifecycle, specifically the process of trust development and the value of trust in supply chain relationships, and particularly between the main contractor and the second-tier subcontractor. The first part of this chapter argued that attempts to study trust in construction project management focusing on projects or series of projects as repeated transactions, short-term profits, and static snapshots of trust, cloud understanding of the dynamics and value of trust. It then argued a 'becoming' ontology and service-dominant logic as the basis of understanding trust in supply chain relationships and offered structuration theory as part of the analytical lens (Giddens 1984) to view the interplay of trust, structures, and process. The second part of this chapter used a case study to explore the issues discussed in the first part. The empirical findings demonstrate that the constitution of trust is an engineered but also emergent process and illustrates five fundamental processes of constituting trust. The interplays of these processes form a sense of familiarity and security upon which trust is constituted. Self-reinforcing cycles of trust and collaboration in the service provision generates the interpretation of trustworthiness, promotes the use of trust relations, and forms efficiency, equity, reciprocity, and bounded solidarity that in return sustain the virtuous cycle. In the process of generating trust, actors and organisations had a higher level of security and were more informative, flexible, and cohesive in operations. Better experiences enabled actors and organisations to achieve higher levels of performance in the project.

On a broader level, this chapter also indicates the influences of organisations and interorganisational networks in the context of multiple levels of service ecosystems. While trust benefits from shared systems, collaborative policy, and common knowledge of the parent organisation, injecting market elements into the hierarchical structure to govern internal relationships weakens the positive effects but also induces transactional interpretations of internal companies and hinders the structuring of strategic relationships to lever value for internal units and organisation as a whole. The shadow of both the past and future between two companies also affects the constitution of trust as actors form interpretations of trustworthiness through learning about the past and future and allocating resources to exploit past experiences and/or explore future opportunities. At the interorganisational network level, collaborative procurement for the main contract encourages the involvement of the client, balances power relations, and constrains the use of power in actor-to-actor interactions.

This chapter presents trust development and value between an internal main contractor–subcontractor relationship, though the organisational structure induced market elements between the two companies. Future research needs to be expanded to external relationships. Such research should explore a wide range of structural influences of macro-level ecosystems. Moreover, longitudinal research will help extend the shadow of the past and future beyond the most recent experiences and provide a more comprehensive view of trust over the course of time. It also facilitates the linking of trust at the project level to the firm level and offers opportunities for exploring the influences of trust on organisations and networks. Trust research needs to enhance knowledge and raise awareness among practitioners and researchers and establish healthy interdependences in the service provision so as to leverage value for end-users and clients but also main contractors and their supply chains as co-creators shaping and realising value propositions.

References

Aarikka-Stenroos, L. and Jaakkola, E. (2012). Value co-creation in knowledge intensive business services: a dyadic perspective on the joint problem solving process. *Industrial Marketing Management* 41: 15–26.

Alderman, N. and Ivory, C. (2007). Partnering in major contracts: paradox and metaphor. *International Journal of Project Management* 25 (4): 386–393.

Bachmann, R. (2001). Trust, power and control in trans-organizational relations. *Organization Studies* 22 (2): 337–365.

Bachmann, R. and Inkpen, A.C. (2011). Understanding institutional-based trust building processes in inter-organizational relationships. *Organization Studies* 32 (2): 281–301.

Ballantyne, D. and Varey, R.J. (2006). Creating value-in-use through marketing interaction: the exchange logic of relating, communicating and knowing. *Marketing Theory* 6 (3): 335–348.

Ballantyne, D., Frow, P., Varey, R.J., and Payne, A. (2011). Value propositions as communication practice: taking a wider view. *Industrial Marketing Management* 40 (2): 202–210.

Bastiat, F., de Huszar, G.B., and Hayek, F.A. (1964). *Selected Essays on Political Economy*. Princeton: Van Nostrand.

Bechky, B.A. (2006). Gaffers, gofers, and grips: role-based coordination in temporary organizations. *Organization Science* 17 (1): 3–21.

Bourdieu, P. (1986). The forms of capital. In: *Handbook of Theory and Research for the Sociology* (ed. J.G. Richardson), 241–258. New York: Greenwood.

Brady, T. and Davies, A. (2004). Building project capabilities: from exploratory to exploitative learning. *Organization Studies* 25 (9): 1601–1621.

Brady, T. and Davies, A. (2014). Managing structural and dynamic complexity: a tale of two projects. *Project Management Journal* 45 (4): 21–38.

Brady, T., Davies, A., and Gann, D. (2005). Creating value by delivering integrated solutions. *International Journal of Project Management* 23 (5): 360–365.

Bresnen, M. and Marshall, N. (2000). Partnering in construction: a critical review of issues, problems and dilemmas. *Construction Management and Economics* 18 (2): 229–237.

Bresnen, M., Goussevskaia, A., and Swan, J. (2004). Embedding new management knowledge in project-based organizations. *Organization Studies* 25 (9): 1535–1555.

Buvik, M.P. and Rolfsen, M. (2015). Prior ties and trust development in project teams–a case study from the construction industry. *International Journal of Project Management* 33 (7): 1484–1494.

Bygballe, L.E., Jahre, M., and Swärd, A. (2010). Partnering relationships in construction: a literature review. *Journal of Purchasing and Supply Management* 16 (4): 239–253.

Carlile, P.R. (2004). Transferring, translating, and transforming: an integrative framework for managing knowledge across boundaries. *Organization Science* 15 (5): 555–568.

Chia, R. (2002). Essai: time, duration and simultaneity: rethinking process and change in organizational analysis. *Organization Studies* 23 (6): 863–868.

Cicmil, S. and Marshall, D. (2005). Insights into collaboration at the project level: complexity, social interaction and procurement mechanisms. *Building Research and Information* 33 (6): 523–535.

Coleman, J.S. (1988). Social capital in the creation of human capital. *American Journal of Sociology* 94: S95–S120.

Cova, B. and Salle, R. (2008). Marketing solutions in accordance with the SD logic: co-creating value with customer network actors. *Industrial Marketing Management* 37 (3): 270–277.

Doloi, H. (2009). Relational partnerships: the importance of communication, trust and confidence and joint risk management in achieving project success. *Construction Management and Economics* 27 (11): 1099–1109.

Dubois, A. and Gadde, L.-E. (2002a). Systematic combining: an abductive approach to case research. *Journal of Business Research* 55 (7): 553–560.

Dubois, A. and Gadde, L.-E. (2002b). The construction industry as a loosely coupled system: implications for productivity and innovation. *Construction Management and Economics* 20 (7): 621–631.

Ebers, M. and Maurer, I. (2016). To continue or not to continue? Drivers of recurrent partnering in temporary organizations. *Organization Studies* 37 (12): 1861–1895.

Edvardsson, B., Gustafsson, A., and Roos, I. (2005). Service portraits in service research: a critical review. *International Journal of Service Industry Management* 16 (1): 107–121.

Egan, J. (1998) The Egan Report – Rethinking Construction. Report of the Construction Industry Task Force to the Deputy Prime Minister. London: HMSO.

Egan, J. (2002) Accelerating Change: A Report by the Strategic Forum for Construction. London: Rethinking Construction.

Elfenbein, D.W. and Zenger, T.R. (2014). What is a relationship worth? Repeated exchange and the development and deployment of relational capital. *Organization Science* 25 (1): 222–244.

Enberg, C. (2012). Enabling knowledge integration in coopetitive R&D projects–the management of conflicting logics. *International Journal of Project Management* 30 (7): 771–780.

Giddens, A. (1984). *The Constitution of Society: Outline of the Theory of Structuration.* University of California Press.

Giddens, A. (1990). *The Consequences of Modernity.* Cambridge, UK: Polity.

Grant, R.M. (1996). Toward a knowledge-based theory of the firm. *Strategic Management Journal* 17 (S2): 109–122.

Grönroos, C. (2008). Service logic revisited: who creates value? And who co-creates? *European Business Review* 20 (4): 298–314.

Gulati, R. (1995). Does familiarity breed trust? The implications of repeated ties for contractual choice in alliances. *Academy of Management Journal* 38 (1): 85–112.

Kadefors, A. (2004). Trust in project relationships—inside the black box. *International Journal of Project Management* 22 (3): 175–182.

Karpen, I.O., Gemser, G., and Calabretta, G. (2017). A multilevel consideration of service design conditions: towards a portfolio of organisational capabilities, interactive practices and individual abilities. *Journal of Service Theory and Practice* 27 (2): 384–407.

Kowalkowski, C., Persson Ridell, O., Röndell, J.G., and Sörhammar, D. (2012). The co-creative practice of forming a value proposition. *Journal of Marketing Management* 28 (13–14): 1553–1570.

Kramer, R.M. and Tyler, T.R. (1996). *Trust in Organizations: Frontiers of Theory and Research*. Sage.

Lanning, M.J. (1998). *Delivering Profitable Value: A Revolutionary Framework to Accelerate Growth, Generate Wealth, and Rediscover the Heart of Business*. Da Capo Press.

Lawson, B., Petersen, K.J., and Cousins, P.D. (2009). Knowledge sharing in interorganizational product development teams: the effect of formal and informal socialization mechanisms. *Journal of Product Innovation Management* 26 (2): 156–172.

Lewis, J.D. and Weigert, A. (1985). Trust as a social reality. *Social Forces* 63 (4): 967–985.

Ling, F.Y.Y., Ning, Y., Ke, Y., and Kumaraswamy, M.M. (2013). Modeling relational transaction and relationship quality among team members in public projects in Hong Kong. *Automation in Construction* 36: 16–24.

Luhmann, N. (1979). *Trust and Power*. Willey.

Luhmann, N. (1988). Familiarity, confidence, trust: problems and alternatives. In: *Trust: Making and Breaking Cooperative Relations* (ed. D. Gambetta), 94–107. Basil Blackwell: Oxford.

Lundin, R.A. and Söderholm, A. (1995). A theory of the temporary organization. *Scandinavian Journal of Management* 11 (4): 437–455.

Lusch, R.F., Vargo, S.L., and Tanniru, M. (2010). Service, value networks and learning. *Journal of the Academy of Marketing Science* 38 (1): 19–31.

Macneil, I.R. (1980). *The New Social Contract: An Inquiry Into Modern Contractual Relations*. Yale University Press.

Manning, S. (2008). Embedding projects in multiple contexts–a structuration perspective. *International Journal of Project Management* 26 (1): 30–37.

Manning, S. and Sydow, J. (2011). Projects, paths, and practices: sustaining and leveraging project-based relationships. *Industrial and Corporate Change* 20 (5): 1369–1402.

Mason, J.R. (2007). The views and experiences of specialist contractors on partnering in the UK. *Construction Management and Economics* 25 (5): 519–527.

Maurer, I. (2010). How to build trust in inter-organizational projects: the impact of project staffing and project rewards on the formation of trust, knowledge acquisition and product innovation. *International Journal of Project Management* 28 (7): 629–637.

Mayer, R.C., Davis, J.H., and Schoorman, F.D. (1995). An integrative model of organizational trust. *Academy of Management Review* 20 (3): 709–734.

Möllering, G. (2005). The trust/control duality: an integrative perspective on positive expectations of others. *International Sociology* 20 (3): 283–305.

Nooteboom, B., Berger, H., and Noorderhaven, N.G. (1997). Effects of trust and governance on relational risk. *Academy of Management Journal* 40 (2): 308–338.

Olson, J.S., Teasley, S., Covi, L., and Olson, G. (2002). The (currently) unique advantages of collocated work. In: *Distributed Work*, 113–135.

Pervan, S.J., Bove, L.L., and Johnson, L.W. (2009). Reciprocity as a key stabilizing norm of interpersonal marketing relationships: scale development and validation. *Industrial Marketing Management* 38 (1): 60–70.

Pettigrew, A.M. (1990). Longitudinal field research on change: theory and practice. *Organization Science* 1 (3): 267–292.

Poppo, L., Zhou, K.Z., and Ryu, S. (2008). Alternative origins to interorganizational trust: an interdependence perspective on the shadow of the past and the shadow of the future. *Organization Science* 19 (1): 39–55.

Portes, A. (1998). Social capital: its origins and applications in modern sociology. *Annual Review of Sociology* 24 (1): 1–24.

Pryke, S. (2009). *Construction Supply Chain Management*. Wiley.

Pryke, S. (2017). *Managing Networks in Project-Based Organisations*. Wiley.

Pryke, S., Badi, S., Almadhood, H. et al. (2018). Self-organising networks in complex infrastructure projects. *Project Management Journal* 49 (2): 18–41.

Rahman, M.M. and Kumaraswamy, M.M. (2012). Multicountry perspectives of relational contracting and integrated project teams. *Journal of Construction Engineering and Management* 138 (4): 469–480.

Ring, P.S. and Van de Ven, A.H. (1994). Developmental processes of cooperative interorganizational relationships. *Academy of Management Review* 19 (1): 90–118.

Rousseau, D.M., Sitkin, S.B., Burt, R.S., and Camerer, C. (1998). Not so different after all: a cross-discipline view of trust. *Academy of Management Review* 23 (3): 393–404.

Saxon, R. (2005). *Be Valuable: A Guide to Creating Value in the Built Environment*, 1. London: Constructing Excellence.

Skålén, P., Gummerus, J., von Koskull, C., and Magnusson, P.R. (2015). Exploring value propositions and service innovation: a service-dominant logic study. *Journal of the Academy of Marketing Science* 43 (2): 137–158.

Smith, J.A. (2004). Reflecting on the development of interpretative phenomenological analysis and its contribution to qualitative research in psychology. *Qualitative Research in Psychology* 1 (1): 39–54.

Smyth, H. (2008). Developing trust. In: *Collaborative Relationships in Construction: Developing Frameworks and Networks* (eds. H. Smyth and S. Pryke), 129. Oxford: Wiley-Blackwell.

Smyth, H. (2015a). *Market Management and Project Business Development*. Routledge.

Smyth, H. (2015b). *Relationship Management and the Management of Projects*. Routledge.

Smyth, H. and Thompson, N.J. (2005). Developing conditions of trust within a framework of trust. *Journal of Construction Procurement* 11 (1): 4–18.

Smyth, H.J., Gustafsson, M., and Ganskau, E. (2010). The value of trust in project business. *International Journal of Project Management* 28 (2): 117–129.

Smyth, H., Fellows, R., Liu, A., and Tijhuis, W. (2016). *Editorial for the Special Issue on Business Development and Marketing in Construction*. Taylor & Francis.

Swärd, A. (2016). Trust, reciprocity, and actions: the development of trust in temporary inter-organizational relations. *Organization Studies* 37 (12): 1841–1860.

Sydow, J. (1998). Understanding the constitution of interorganizational trust. In: *Trust Within and Between Organizations: Conceptual Issues and Empirical Applications*, 31–63.

Sydow, J. (2017). Managing inter-organizational networks: governance and practices between path dependence and uncertainty. In: *Networked Governance* (eds. B. Hollstein, W. Matiaske and K.U. Schnapp), 43–53. Cham: Springer.

Tuli, K.R., Kohli, A.K., and Bharadwaj, S.G. (2007). Rethinking customer solutions: from product bundles to relational processes. *Journal of Marketing* 71 (3): 1–17.

van der Valk, W. and Wynstra, F. (2012). Buyer–supplier interaction in business-to-business services: a typology test using case research. *Journal of Purchasing and Supply Management* 18 (3): 137–147.

Vargo, S.L. and Lusch, R.F. (2004). Evolving to a new dominant logic for marketing. *Journal of Marketing* 68 (1): 1–17.

Vargo, S.L. and Lusch, R.F. (2008). Service-dominant logic: continuing the evolution. *Journal of the Academy of Marketing Science* 36 (1): 1–10.

Vargo, S.L. and Lusch, R.F. (2016). Institutions and axioms: an extension and update of service-dominant logic. *Journal of the Academy of Marketing Science* 44 (1): 5–23.

Walter, A., Ritter, T., and Gemünden, H.G. (2001). Value creation in buyer–seller relationships: theoretical considerations and empirical results from a supplier's perspective. *Industrial Marketing Management* 30 (4): 365–377.

Wolstenholme, A., Austin, S.A., Bairstow, M. et al. (2009). *Never Waste a Good Crisis: A Review of Progress Since Rethinking Construction and Thoughts for Our Future.* Constructing Excellence.

Wong, W.K., Cheung, S.O., Yiu, T.W., and Pang, H.Y. (2008). A framework for trust in construction contracting. *International Journal of Project Management* 26 (8): 821–829.

Woolthuis, R.K., Hillebrand, B., and Nooteboom, B. (2005). Trust, contract and relationship development. *Organization Studies* 26 (6): 813–840.

Yin, R.K. (2009). *Case Study Research: Design and Methods.* Sage Publications.

Zaheer, A., McEvily, B., and Perrone, V. (1998). Does trust matter? Exploring the effects of interorganizational and interpersonal trust on performance. *Organization Science* 9 (2): 141–159.

16

Summary and Conclusions

Stephen Pryke

The purpose of this summary and conclusions chapter is to attempt to define the problem that we are trying to address in this book and synthesise the key themes emerging from the work of the chapter authors. Finally, this chapter aims to draw some conclusions in both theoretical and applied terms and consider what might be appropriate next steps for both the research agenda and practice.

16.1   Context – What's the Problem?

During the late 1990s in the UK, pressure for reform was placing more emphasis on avoiding adversarial relationships and shifting the focus to relational approaches to contracting (Pryke 2009) – the sort of arrangements that established project coalitions concentrating on problem-solving and using long-term relationships to engineer continuous improvements for both clients and suppliers. Prime Contracting (Holti et al. 2000) enjoyed some interest from both the public and private sectors, providing structure to the application of supply chain management through its seven key principles. We see examples of elements of these principles being applied in industry but the recessionary years in the UK turned contractors and client interests towards price reduction through fierce financial competition. More recently there has been a resurgence of interest in the principles of the Prime Contracting initiative and those seven key principles in the rail infrastructure sector. Some of the quite specific aspirations that are appropriate to this area of supply chain management are:

- An increased focus on value as a means of selecting suppliers and as a basis of competition between firms in the supply chain.
- Maximisation of innovation, learning, and efficiency.
- Development of suppliers and the measurement of their performance over time.

Successful Construction Supply Chain Management: Concepts and Case Studies, Second Edition.
Edited by Stephen Pryke.
© 2020 John Wiley & Sons Ltd. Published 2020 by John Wiley & Sons Ltd.

- Smoothing supplier workload to provide certainty and reduce risk for both clients and suppliers.
- Capturing innovations through the use of standardised systems and components.

(Holti et al. 2000)

These aspirations remain primarily an agenda for action, and have yet to be widely adopted by the construction industry in any kind of formal or strategic manner.

16.2   A Summary of the Contributions

16.2.1   IT, Digital, and BIM

The book opens with Papadonikolaki's chapter – *The Digital Supply Chain: Mobilising Supply Chain Management Philosophy to Reconceptualise the Digital and Building Information Modelling (BIM)*. Papadonikolaki looks at how the transition from analogue to digital processes (or digitisation) affects construction. Construction information flows, with high levels of risk flowing from the business environment, have a real need for rapid implementation of digitisation. Papadonikolaki suggests that construction focuses too heavily on *projects* and places insufficient emphasis on *products*. The overemphasis on the project is a common theme throughout the book. Inevitably, clients and end-users focus on projects, but better projects are delivered through focus by the supply chain on the products of the supply chain.

Over the long-term, supply chain management within the context of real collaboration is important. Establishing appropriate standards in a digital built environment, through Industry Foundation Classes, is important to provide structure for the ongoing maturity of BIM. Papadonikolaki ponders how small and medium enterprises fair alongside the small number of very large organisations that the construction industry comprises. Finally, the chapter considers the way in which digital technology might usefully deal with processual, technical, and relationship complexities.

16.2.2   Self-Organising Networks in Supply Chains

Almadhoob's chapter – *At the Interface – when Social Network Analysis and Supply Chain Management Meet,* builds some theory linking supply chains and networks based upon a large infrastructure case study, researched over a period of four years. Almadhoob explores the proposition that supply networks in construction are complex adaptative systems. Like Papadonikolaki, she argues that complexity theory can be explored through the use of network theory in a supply chain context, and refers to the 'nonlinear', self-organising, and emergent properties. It is argued that it is useful to compare, for the purpose of the analysis, formal organisational forms with the possibly larger informal organisations that are essentially self-organising and most fundamentally, constitute the system of relationships that *delivers the project.*

The importance of dealing with interdependence (see Tavistock 1966) when studying construction supply chains is emphasised. The author finally reflects upon the need for the management of supply chain relationships to focus upon complexity and relationships in the context of a nonlinear and dynamic network environment where autonomy of action is fundamental to success in supply chain activity.

16.2.3 Green Issues

Murtagh and Badi look at sustainability in *Green Supply Chain Management in Construction: A Systematic Review*. This chapter on green supply chains deals with green design, green manufacturing, green transportation, waste management, green operation, and end of life management – reverse logistics and recycling of building components.

Murtagh and Badi conclude that further work (research and practice) needs to be done on:

- Complex decision-making in practice.
- The gap between academic and practitioners.
- 'Hot-spots' in supply chains – which are high impact in sustainability terms.
- Innovation in sustainability through contract conditions.
- The unique nature of the construction industry in green terms.
- The green agenda in the context of each of the interdependency of roles in the supply chain.
- Industry bodies – their roles and effectiveness.

16.2.4 Demand Chains and Supply Chains

Addyman provides *Connecting the 'Demand Chain' with the Supply Chain: (Re)creating Organisational Routines in Life Cycle Transitions*. Addyman, in common with a number of other chapter authors, looks at the ongoing interdependent actions that comprise the activities of both the supply chain and their corresponding *demand chains* and the need within the context of a project-based approach for these interdependent relationships to be *recreated at the commencement of each project undertaken*. This is not necessary where supply chains or demand chains are 'standing' or identical over time.

Addyman's chapter delivers a descriptive case study (alongside some theory development) based upon a live (at the time of going to press) transport infrastructure supply chain.

16.2.5 Lean

Koskela's chapter with Vrijhoef and Broft looks at *Construction Supply Chain Management Through a Lean Lens*. This chapter sets out to distinguish the subject of lean from supply chain management and to explain the relationship between the two topics. The chapter starts with an exploration of the concept of *production management*. The authors make the link between production of the end product and the production of waste, discussing the 'just-in-time' and Toyota Production System along the way. There is a link to Addyman's discussion of production as a *value generation function*. Koskela, Vrijhoef, and Broft identify the production, the economic and organisational, and the social perspectives of supply chains and their management.

16.2.6 Power Structures and Systemic Risk

Edkin's chapter – *Supply Chain Management and Risk Set in Changing Times: Old Wine in New Bottles?* – is the first of two chapters that reference risk in the supply chain.

Edkins starts with the premise that achieving client satisfaction must essentially involve more than simply *sourcing and procuring products and services*. Edkins makes the point that for every interorganisational relationship there is a power dynamic which depends on the relative prominence of the individual actors (firms) both within an individual supply chain and the market for a particular product or service within the market place – see also Cox and Townsend's work on leverage in supply chains (Cox and Townsend 1998). Finally, Edkins concludes that supply chain actors need to recognise the systemic risk associated with power structures. These systems comprise diverse organisational entities which are driven by different forces. The importance of careful selection of supply chain members, coupled with appropriate governance through contract and the management of relationships, are stressed.

16.2.7 Decision-Making Maturity

Arthur's chapter – *Linkages, Networks, and Interactions – Exploring the Context for Risk Decision Making in Construction Supply Chains* considers the socially constructed nature of risk and its assessment and management: objectively and subjectively biased risk and socially mediated and transformed risk. Arthur argues that where the client creates a context with high levels of risk being transferred, legitimately or otherwise, from the client to the supply chain actors, supply chain actors set about transferring some or all of this risk to other supply chain actors.

16.2.8 Culture

Fellows' and Liu's chapter entitled: *Culture in Supply Chains* argues essentially that *success* in supply chain management hinges around the ability to create and maintain a *culture of value creation in the supply chain*. The authors conclude by recognising that individuals bring culture to the supply chain and they suggest that the project perhaps has a culture of its own, based on history and context.

Part B consists of chapters that are essentially presenting case study matter.

16.2.9 Lessons from Megaprojects

Denicol's chapter *Managing Megaproject Supply Chains: Life After Heathrow Terminal 5*, looks at a group of megaprojects which used lessons learned from Heathrow Terminal 5. The four megaprojects the chapter draws on were:

London Olympics 2012 – A programme of projects delivering the facilities required to host the 2012 Olympics in London.

Crossrail (the Elizabeth Line) – A 118-km long railway line in London and the counties of Berkshire, Buckinghamshire, and Essex, UK.

Thames Tideway Tunnel – A 25 km tunnel running under the River Thames through central London, which captures, stores and disposes of the sewage and rainwater discharges that currently overflow into the river at various points along its length.

High Speed Two (HS2) – At the time of writing this was the largest infrastructure project in Europe – a new high speed overground railway connecting London to Birmingham in excess of 200 km long (with subsequent phases to follow).

Denicol was able to conclude, after his study of these four massive projects, that:

- Frequency and consistency of demand from clients to construction supply chains were significant factors influencing the ability to exploit good supply chain management practice.
- Although we see megaproject clients recognising the need to manage supply chains and employing organisations to act as supply chain management on their behalf, on much of the remainder of the workload of the construction industry, clients do not necessarily recognise the need or allocate the resources to supply chain management.
- The construction sector has been resistant to change in a way that much newer industries tend not to be.
- The metrics and indicators to support effective supply chain management have been relatively immature in the UK construction industry.
- There is a need to provide structure and incentives to support supply chain management activities.

16.2.10 Collaboration and Integration

In their chapter *Anglian Water @one Alliance: A New Approach to Supply Chain Management,* Mills, Evans, and Candlish contend that within the construction management literature, supply chain management is frequently seen as a function performed by the *client* or *contractor*. They stress the importance of a *relational approach* to supply chain management that engages and aligns the business goals of partners. This approach sets both long- and short-term targets for delivery, collaboratively manages supply chain risk, measures success, and provides both incentives and penalties for non-delivery when viewed against the client's business outcomes.

The case study provides evidence of early supplier engagement, collaborative planning, and production. Mills et al. look at the evaluation of alliance supply chain management *value*. The Anglian Water case study used a project supplier work cluster and involvement map indicating planning, design, and delivery of integrated solutions. A long-term relational approach provided a vehicle, in this case study, for better management of risk and a range of measures of value that align with the business goals of the client.

16.2.11 Lesson Learned and Findings from Tier 1 Contractors

Manu and Knight provide *Understanding Supply Chain Management from a Main Contractor's Perspective* which focuses on the role of the Tier 1 contractor in managing the supply chain. The authors report on the following from industry:

- The audit of supply chain actors' performance measurement across projects.
- The use of collaborative IT systems.
- Engagement in continuous performance measurement.
- Motivation and incentivisation of the supply chain.
- Prioritisation and promotion of long-term supply chain relationships with best performers.

The research found that where a main contractor achieves collaborative long-term relationships with a select group of Tier 2 contractors, the benefits include:

- Support for the Tier 2 contractor in establishing higher levels of certainty about future work load.
- The Tier 2 contractors provided support for Tier 1 during the bid stage for projects through pricing, specification development, and design contributions and planning.
- The supply chain management practices seemed to promote integration of the supply chain and achieve mutual competitive advantage.
- Visible supply chain management champions within individual supply chain firms were valuable in promoting the values of supply chain management across the firm and the supply chain.

The chapter ends with a discussion about the damaging effects of unethical behaviour in supply chain management and the importance of openness and trust.

16.2.12   Lean Practices in The Netherlands

Broft's chapter, *Lean Supply Chain Management in Construction: Implementation at the Lower Tiers of the Construction Supply Chain* focuses on Tier 1 contractors and their role in achieving outstanding performance through management of the supply chain. Once again, the project focus is seen as overemphasised and a limiting factor in the development and evolution of supply chain management in construction and it is argued that supply chain management in construction lacks maturity. There is some discussion about the existence of *fragmentation* in construction. The value stream is identified as supporting and underpinning the entire processes of product or service delivery, emphasising the point that value is created in the most unlikely places. It is argued that a project mindset tends to focus upon the *upper tiers of the supply chain*, whereas a lot of highly specialised knowledge and expertise is located *lower in the supply chain* (i.e. with specialist subcontractors). Complexity drives knowledge to be more specialised and tends to lead to the evolution of highly specialised subcontractors and consultants.

16.2.13   Knowledge Transfer

Smyth and Duryan, in *Knowledge Transfer in Supply Chains,* suggest that interdependence might be regarded as a *benefit* to construction. Dense linkages exist between supply chain actors and those linkages enable knowledge capture and management, rather than constituting a problem.

The case study material for this chapter was drawn from an extensive research project conducted while one of the authors was working with one of the UK's largest rail infrastructure providers. In summary, the findings were:

- Organisations and their supply chains rely on individuals to take the 'lessons learned' initiative, rather than organise dissemination to supply chain actors.
- There was quite frequently a disconnect between knowledge captured and its transfer usefully to other supply chains.
- Most fundamentally in the context of this book on supply chain management, Smyth and Duryan found that where knowledge management was successfully implemented it was done so with a *project* focus, rather than a *supply chain* focus. This inhibits the value of knowledge management and perhaps fosters an emphasis on *operational and technical issues* rather than *long-term strategic issues*.

- A number of barriers to effective knowledge management in the supply chain were detected, both internal and external.

Finally, Part B is brought to a close with Xu's contribution.

16.2.14 The Role of Trust in Managing Supply Chains

Xu argues in *Understanding Trust in Construction Supply Chain Relationships* that trust sustains institutional, social, and organisational life. The findings of the case study-based research were that:

- Trust in supply chains is emergent but needs to be fostered and promoted.
- Self-reinforcing cycles of trust and collaboration are useful in relation to establishing efficiency, equity, reciprocity, and 'bounded solidarity'.
- Through the process of creating a trusting environment, supply chain actors created higher levels of security, were more informative, flexible, and cohesive.
- Better experiences in relation to trust tended to enable better performance overall.

16.3 Key Themes and Agendas for Research and Practice

16.3.1 Complexity and Interdependence

Projects and their systems are becoming more complex and this influences the location of specialist knowledge. It affects the interdependence of roles and the manner in which communications occur. Increasing complexity drives the location of highly specialised knowledge into lower tiers of the supply chain and further from the client. Complexity brings higher levels of interdependence. Industry needs to recognise the disadvantages of poor communications between lower tiers of the supply chain, client, and Tier 1. Clearly, there is a research agenda looking at the impact of increasing systems complexity and interdependence of actor roles. Papadonikolaki and Amadhoob's chapters (2 and 3) provide a suitable framework to further these agendas.

The work of the Tavistock Institute (1966) is still resonant today: interdependence in roles and uncertainty in the environment. Dealing with interdependencies and uncertainties are two key tasks for the construction supply chain.

16.3.2 Work Packages

Projects tend to reflect classifications by trade or element. Whether this way of packaging projects into subcontracts is valuable to subcontractors is a moot point among those subcontractors. Subcontractors often complain about the lack of consideration given to work packaging and the effect this has on logistics and the total cost to the client. Fundamentally, the delineation of the subcontract packages affects the distribution of risk between project actors. Work needs to be done to assess the impact of package decisions on the ability to integrate design and production systems. There is some relevant discussion in chapters provided by Edkins, Arthur, and Denicol (Chapters 7, 8, and 10). The uncertainty associated with risk in construction is as real today as it was 50 years ago. The context of uncertainty reduces investment in systems and is destructive to supply chain management development.

Arguably, there are too many uncertainties, complexities, and interdependencies in construction to form 'complete' lump sum contracts (and subcontracts). We therefore need to establish emergent and adaptive networks that self-organise around the evolving needs of clients, end-users, and supply chain functions.

16.3.3 Resistance to Change

Examples of firms entering construction markets showing a tendency to innovate and diversify partly as a strategy for long-term survival are given later. Yet the average supply chain actor in construction shows very little tendency to diversify even into other trades. Work needs to be carried out to understand the nature of competition seen in new market entrants. There is material relevant to this agenda item in chapters provided by Papadonikolaki, Arthur, Fellows and Lui, Edkins and Denicol (Chapters 2, 7, 8, 9, and 10).

Supply chain actors come to their relationships inevitably with unequal power relationships – this inequality is affected by, and affects, the networks that can be established and maintained amongst supply chain actors. There is a need for more conceptual research to better understand how novel organisational forms and governance structures between owners, operators, sponsors, clients, delivery partners, and suppliers are being developed to improve performance.

Construction is slow to change mostly because a lot of competition is purely price based – quite unlike motor manufacturing where much of the competition is innovation based. This means lots of inertia for firms and individuals. New entrants to markets, however, quickly develop accurate and valuable measures of value when compared with existing actors in the market place. For example, Drogados from Spain won the contract which formed the case study for Simon Addyman's chapter; IKEA are penetrating the UK hospitality development market – building hotels.

16.3.4 Risk

Much of what is taught about the management of risk is based upon an assumption of rationality – an analytical, possibly statistical, basis for decision making. The distribution of risk throughout the supply chain fundamentally affects the efficiency of the operation of the supply chain. In particular, the unfair and opaque transfer of risk, perhaps through packaging of work packages, or through decisions to redistribute risk post-contract, has a negative effect on communications, trust, and the prevailing culture. The work of Arthur, Edkins, Fellows and Lui, and Smyth and Duryan provide weight for further analysis and discussion (Chapters 7, 8, 9, and 14).

Risk quantification and management impacts upon decision making in networks. Risk internalisation is often eschewed in supply chains yet is frequently the cheapest and an effective means of dealing with risk – especially the type of highly complex, specialised, and interdependent risk that needs a lot of knowledge and experience to deal with. Mutual trust borne out of knowledge management and the development of a culture focused upon value are important here. The identification, development, and delivery of accurately defined value is a key feature of supply chain management in construction. A project focus encourages supply chain actors to put too much attention upon 'my project' – what personally am I trying to achieve here – how do I personally define the project?

16.3.5   Communications and Integration of Systems and the Green Agenda

A great deal of work needs to be performed to further the green agenda in construction. Green issues will not be resolved by focusing on projects and their performance based upon an, arguably, irrelevant set of metrics. Murtagh and Badi (Chapter 4) recognise that a better understanding of complex decision making in practice would lead to improvements in these important areas.

Developing the self-organising abilities in the supply chain is more important than trying to design systems that are long-lived and regarded as robust. Perhaps there is an issue of ownership here. Systems that are designed by the network are owned and supported by the network. This is in contrast to systems designed by a third party and imposed upon a group of network actors.

Clients want their buildings to be green in design, manufacture (through transport and logistics), and the management of waste and demolition/recycling. The green issues are highly interdependent but tend to be dealt with separately in construction supply chains.

There is a lot of emphasis on the initiatives of individuals in establishing and maintaining networks and knowledge transfer within the supply chain. It is argued that the former is inevitable, but the latter needs more infrastructure. *Dissemination* is a very important part of knowledge management. Knowledge management and transfer tends to be *project* focused rather than *supply chain* focused (see Smyth and Duryan, Chapter 14).

16.3.6   The Role of the Contractor

Mills, Evans, and Candlish and Manu and Knight consider the role of the main contactor in managing the supply chain (Chapters 11 and 12). Mills et al. deal with the fundamentally important issue of value definition and delivery: the alignment of value definitions with the long-term business objectives of the client organisation. Manu and Knight see the management of the supply chain as a logical role for the main contractor to adopt. They see the main contractor as a 'safe pair of hands' to deal with performance measurement, integration of supply chain IT systems, and incentivisation of the supply chain through the prioritisation and promotion of long-term relationships.

16.3.7   The Role of the Client

Property developers, as construction clients, tend to manage the supply chain quite intensively, frequently dispensing with the Tier 1 main contractor altogether (by using construction management type procurement approaches). In these cases, the client has short communication path lengths with Tier 2 and below and very high levels of knowledge of the production processes are acquired by the client actors. Elsewhere, clients insulate themselves from the production part of the supply chain through the use of consultants. In these cases, clients are inclined to repeat mistakes, and knowledge management between client and the remainder of the supply chain is weak.

16.3.8   Lean Construction

Koskela, Vrijhoef, and Broft make a link with Addyman's discussion of the value generation function (Chapters 6 and 5). They look at the production, economic, and organisational perspectives of supply chains and their management. Broft's solo chapter (13) laments the overemphasis on the management of projects and links to Manu and Knight's chapter (12) with the main contractor as supply chain manager. Broft's contention is also that the main contractor is a good location for the function of supply chain management. Broft found that main contractors are able to identify and share knowledge and locate and realise value of relevance to the client.

16.3.9   Collaborative Behaviour and Quality of Relationships

Our organisational systems make certain assumptions about the nature of the human being – rational, analytical, being perfectly well-informed, and mentally fit among others. Yet we seem surprised when our systems disappoint. This is because we work within a framework of trust (Xu, Chapter 15) and our behaviours in any organisation or setting are influenced by the level of, and the nature of, trust. Culture within project supply chains and trades affect the expectations of individuals embedded within the various cultures (Fellows and Lui, Chapter 9). Collaborative behaviour is affected by culture and trust. Mills, Evans, and Candlish (Chapter 11) looked at a framework designed to align client business goals with supply chain objectives and emphasised the importance of incentivising appropriate behaviour.

Some effort has been made in this last chapter to draw together key themes within individual chapters and provide sense and meaning. Essentially, however, all authors worked independently to devise and produce their chapters.

16.4   Final Remarks

Finally, I return to the issue of the way in which we conceptualise activity. The point is often made that managing the project focuses effectively on the management of cost, time, and quality within a given project. Yet for all but the smallest and simplest projects, experience shows that the management of the project *does not* lead to the effective management of cost, time, and quality. This is a fundamental issue that undermines the effective exploitation of supply chain management. While we continue to cling to the project as the focus of our management analysis and providing the research unit of analysis, we will not realise the potential that effective long-term relationships coupled with supply chain management can offer.

For a long time there has been too much focus on projects and the management of the project. To achieve further exploitation of supply chain management principles, however, the focus needs to change to the management of the industry and the management of supply chains. This has been clearly illustrated by the work of key property developers and major public sector clients, some of which were discussed in this book. Good projects are delivered through effective management of the supply chain.

References

Cox, A. and Townsend, M. (1998). *Strategic Procurement in Construction: Towards Better Practice in the Management of Construction Supply Chains*. Thomas Telford Publishing.

Holti, R., Nicolini, D., and Smalley, M. (2000). *The Handbook of Supply Chain Management.* London: CIRIA.

Pryke, S.D. (ed.) (2009). *Construction Supply Chain Management: Concepts and Case Studies*, vol. 3. Wiley.

Tavistock Institute of Human Relations (1966). *Interdependence and Uncertainty: A Study of the Building Industry: Digest of a Report from the Tavistock Institute to the Building Industry Communication [s] Research Project.* Tavistock.

Index

a

acculturation 180
achievement (motivational types of values)
    185
action research   71, 75, 78, 92, 295
actors   17, 21, 24, 26, 31, 32, 46, 51, 67, 92,
        143, 149, 190, 191, 195, 228, 280,
        292, 299, 307, 310–313, 319–325,
        328, 338, 341–343
    network role   49
adaptive capacity of a network   48
    collective behaviour   116
    systems   336
Advanced Supply Chain Management
        Group   8
adversarial relationships   43, 196, 199, 219,
        220, 228, 238, 239, 273, 274, 290, 335
AECOM   225
affective heuristics   157, 161
Africa   75
agency   22, 89, 95, 173, 191, 198, 312
    principal-agent   187, 188
agile   18
alliances   23, 73, 188, 192, 226, 237–239,
        241, 244–246
    strategic   192
altruism   187
Amey   226
analogue   15
Anglian Water   8
    @one Alliance   238, 240, 242, 245, 339
architect   49, 143
architecture, engineering and construction
        industry (AEC)   15, 17, 18, 20, 31
arctic ice sheets   63

b

Aristotelianism   121
artefacts   89, 96, 216
artificial intelligence (AI)   31, 128
assemble-to-order (ATO)   215, 217, 219,
        230
asset owners   15
    specificity   116
assuming   100
Australia   23, 28
Autodesk Revit   26
autonomy   3

Bachy Soletanche   226
Balfour Beatty Group Limited   226
BAM Nuttall Ltd   226
barrier to entry   55, 195, 273
Bazalgette Tunnel Limited   226
Bechtel   225
behaviour   193, 194, 228, 344
    modification   183
behavioural   92, 200
    science   156
benchmarking   257
benevolence   185
Bentley Microstation   26
bills of quantities   30
BIM execution plans (BEP)   27
biodiversity   64
Blockchain   31, 132, 138, 266
Boston Big Dig   222
boundary objects   320
bounded rationality   194
    solidarity   321, 325, 328, 341
Bouygues   1

Successful Construction Supply Chain Management: Concepts and Case Studies, Second Edition.
Edited by Stephen Pryke.
© 2020 John Wiley & Sons Ltd. Published 2020 by John Wiley & Sons Ltd.

Brazil 213
BREEAM 64, 145
buildability 168, 239
Building Control 168
Building Down Barriers 254, 255
Building Information Management 27
building information modelling (BIM) 2, 15, 19, 20, 24–29, 31, 32, 53, 129, 132, 135, 137–139, 145, 227, 230, 262, 336
building product model (BPM) 25, 27
BuildingSMART 26
business goals 344
  intelligent 225
  representatives 144
  requirements 275
  satisfaction 5, 49, 337
  specification 252
business model 77
  plan 314

C

California 64
Cambridge Water Recycling Centre 240, 244
capabilities 97, 238–240, 274, 277, 289, 296, 301, 310, 324
capital investment 89
carbon emissions 63, 240
Carillion plc. 6, 129–132, 137, 139, 147, 267
cash flow 218, 300
(de)centralised design and engineering control 18
centrality 56, 295, 298
change management 227
Channel Tunnel 222, 223
chemistry 323
China 213
CH2M 225
circular economy 75, 134, 135
  supply chains 132, 167
client(s) 1, 5, 6, 8, 11, 21–23, 73, 74, 77, 88, 90, 93, 115, 119, 134, 143, 146, 147, 167, 168, 186, 187, 196, 213, 218, 220, 221, 223–228, 237, 238, 240, 245, 246, 251, 255–257, 259, 280, 289, 291, 294, 298–300, 302, 309, 311, 321, 335, 336, 338, 339, 342–344
climate change/global warming 135
cluster 21, 57, 72, 238, 240–242, 246, 256, 295, 339
  leader 49
coalitions 119
codes of conduct 186
co-evolution 47
cognitive dissonance 195
  mapping 295, 296, 299
collaboration 2, 8, 11, 27, 28, 31, 48, 54, 56, 73, 79, 113, 116, 188, 197, 219, 220, 231, 237–240, 246, 255, 257, 259, 260, 265, 273, 274, 289, 291, 293, 294, 296, 298–300, 307, 308, 311, 320–322, 324, 325, 328, 336, 339, 341
collaborative IT systems 9
  customer focus 272
collaborative relationships 1, 2, 18, 19, 30, 31, 48, 88, 184, 197, 226, 251, 254, 274, 290, 299, 300, 339, 344
  collocated 226
  planning 339
  production 339
co-location 27
co-makership 78
common data environment (CDE) 27
communication 28, 90, 116, 143, 146, 148, 159, 171, 188, 189, 196, 198, 217, 255, 259, 307, 313, 316, 319, 320, 322, 326, 341–343
  interpersonal 52, 53
communities of practice 289
compartmentalisation 187
competencies 310, 316, 320
competition 48, 189, 195, 335
competitive advantage 1, 30, 47, 74, 252, 254, 257, 265, 282, 340
competitiveness 1, 74, 282
complex adaptive systems 3, 43, 46, 48–50
complex decision making 4
complexity 1, 3, 17, 18, 49, 56, 73, 76, 96, 169, 173, 222–226, 252, 274, 275, 279, 300, 307, 340, 341
  in decision-making 337

processual   19
relational   19, 336
technical   19
theory   3, 43, 45, 46, 48, 336
complex systems   54, 151
decision-making   78
transient project environment
   44
computer aided architectural design
   (CAAD)   15, 23–25, 29
computer-aided design (CAD)   15
computing infrastructure   2
conflict management   181, 323
conformity   185
Confucian dynamism   174
congruence   198
connaught   147
consenting   98
constructing   98, 168
Construction Industry Review Committee
   189
construction manager   258
Construction Operations Building
   information exchange (COBie)   26,
   29
Construction (Design and Management)
   Regulations   133
Constructive Objects and the Integration of
   Systems (COINS)   26
constructivists/interpretivists
   173
consultant/professional service firm   146,
   167, 186, 215, 256, 289, 295, 299,
   300, 340, 343
continuous improvement   1, 117, 119, 229,
   255, 257, 261, 262, 264, 267, 272,
   279, 335, 339
contract(s)   3, 23, 27, 77–79, 91, 94, 102,
   131, 138, 168, 192, 200, 215, 216,
   224, 226, 227, 239, 240, 242, 274,
   282, 317, 318, 320, 321,
   327, 338
administration   239
disputes   11, 222, 259
lump sum   222, 342
management   266
sustainability   337

contractor(ing)/constructor   8, 29, 65, 72,
   73, 77, 78, 80, 88, 90, 98, 131, 134,
   146, 167, 237, 238, 251–267, 291,
   295, 335, 339, 343
contractual relationships   3, 23, 31, 48, 50,
   90, 140
controlling   320, 324, 325
control theory   20
converting requirements   112
cooperative relationships   48, 88, 197, 256,
   290–292
coordinated behaviour   48
coordination   31, 48, 53, 56, 88, 90, 95, 116,
   146, 154, 155, 189, 196, 214, 251,
   252, 254, 271, 274, 282, 290
core competencies   274
corporate social responsibility (CSR)   187,
   188
cost(s)   1, 239, 281, 289–292, 294, 300
Costain Ltd   226
cost-plus   77
cost, time and quality   168, 242, 283, 344
benefit analysis   186
control   239
savings   56, 257, 299
target   255
co-working   27
Crossrail (the Elizabeth Line)   7, 213, 223,
   225–227, 338
culture   6, 7, 11, 31, 78, 117, 167–201, 239,
   255, 289, 293, 302, 312, 338, 342, 344
adhocracy   179
clan   179
collectivist   176, 194
cultural norms   301
distance   193, 194, 197, 198
hierarchy   179, 196
market   179
national   7, 174, 189, 193, 197
organisational   177, 178, 190, 229
person   178
power   177
project   173
role   177
task   178
'customer delight,'   53
customisation   218–220

cybernetics   20
  complexity   343
  decentralisation of   47, 53, 237,
    312
  decision making   92, 156, 191, 223, 227,
    231, 298, 342
  intuitive   157, 160
  maturity   6
  support   72

d

Decision Explorer software   295
decomposition   5, 110, 119
defensiveness   289, 292, 295
delegated professional roles   50
delegation   5, 47
delivery   103
  partner   225, 226, 342
demand chain   4, 20, 87, 88, 90, 93, 95, 101,
    103, 337
The Department for Transport (DfT)   225,
    226
derogation   298
design(ing)   98, 112, 128, 131,
    134, 155, 168, 189, 239, 320,
    322, 340
  manager   49
design chain   238
development partner   226
dialogical   87, 101, 104
diffusion theory   28
Digital Built Britain   27
digitalisation   2, 16, 26, 30
digital processes   15, 336
digital supply chain   15, 16
  technology   2, 15, 20
  thread   2
digitisation   16, 23, 28–34
diversification   342
division of labo(u)r   90, 307
Drogados (contractor from Spain)
    342
Dubai   147
duty of care   133, 186
dyadic   19, 23, 24, 77, 251, 266

e

early supplier engagement   238–240,
    244, 247, 265, 272, 300, 310, 318,
    339
(e-business)   18
economic performance   75
  reciprocity   321
economies of scale   216, 220
edge of chaos   46, 48, 56
efficiency   146, 159, 167, 238–240, 256,
    257, 273, 274, 282, 283, 290, 299,
    307, 310, 317, 328, 335, 341, 342
Egan Report   29, 88, 145, 214, 238, 256,
    290, 307
electronic data interchange (EDI)   24
electronic document management (EDM)
    24
empathy   117, 181, 325
employers information requirements
    27
enacting   4, 5, 101
end-of-life management   3, 4, 76,
    132, 337
end-users   1, 6, 15, 155, 254, 272, 308, 311,
    329, 336, 342
energy   64
engineer-to-order (ETO)   215, 217–220,
    230, 276
enterprise resource planning (ERP)-systems
    18
environment   168
  natural   17, 65
environmental footprinting   73
  impact   74–76, 134
epistemology   110, 111, 113
equilibrium   48
equitable exchange   311
equity   308, 328, 341
  in the exchange   319
ethical practices   9, 132, 139, 145, 167
ethics   185, 186, 259, 267
Europe   28
exchange power   117
external uncertainties   17
extranets   24

f

face-to-face interactions 52
facilities managers 155, 253, 342
fast-track scheduling 28
fatal injuries 134
Ferrovial Agroman UK Ltd 226
Fifth Intergovernmental Panel on Climate
    Change Assessment Report 63
final account 319
financial incentives 48, 73
    management 289, 292, 302, 320
Fire of London 144
flexibility 252, 322
flow 51, 90, 111, 115, 132, 169, 215, 252,
    254, 272, 274, 281, 282, 314, 336
    material 30, 272
    model of production 114
    theory 20
    of time 87
    upstream/ downstream 239
Ford Production System 279, 280
formal/informal organisation 49
Forum for Supply Chain Innovation
    127
fragmentation 1, 6, 9, 17, 19, 32, 34, 76,
    119, 145, 146, 170, 196, 214, 215,
    217, 219, 220, 242, 273, 274, 282, 340
framework agreements 23, 251, 256
(transaction) frequency 116, 216, 217,
    220, 229
front end 93, 97, 101, 213, 217, 222, 227,
    298, 300, 301, 310

g

gate keeper 215
gigaprojects 221
global positioning system (GPS) 30
Global Supply Chain Forum 254
GLOBE (methodology) 176
goal alignment 48, 259, 299
good practice 298
goods-dominant logic 308–310, 317
governing (ance) 98, 102, 103, 116, 237,
    240, 307, 338, 342
Graphisoft Archicad 26

Greenacre Homes Ltd 147
Green Building Council 64
green design 3, 74, 76, 77, 337, 343
    innovation 78
    logistics 343
    manufacturing 3, 4, 76, 78, 337, 343
    operation 3, 76, 337
    procurement 73, 74, 78
    supply chain management 3, 63, 65, 72,
        74, 75, 77, 79, 80, 337, 343
    transportation 3, 76, 337, 343
greenhouse gas emissions 64, 79
grounded theory 23
groundworks 11

h

Halcrow Ltd. 225
Health and Safety at Work Act 133, 168,
    252, 258, 259, 261, 262, 294, 319
Health and Safety Executive 134
Heathrow Terminal 5–7, 213, 222–225,
    227, 338
hedonism (motivational types of values)
    185
hierarchies 168
    organisational structure 44, 56
High Speed Two Ltd (HS2) 7, 213, 223,
    226, 227, 338
hold-up 57
homeostatic and callibrationist (risk
    ideologies) 150, 151, 160
honesty 322
Hong Kong 73, 197
horizontal integration 17
hospitals 130
hotels 342
human resource management 115, 302
    practises 289, 299

i

IKEA 342
incentives 8, 9, 48, 73, 77, 102, 131, 132,
    192, 221, 223, 226, 237, 242, 256,
    257, 261, 262, 265, 266, 289, 294,
    298, 339, 343, 344

incomplete information   88, 94, 100–102
India   213
individualism/collectivism (in culture)
     174
induction   112
Industry 4.0   16
Industry Foundation Classes (IFC)   26, 29,
     336
informal relationships   3, 48, 52, 336
information   25, 26, 32, 56, 103, 190, 195,
     215, 239, 254, 257, 259, 274, 282
information exchange networks
     53, 57
information flows   2, 15, 16, 21, 31, 94,
     128, 190, 257
information management   24, 109
information processing   128
information theory   20
informing   4, 100
   and assuming   4
infrastructure   3, 4, 10, 88–90, 294, 335,
     340
   Infrastructure and Projects Authority
     (IPA)   91
   Infrastructure Client Working Group
     (CWG)   91
   projects   31, 154, 226, 301, 336, 337
innovation   79, 119, 220, 223, 224, 227,
     237–240, 242, 245, 247, 256, 257,
     259, 272, 274, 291, 298, 335, 336, 342
   in sustainability   4
insurance   131, 147
integrated project delivery (IPD)   19
integration   8, 9, 18, 21, 53, 56, 77–79, 146,
     167, 168, 189, 195, 215, 216, 226,
     228, 237–240, 246, 254, 255, 257,
     259, 265, 274, 281, 282, 292, 307,
     317, 339–341
integrity   307
interactions   48
interdependence   3, 4, 10, 17, 53, 90, 102,
     168, 169, 175, 188, 195, 199, 223,
     246, 273, 290, 292, 296, 318, 329,
     336, 337, 340–342
   patterns of action   88
interdisciplinary clusters   57
   interfirm relationships   54

International Alliance for Interoperability
     (IAI)   26, 27
International Journal of Project
     Management   213
internet of things (IoT)   16
interpersonal communications   52, 53
interpersonal relationships   54
interpretive   313
Interserve   137
Iran   147
iterative information exchange relationships
     56
IT, ICT   2, 10, 15, 16, 18, 23–25, 31, 129,
     257, 260, 262, 263, 265, 266, 293,
     302, 336, 339, 343

j
John Laing Group plc   147
joint venture   170, 188, 192–194, 197, 226,
     227, 242
   Activities   320
Journal of Management in Engineering
     213
just-in-time   5, 218–220, 228, 337

k
key performance indicators (KPIs)   257
knowledge   1, 25, 111, 180, 188, 190, 198,
     238, 242, 245, 274, 275, 310, 319,
     320, 324, 326, 340–344
   capture   340
   management   10, 341, 342
   platonic   114
   sharing   289, 294, 301, 302, 311
   tacit   293, 294, 302
   transfer   10, 54, 289–302, 310, 340, 343

l
Laing O'Rourke   225, 226
land fill sites   135, 136
Latham report   22, 29, 88, 189, 214, 237,
     238, 290
leadership   246
Leadership in Energy and Environmental
     Design (LEED)   64
Lean   5, 9, 18, 64, 65, 73, 109–125,
     271–283, 337, 340, 343

learned helplessness   289, 293, 301, 302
learning   28, 30, 116, 117, 238, 239, 256,
      275, 291, 302, 311, 316, 321–325, 335
   interorganisational   28, 30
legal obligations   52
lessons learned   296, 298, 340
leverage   338
life cycle   4, 87, 92–97, 102–104, 132, 213,
      217, 225, 313, 325, 327
   transitions   337
life cycle analysis (LCA)   71, 72
linkages   143, 159, 251
literature review   3, 63
lock-in   57
logistics   21, 109, 145, 146, 159, 254, 256,
      272, 273, 275, 276, 318, 341, 343
   integration   20
   reverse   337
London Olympics 2012   7, 213, 222, 223,
      225, 226, 338
London Underground extension of the
      Jubilee line   222
longitudinal study   314

m
Mace   225
main contractor   1, 8, 49, 88, 145, 214, 215,
      223, 230, 271–283, 294, 299, 308,
      310, 311, 313, 314, 317, 318, 321,
      323–325, 328, 329, 339, 344
make-or-buy   224, 238, 273, 282, 290
make-to-order (MTO)   218
make-to-stock (MTS)   215, 217, 218
management   4, 5, 24, 109, 111, 112, 115,
      275, 276, 279, 281, 283, 337
   system   215, 279
   theory   110–113, 276, 277
manufacturing   2
markets   168
masculinity/femininity (in culture)   174
maturity   6, 9, 22, 26, 27, 29, 73, 215, 228,
      272, 336, 338, 340
measurement   9, 257, 259, 260, 262, 263,
      265, 335, 339
mechanisation and automation   18
megaprojects   7, 93, 96, 213, 219, 221–225,
      227, 228, 338, 339

metaphysics   112
milestone date   101
modularisation   219, 246
monitoring   320
morals   186
Morgan Sindall plc   226
motivated reasoning   190
motivation   9, 184
multidisciplinary   52
multiplicity of authority   48

n
NASA   222
National Audit Office   130
National BIM Standards (NBIMS)   27
National Infrastructure and Construction
      Pipeline   90
natural environment   63
near field communication (NFC)   18
negative uncertainties   5
negotiated dialogue   103
The Netherlands   23, 340
network(s)   3, 16, 20, 21, 23, 31, 48, 52, 54,
      65, 115, 117, 143, 146, 154, 156, 159,
      167–170, 197, 214, 220, 239, 252,
      254, 266, 273, 280, 281, 292, 312,
      313, 327–329, 336, 338, 342
   analysis   336
   cohesiveness   341
   coordination   20
   environment   52
   managers   49, 57
   roles   50
   sampling   55
   science   44
   self-organising   343
   single layer   56
   supply networks   46, 56
   theory   56
newness   89
Newtonian paradigm   48
nexus of contracts   6
Nichols   225
nodes   295
non-blaming   322
nonhierarchical and nondyadic
      representation   53

nonlinear and complex iterative processes 53, 336
normative 313
norms 199, 312
  relational 322
not-for-profit organisations 169

o

obligations 175
off-site production 21, 30, 64
oligopolistic organisations 193
Olympic Delivery Authority 226
One Alliance 8
ontology 110–113
  becoming 312, 328
openness 9, 307, 322, 326
operational 21
operations and logistics research 20
operations management (OM) research 18
operations research (OR) 16
opportunistic behaviour 9, 48, 91, 167, 192, 194, 196, 197, 199, 273, 274, 313, 321, 323, 325
opportunity cost 128
opportunity management 128, 149, 155, 237
optioneering 100
organisational capabilities 88
  climate 169, 170, 172, 182
  culture 172
  effectiveness 52
  national 172
organisational citizenship behaviour (OCB) 187, 188, 197
organisational structure 44, 56
organising 98, 103
ostensive 96
owners 239

p

4.0 Paradigm 20
parametric intelligence 15
parametrisation 18
partnering 23, 24, 30, 79, 145, 188, 251, 307, 308, 328
  partnership 19

pattern recognition 20
patterns of action 87, 88, 95, 96, 102
peer-reviewed papers 3
penalties 8
performance 78, 96, 195, 271, 272, 281, 342
personality 199
personal shielding 187
piling 314, 318, 319, 322
Plan-Do-Check-Act Cycle 112
Platonism 112, 121
policy makers 15
positivist/nonpositivist paradigm 22, 32, 199, 200
power 5, 321, 327, 337, 338, 342
  coercive 323
  distance (in culture) 174, 176
  dominance 266
  and influence 173, 251, 313
  and motivational types of values 185
  and performance 54, 198
  structures 5, 132, 134, 135, 138, 173, 195, 298, 311, 338
practice 13, 211
preconstruction agreement (PCA) 317
prefabricated components 18, 281
prequalification questionnaires 257, 260–262, 265, 294, 299, 302
Prime Contractor 219, 220, 251, 335
Private Finance Initiatives (PFI) 79, 130
probability 150, 159, 191
problematisation methodology 16
problem solving 335
process-based view 311, 313
processing 89, 128, 191
process ontology 87, 88, 92, 95, 96, 114, 312
processual 16, 33, 34
procurement 5, 18, 20, 22, 23, 49, 56, 79, 88, 90, 91, 102, 103, 119, 127, 146, 147, 214, 226, 237, 238, 240, 251, 255, 256, 262, 274, 289, 294, 302, 314, 317, 318, 320, 327, 337, 343
  strategies 49, 56, 102
production 112, 114, 118–120, 127
productivity 22, 240, 246
  paradox 222

product modularisation 18
  standardisation 25
products 2, 18, 25, 180, 336, 340
  specification 112
professionalism 17
professional service providers 15
profit (ability) 2, 191, 221, 240, 246, 274,
    300, 308, 314, 317, 318
programme management 215, 226, 237,
    289, 296, 301
project(s) 2, 10, 15, 18, 80, 118, 119, 128,
    146, 161, 167, 216, 217, 222, 238,
    240, 265, 275, 279, 280, 289, 311,
    313, 342–344
  actor roles 56
  atmosphere 172, 173, 182,
    197, 298
  based organisations 307
  chemistry 196
  coalitions 335
  culture 183
  databases 24
  delivery 128, 154, 155, 193, 238, 251,
    257, 273, 307, 310, 311, 314, 319,
    320, 322, 336, 339
    partner 225
  ecosystem 292, 326, 328
  functionality 43
  governance 54, 100
  interorganisational 325
  management 22, 91–95, 145, 168, 170,
    197, 215, 221, 262, 300, 308, 311,
    312, 317, 344
  performance 259, 274, 308, 343
  planning 189, 252, 340
  portfolio 200, 220, 222
  realisation 168, 195
  software 200
  specific approach 273
  sponsor 227
  supply networks 43
  unique (ness) 271, 278, 300
project-focused logic 308, 310
Project Management Journal 213
project-related boundaries 52
property developers(ment) 31, 73, 77, 193,
    343, 344

q
quantitative method 55, 78
  analysis 18
quantity surveyor 49, 144, 258, 323
quasi-firm 168

r
radiofrequency identification (RFID) 18,
    30
realising 4
reciprocity 308, 323, 325, 341
recycling building components 337
reflective practitioners 299
relating 320, 324, 325
relational 16, 21–23, 30–34, 79, 88–91, 94,
    102, 104, 117, 119, 121, 159, 239,
    274, 290, 312, 339
  contracting 307, 335
  governance 100, 143, 145, 307
  management 100, 143, 145
  marketing 307
  norms 308, 313, 325
relationship(s) 6, 17, 39, 78, 88, 102, 117,
    127–140, 143, 146, 151, 156, 167,
    169, 193, 198, 199, 207, 228, 237,
    239, 245, 254, 262, 271–275,
    310–314, 316, 318–321, 324, 325,
    337, 338, 341, 344
  benefits 317
  cohesion 324
  informal 145
  long term 23, 34, 88, 91, 117, 194, 214,
    218, 220, 237, 239, 241, 242, 255,
    261, 262, 264, 265, 273, 275, 281,
    282, 335, 339, 343, 344, 420
  management 272, 290, 338
  stability 323
  supplier 255
  webs of 54
repetition/repeatability 89, 219, 220, 273,
    276, 278, 279, 283
reputation 240, 246
resistant to change 8, 339, 342
revenue maximisation 191
risk 1, 2, 6, 77, 127–140, 143–161, 193,
    195, 197, 223, 224, 227, 240, 242,
    267, 274, 275, 290, 302, 317, 319,
    321, 336, 338, 339, 341, 342

risk (*contd.*)
  aversion   293
  generation   154
  inertalisation   342
  management   54, 147, 150, 156, 157,
    159–161, 342
  shared   23, 235, 237, 239
  socially constructed   148, 149, 338
  systemic   5, 337
  transfer   338, 342
Rok   147
role(s)   226, 237
  uncertainty   341
routines   4, 87–93, 95–98, 102, 103, 190,
    193, 312, 316, 320, 325, 337
  routinising   324, 325
Russia   147

s
Sankey Diagrams   72
Scandinavia   23
scenario analysis   100
schools   130
Science for Environment Policy
    135
security   11
  (Motivational types of values)
    185
self-direction (motivational types of values)
    185
self-organising   3, 31, 43, 45, 47, 50, 57,
    336, 342
  protectionism   181
  subjugation   189
sensemaking   189, 190
service delivery   132
  ecosystems   310, 311
  provider   310
service-dominant logic   308, 310, 311, 313,
    328
Shanghai   147
shared goals   46
  learning   230
sharing   322
Singapore   64
single-layer networks   56
SIR COPE   189

site production   146
situation, task, intent, concerns, and
    calibration (STICC)   189
Slough Estates plc.   238
small and medium enterprises (SMEs)   17,
    29, 273, 336
social actors   51
social capital   197
social network analysis   3, 43, 49–51, 54,
    56, 94, 336
social norms   175
social position   54
social system   50
The Society of Engineers   145
South America   75
Specialist expertise   290
stakeholders   31, 32, 64, 72–76, 93, 101,
    169, 184, 217, 222, 254, 300, 312
standardisation   216, 220, 239, 276, 281
standing supply chains   4, 337
statistics   20
stimulation (motivational types of values)
    185
strategic purchasing   20
strategy   10, 217
structurationist   313, 315
structuration theory   308, 312, 313, 328
structures   49, 87, 92, 94, 100, 103, 311
(organisational) structures   46, 56, 87, 88,
    93–95, 143, 159, 200, 214–216, 227,
    228, 238, 279, 312–314, 316, 317,
    325, 328, 336, 339, 342
subcontracting   1, 119, 131, 251, 291
subsystems   152
supplier integration   73
  relationship   262
supply chain champions   340
supply chain manager   6, 258, 344
  integrator   281
sustainability   3, 129, 132, 167, 259, 294,
    337
sustainable development goals   129, 132,
    133, 139
sustainable innovation   79
Sydney Opera House   223
systematic review of literature   66
systems   312

systems thinking   17, 31, 140, 151,
     154–156, 160, 178, 200, 214, 221,
     237, 256, 277, 280, 282, 293, 296,
     320, 341, 343, 344
  integration   79, 223, 224, 226, 241, 292,
     343
  subsystems   152
Systra   225

t

Tavistock Institute   10, 90, 94, 196, 290,
     336, 341
tax relief   74
teamwork   188, 189, 195, 196, 198,
     237, 239
teleology   186, 188
temporal boundaries   87
  and spatial frame   101
temporary project team   48, 96
  organisation(ing)   88, 89, 92–94, 102,
     121, 170, 195, 213, 216–219, 225,
     227, 276–278, 310
tenders   78, 103, 265
Thames Tideway Tunnel   7, 213, 223, 226,
     227, 338
Thames Water   226
thematic analysis   258
theory   13, 211
Tier 1   1, 8, 146, 214, 215, 223, 225–228,
     237, 251–253, 266, 289, 291, 298,
     339, 340, 343
Tier 2   1, 9, 80, 146, 214, 251–253, 266,
     289, 291, 298, 308, 314, 328,
     339–341, 343
Tier 3 and 4   228, 252, 253, 266,
     342
time   88, 89, 94, 97, 112
  cost and quality   237
topological space   51
Town Planning   168
Toyota Production System   5, 9, 18, 109,
     110, 272, 279, 280, 337
tradition (motivational types of values)
     185
traditional buyer–supplier dyads   43
  development   256
training   1

transaction(al)   88–91, 94, 102, 104, 117,
     119, 184, 192, 215, 289, 290, 298,
     302, 308–310, 316–318, 328
  frequency   216, 336, 339
transaction costs   246, 307
  economics (TCE)   115, 237, 282
transformation   110, 114, 155, 215, 281,
     290, 292, 295, 302
transient(ence)   56, 276
transitioning   87, 89, 91, 98–100,
     102, 104
transition rituals   98
Transport and Works Act Order (TWAO)
     100
Transport for London (TfL)   225
triadic   266
trust   9–11, 79, 116, 117, 193, 196, 197,
     259, 265, 273, 274, 281, 292, 298,
     307–329, 341, 342, 344
turning and preparing   4, 5, 100
two-stage   318, 327
  tendering   11, 317

u

uncertainty   102, 116, 149, 159, 191, 193,
     215, 307, 310, 341
  avoidance (in culture)   174
understanding   322
unfairness   316, 340
uniqueness   310
United Arab Emirates   64
universalism   185, 222
USA   23, 28, 29, 213
USA Air force   25

v

validating   100
value   1, 7, 21, 77, 78, 88, 103, 111, 112,
     115, 117, 167, 171, 178, 180, 183, 184,
     194–196, 215, 238, 246, 247, 251,
     254–256, 265, 271–273, 279–283,
     292, 293, 309–311, 313, 320, 324,
     327, 329, 335, 337, 339, 340, 342,
     343
  added   168, 279, 290
  chains   169
  co-creation   311, 322, 323, 338

competing   196
for money   240, 290
proposition   310
reciprocal   321
stream   9
value-in-exchange   309
value-in-use   311
Vinci Construction Grands Projets
     1, 226
Virtual Design and Construction (VDC)
     27
virtual reality   228

w

warranties   78
waste   4, 5, 64, 115, 117, 135–137, 145, 214,
     215, 239, 242, 246, 259, 279–281, 337
   to energy   135, 137
   management   3, 76, 240, 337
Wembley Stadium project   222
whole life cost   237, 255
win-win   238, 245, 246
Wolstenholme Report   90
workload management   256
work packages   341, 342